ISNM
International Series of Numerical Mathematics
Volume 151

Trends and Applications in Constructive Approximation

Detlef H. Mache
József Szabados
Marcel G. de Bruin
Editors

Birkhäuser Verlag
Basel · Boston · Berlin

Editors:

Detlef H. Mache
University of Applied Science Bochum
FB3, Applied Mathematics
(Constructive Approximation)
Herner Str. 45
D-44787 Bochum

Marcel G. de Bruin
Department of Applied Mathematics
Delft University of Technology
P.O. Box 5031
2600 GA Delft
The Netherlands

József Szabados
Alfréd Rényi Institute of Mathematics
Hungarian Academy of Sciences
P.O. Box 127
H-1364 Budapest

2000 Mathematics Subject Classification 41Axx, 65Dxx

A CIP catalogue record for this book is available from the Library of Congress, Washington D.C., USA

Bibliographic information published by Die Deutsche Bibliothek
Die Deutsche Bibliothek lists this publication in the Deutsche Nationalbibliografie; detailed bibliographic data is available
in the Internet at http://dnb.ddb.de.

ISBN 3-7643-7124-2 Birkhäuser Verlag, Basel – Boston – Berlin

© 2005 Birkhäuser Verlag, P.O. Box 133, CH-4010 Basel, Switzerland
Part of Springer Science+Business Media
Printed on acid-free paper produced of chlorine-free pulp. TCF ∞
Printed in Germany
ISBN-10: 3-7643-7124-2
ISBN-13: 978-3-7643-7124-1

9 8 7 6 5 4 3 2 1   www.birkhauser.ch

Table of Contents

Preface

During the last years, constructive approximation has reached out to encompass the computational and approximation-theoretical aspects of different fields in applied mathematics, including multivariate approximation methods, quasi-interpolation, and multivariate approximation by (orthogonal) polynomials, as well as modern mathematical developments in neuro fuzzy approximation, RBF-networks, industrial and engineering applications.

Following the tradition of our international Bommerholz conferences in 1995, 1998, and 2001 we regard this 4th IBoMAT meeting as an important possibility for specialists in the field of applied mathematics to communicate about new ideas with colleagues from 15 different countries all over Europe and as far away as New Zealand and the U.S.A. The conference in Witten Bommerholz was, as always, held in a very friendly and congenial atmosphere.

The IBoMAT-series editor Detlef H. Mache (Bochum) would like to congratulate Marcel de Bruin (Delft) and József Szabados (Budapest) for an excellent editing job of this 4th volume about *Trends and Applications in constructive approximation.*

After the previous three published books in *Akademie Verlag* (1995) and *Birkhäuser Verlag* (1999 and 2003) we were pleased with the high quality of the contributions which could be solicited for the book. They are refereed and we should mention our gratitude to the referees and their reports.

At this point we also thank the *Deutsche Forschungsgemeinschaft (DFG, Bonn)* for providing the majority of the financial support of this 4th conference and the publisher *Birkhäuser / Springer Publishing Group* for accepting the proceedings into its *International Series of Numerical Mathematics.*

Also we would like to thank all participants for their efforts in making IBoMat 2004 a more than successful conference.

Finally, we would like to express our special thanks to Petra Mache who assisted in the preparation of this book and Jennifer Meyer who produced the manuscript in a camera-ready form. We wish to express our appreciation of their friendly assistance and wonderful cooperation.

The very positive resonance of the IBoMAT conferences in the years 1995, 1998, 2001 and 2004 will encourage us to continue this successful series in Witten-Bommerholz with new international developments in applied mathematics and applications in constructive approximation.

Witten-Bommerholz, January 2005   Marcel de Bruin (Delft)
                                         Detlef H. Mache (Bochum)
                                         József Szabados (Budapest)

Participants of the IBoMAT Conference in Witten-Bommerholz

D.H. Mache and E. Becker

List of Participants

- **Adell, Jose A.**, Universidad de Zaragoza, Dept. de Metodos Estadisticos, Facultad de Ciencias, E-50009 Zaragoza, Spain;
  e-mail: adell@unizar.es

- **Altomare, Francesco**, University of Bari, Department of Mathematics, Campus Universitario, Via E. Orabona 4, 70125 Bari, Italy;
  e-mail: altomare@dm.uniba.it

- **Beatson, Rick**, University of Canterbury, Department of Mathematics and Statistics, Private Bag 4800, Christchurch, New Zealand;
  e-mail: rick.beatson@canterbury.ac.nz

- **Berrut, Jean-Paul**, Université de Fribourg, Départment of Mathématiques, CH-1700 Fribourg / Pérolles, Switzerland;
  e-mail: jean-paul.berrut@unifr.ch

- **Braess, Dietrich**, Ruhr-University Bochum, Faculty of Mathematics, Universitätsstr. 150, D-44780 Bochum, Germany;
  e-mail: braess@num.ruhr-uni-bochum.de

- **Bruin, Marcel G. de**, Delft University of Technology, Faculty of Electrical Engineering, Mathematics and Computer Science, Department of Applied Mathematics, Mekelweg 4, 2628 CD Delft, The Netherlands;
  e-mail: m.g.debruin@EWI.TUDelft.nl

- **Bucci, Anthony**, Brandis University, Computer Science Department, Dynamical & Evolutionary Machine Organization, Waltham MA, U.S.A.;
  e-mail: abucci@cs.brandis.edu

- **Campiti, Michele**, University of Lecce, Department of Mathematics "E. De Giorgi", P.O. Box 193, I-73100 Lecce, Italy;
  e-mail: michele.campiti@unile.it

- **Catinas, Emil**, Romanian Academy, Institute of Numerical Analysis, P.O. Box 68-1, RO-3400 Cluj-Napoca, Romania
  e-mail: ecatinas@ictp.acad.ro

- **Catinas, Teodora**, Babes Bolyai University, Faculty of Mathematics and Computer Science, Department of Applied Mathematics, str. M. Kogalniceanu 1, RO-3400 Cluj-Napoca, Romania
  e-mail: tcatinas@math.ubbcluj.ro

- **Chiv, Henning**, University of Dortmund, Institute for Applied Mathematics, Vogelpothsweg 87, D-44221 Dortmund, Germany;
  e-mail: henning.chiv@web.de

- **Davydov, Oleg**, University of Giessen, Mathematical Institute, Arndtstr. 2, D-35392 Giessen, Germany;
  e-mail: Oleg.Davydov@math.uni-giessen.de

- **Dette, Holger**, Ruhr-University Bochum, Mathematical Institute, Universitätsstr. 150, D-44780 Bochum, Germany;
  e-mail: holger.dette@ruhr-uni-bochum.de

- **Fredebeul, Christoph**, Kemna-Berufskolleg-School, Kemnastr. 11, D-45657 Recklinghausen, Germany;
  e-mail: Christoph.Fredebeul@math.uni-dortmund.de

- **Gellhaus, Christoph**, University of Applied Sciences Bochum, FB 3, Herner Str. 45, D-44787 Bochum, Germany;
  e-mail: gellhaus@TFH-Bochum.de

- **Giefing, Gerd J.**, University of Applied Sciences Bochum, FB 3, Herner Str. 45, D-44787 Bochum, Germany;
  e-mail: giefing@TFH-Bochum.de

- **Gonska, Heinz H.**, University of Essen-Duisburg, Fachbereich 11 / Informatik I, Lotharstr. 65, 47057 Duisburg, Germany;
  e-mail: gonska@informatik.uni-duisburg.de

- **Igel, Christian**, Ruhr-University Bochum, Chair of Theoretical Biology, Institute for Neuro-Informatics, Universitätsstr. 150, D-44780 Bochum, Germany;
  e-mail: Christian.Igel@neuroinformatik.ruhr-uni-bochum.de

- **Kacso, Daniela**, University of Essen-Duisburg, Fachbereich 11 / Informatik I, Lotharstr. 65, 47057 Duisburg, Germany;
  e-mail: kacso@math.uni-duisburg.de

- **Le Méhauté, Alain**, Départment de Mathématiques, Université de Nantes, 2 rue de la Houssinière, F-44072 Nantes Cedex, France;
  e-mail: alm@math.univ-nantes.fr

- **Mache, Detlef H.**, University of Applied Sciences Bochum,
  FB 3 - Applied Mathematics (Constructive Approximation),
  Herner Str. 45, D-44787 Bochum, Germany;
  e-mail: Mache@math.uni-dortmund.de or *e-mail: Mache@TFH-Bochum.de*

- **Mache, Petra**, University of Hagen, LG Numerical Mathematics, Lützow-str. 125, D-58084 Hagen, Germany;
  e-mail: Mache-Witten@t-online.de

- **Mazure, Marie-Laurence**, Université Joseph Fourier, LMC - IMAG, BP 53,
  38041 Grenoble Cedex 9, France;
  e-mail: Marie-Laurence.Mazure@imag.fr

- **Meyer, Jennifer**, Ruhr-University Bochum, Chair of Theoretical Biology,
  Institute for Neuro-Informatics, Universitätsstr. 150, D-44780 Bochum,
  Germany;
  e-mail: Jennifer.Meyer@neuroinformatik.ruhr-uni-bochum.de

- **Mhaskar, Hrushikesh N.**, California State University, Department of
  Mathematics, Los Angeles, CA 90032, U.S.A.;
  e-mail: hmhaska@calstatela.edu

- **Michels, Kai**, Business Development Power Supply & Automation Systems,
  Fichtner Consulting, Sarweystr. 3, D-70191 Stuttgart, Germany;
  e-mail: michelsk@fichtner.de

- **Obermaier, Josef**, GSF-National Research Center of Enviroment and
  Health, Institute of Biomathematics and Biometry, Ingolstädter Landstr. 1,
  D-85764 Neuherberg, Germany;
  e-mail: josef.obermaier@gsf.de

- **Oberrath, Jens**, University of Applied Sciences Bochum, FB 3, Herner Str.
  45, D-44787 Bochum, Germany;
  e-mail: jens.oberrath@t-online.de

- **Platte, Frank**, University of Dortmund, Institute for Applied Mathematics,
  LS III, Vogelpothsweg 87, D-44221 Dortmund, Germany;
  e-mail: frank.platte@math.uni-dortmund.de

- **Prestin Jürgen**, University of Lübeck, Mathematical Institute, Wallstr. 40,
  D-23560 Lübeck, Germany;
  prestin@math.uni-luebeck.de

- **Rasa, Ioan**, Technical University Cluj-Napoca, Mathematical Department,
  RO-3400 Cluj-Napoca, Romania;
  e-mail: Ioan.Rasa@math.utcluj.ro

- **Revers, Michael**, Universität Salzburg, Mathematisches Institut, Hellbrunnerstr. 34, A-5020 Salzburg, Österreich;
  e-mail: Michael.Revers@sbg.ac.at

- **Sablonnière, Paul**, INSA de Rennes, 20 Avenue des Buttes de Coesmes, CS 14315, F-35043 Rennes cédex, France;
  e-mail: Paul.sablonniere@insa-rennes.fr

- **Sanguesa, Carmen**, Universidad de Zaragoza, Department of Statistical Methods, Pedro Cerbuna, 12, E-50009 Zaragoza, Spain;
  e-mail: csangues@unizar.es

- **Strauß, Hans**, Universität Erlangen-Nürnberg, Institut für Angewandte Mathematik, Martensstr. 3, D-91058 Erlangen, Germany;
  e-mail: strauss@am.uni-erlangen.de

- **Szabados, József**, Alfréd Rényi Inst. of Mathematics of the Hungarian Academy of Sciences, Reáltanoda u. 13-15, H-1053 Budapest, Hungary;
  e-mail: szabados@renyi.hu

- **Szwarc, Ryszard**, Wroclaw University, Institute of Mathematics, pl. Grunwaldzki 2/4, PL-50-384 Wroclaw, Poland;
  e-mail: szwarc@math.uni.wroc.pl

- **Toussaint, Marc**, Ruhr-University Bochum, Chair of Theoretical Biology, Institute for Neuro-Informatics, Universitätsstr. 150, D-44780 Bochum, Germany;
  e-mail: Marc.Toussaint@neuroinformatik.ruhr-uni-bochum.de

- **Van Assche, Walter**, Katholieke Universiteit Leuven, Department of Mathematics, Celestynenlaan 200 B, B-3001 Leuven, Belgium;
  e-mail: Walter.VanAssche@wis.kuleuven.ac.be

- **Vértesi, Peter**, Mathematical Institute of the Hungarian Academy of Sciences, Reáltanoda u. 13-15, H-1053 Budapest, Hungary;
  e-mail: veter@renyi.hu

Scientific Program

Sunday, February 15, 2004:

1st Afternoon Session: Chair: *D.H. Mache (Bochum)*

15:00   Welcome remarks and opening address

15:15–16:15   **C. Igel** (Bochum)
Evolutionary Optimization of Neural Systems

2nd Afternoon Session: Chair: *H. Dette (Bochum)*

16:50–17:20   **M. Toussaint** (& C. Igel and Wan Weishui) (Bochum)
Rprop Using the Natural Contravariant Gradient vs.
Levenberg-Marquardt Optimization

17:25–17:55   **J. Meyer** (& D.H. Mache) (Bochum)
Finding relevant Input Arguments with Radial Basis
Function Networks

18:00–18:30   **K. Michels** (Stuttgart)
T-S Controller – The Link between Fuzzy Control Theory
and Classical Control Theory

Monday, February 16, 2004:

1st Morning Session: Chair: *D.H. Mache (Bochum)*

9:00   **E. Becker** (Dortmund)
Greetings by the Rector of the University of Dortmund

9:15–10:15   **R. Beatson** (Christchurch)(& J. Levesley & W.A. Light)
Fast Evaluation on the Sphere

2nd Morning Session: Chair: *M. de Bruin (Delft)*

10:40–11:40   **F. Altomare** (Bari)
Some Simple Methods to Construct Positive Approximation Processes Satisfying Prescribed Asymptotic Formula

11:45–12:15   **I. Rasa** (Cluj-Napoca)
Semigroups Associated to Mache Operators (II)

1st Afternoon Session: Chair: *J. Obermaier (Neuherberg)*

15:15–16:15 **J.B. Berrut** (Friburg) (& H.D. Mittelmann)
Adaptive Point Shifts in Rational Interpolants with
Optimized Denominator

2nd Afternoon Session: Chair: *H. Strauß (Erlangen-Nürnberg)*

16:50–17:20 **M. Campiti** (Lecce)
Approximation of Fleming-Viot Diffusion Processes

17:25–17:55 **E. Catinas** (Cluj-Napoca)
Approximation by Successive Substitutions: The
Fast Trajectories and an Acceleration Technique

18:00–18:30 **T. Catinas** (Cluj-Napoca)
The Combined Shepard-Lidstone Bivariate Operator

Tuesday, February 17, 2004:

1st Morning Session: Chair: *M.-L. Mazure (Grenoble)*

 9:00–10:00 **J. Szabados** (Budapest)
Weighted Approximation by Generalized Bernstein
Operators

2nd Morning Session: Chair: *W. Von Assche (Leuven)*

10:30–11:30 **R. Szwarc** (Wroclaw)
Strong Nonnegative Linearization of Orthogonal
Polynomials

11:35–12:05 **J. Obermaier** (Neuherberg)
A Continuous Function Space with a Faber Basis

Wednesday, February 18, 2004:

1st Morning Session: Chair: *A. Le Méhauté (Nantes)*

 9:00–10:00 **M. Mazure** (Grenoble)
Subdivision Schemes and Non Nested Grids

2nd Morning Session: Chair: *J. Szabados (Budapest)*

10:30–11:30 **D. Braess** (Bochum)
Approximation on Simplices with Respect to Weighted
Sobolev Norms

11:35–12:05 **M. de Bruin** (Delft) (& D.H. Mache (Bochum))
(0,2) Pál Type Interpolation: A General Method for
Regularity

1st Afternoon Session: Chair: *J.P Berrut (Fribourg)*

15:15–16:15 **W. Van Assche** (Leuven)
Quadratic Hermite-Pade Approximation to the
Exponential Function: A Riemann-Hilbert Approach

2nd Afternoon Session: Chair: *D. Braess (Bochum)*

16:50–17:20 **P. Vertesi** (Budapest)
Paraorthogonal Polynomials

17:25–17:55 **J.A. Adell** (& A. Lekuona (Zaragoza))
A differential Calculus for Positive Linear Operators
and its Application to Best Poisson Approximation

18:00–18:30 **C. Sanguesa** (Zaragoza)
Approximation by B-Spline Convolution Operators
A Probabilistic Approach.

Thursday, February 19, 2004:

1st Morning Session: Chair: *R. Beatson (Christchurch)*

9:00–10:00 **A. Le Méhauté** (Nantes)
Multivariate Polynomial Splines and Rational Splines
via Finite Elements

2nd Morning Session: Chair: *F. Altomare (Bari)*

10:30–11:30 **P. Sablonniére** (Rennes)
Some Recent Results on Univariate and Multivariate
Polynomial or Spline Quasi-interpolants

11:35–12:05 **J. Prestin** (Lübeck)
Multivariate Periodic Shift-invariant Spaces

1st Afternoon Session: Chair: *J. Prestin (Lübeck)*

15·15–16:15 **H.N. Mhaskar** (Los Angeles)
Local Polynomial Operators

2nd Afternoon Session: Chair: *D.H. Mache (Bochum)*

16:50–17:20 **F. Platte** (& C. Fredebeul (Dortmund)
Numerical and Experimental Studies of Instationary
Fixed-bed-Reactors Exhibiting Steep Gradients and
Shocks

17:25–17:55 **O. Davydov** (Giessen)
Efficient Fitting of Scattered Data by Direct Extension
of Local Approximations

18:00   Closing of the conference and final remarks

During the first Lecture in the Morning Session

Opening by the Rector of the University of Dortmund E. Becker

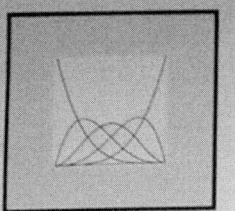

Witten-Bommerholz (Germany)

4 th International Bommerholz - Meeting on

Constructive

Approximation

Feb. 15-20, 2004

Scientific Committee: **M.G.de Bruin – D.H. Mache – M. Revers – J. Szabados**

Invited Speakers

Altomare, F. (Italy)
Assche, W. van (Belgium)
Beatson, R. (New Zealand)
Igel, C. (Germany)
LeMehaute, A. (France)
Mazure, M.-L. (France)
Mhaskar, H. N. (U.S.A)
Schumaker, L.L. (U.S.A)
Szabados, J. (Hungary)
Szwarc, R, (Poland)

Conference Topics

Multivariate Approximation Methods
Constructive Approximation
Approximation by Orthogonal Polynomials
Special Functions and Applications
Quasi-Interpolation & Interpolation
Neuro Fuzzy Methods & RBF-Networks
Industrial and Engineering Applications

Contact-Address: **D.H. Mache** (Bochum, Germany)

eMail: mache-witten@t-online.de

University of Applied Sciences Bochum
FB 3, Applied Mathematics (Constructive Approximation), Herner Str. 45, D-44787 Bochum (Germany)

Conference-Homepage: http://www.sbg.ac.at/mat/events/IDoMAT2004/IDoMAT01.htm

Trends and Applications in Constructive Approximation
(Eds.) M.G. de Bruin, D.H. Mache & J. Szabados
International Series of Numerical Mathematics, Vol. 151, 1–12

Best Poisson Approximation of Poisson Mixtures.
A Linear Operator Approach

José Antonio Adell and Alberto Lekuona

Abstract. We provide closed form solutions, in an asymptotic sense, to the problem of best choice of the Poisson parameter in Poisson approximation of Poisson mixtures, with respect to the Kolmogorov and the Fortet-Mourier metrics. To do this, we apply a differential calculus based on different Taylor's formulae for the Poisson process, either in terms of the forward differences of the function under consideration or in terms of the Charlier polynomials, which allows us to give simple unified proofs. This approach also shows that the zeros of a suitable linear combination of the first and second Charlier polynomials play a key role in determining the leading coefficient of the main term of the approximation.

1. Introduction

It is known that a large class of one-dimensional positive linear operators L_0, usually considered in the literature of approximation theory, allow for a probabilistic representation of the form

$$L_0\phi(t) = E\phi(Z_0(t)), \qquad t \in I, \tag{1}$$

where I is a subinterval of the real line, $Z_0 := (Z_0(t),\ t \in I)$ is a stochastic process of I-valued random variables and $\phi : I \to \mathbb{R}$ is any measurable function such that $L_0|\phi|(t) < \infty$, $t \in I$. In such a case, L_0 is said to be represented by Z_0. Typical examples are the following. The classical nth Bernstein polynomial is represented by

$$B_n\phi(t) := \sum_{k=0}^{n} \phi\left(\frac{k}{n}\right)\binom{n}{k} t^k (1-t)^{n-k} = E\phi\left(\frac{S_n(t)}{n}\right), \qquad t \in [0,1],$$

where $S_n(t) := \sum_{k=1}^{n} 1_{[0,t]}(U_k)$, $(U_k)_{k\geq 1}$ is a sequence of independent and on the interval $[0,1]$ uniformly distributed random variables, and 1_A stands for the indicator function of the set A. Also, for any $r > 0$, the Szász operator S_r is represented by

$$S_r\phi(t) := e^{-rt} \sum_{k=0}^{\infty} \phi\left(\frac{k}{r}\right) \frac{(rt)^k}{k!} = E\phi\left(\frac{N(rt)}{r}\right), \quad t \in [0, \infty), \qquad (2)$$

where $(N(t),\ t \geq 0)$ is the standard Poisson process. Many other examples can be found in Adell and de la Cal [2] and Altomare and Campiti [7, Chap. 5].

Concerning operators of the form (1), an interesting question is to determine which properties of ϕ are retained by $L_0\phi$. It is known that many usual operators preserve properties of first order, such as monotonicity, moduli of continuity, Lipschitz constants or φ-variation (cf. Kratz and Stadtmüller [18], Khan and Peters [17], Altomare and Campiti [7, Chap. 5] and Adell and de la Cal [3]), as well as properties of second order, such as convexity and moduli of continuity and Lipschitz classes of second order (cf. Cottin and Gonska [10], Adell et al. [1] and Zhou [25]). Here, we are mainly interested in preservation of convex functions of arbitrary order, because this property is closely connected with a differential calculus for positive linear operators. Indeed, it is shown in Adell and Lekuona [5, Theorem 3] that, under appropriate integrability assumptions on $\mathbb{Z}_0$, the operator L_0 in (1) preserves generalized convex functions of order n if and only if there are positive linear operators L_k represented by I-valued processes $\mathbb{Z}_k := (Z_k(t),\ t \in I)$, $k = 1, \ldots, n$, having right-continuous nondecreasing paths, satisfying the following Taylor's formula of order n for any smooth enough function ϕ

$$L_0\phi(t) - L_0\phi(s) = \sum_{k=1}^{n-1} L_k\phi^{(k)}(s)\Pi_k(C_k(s,t])$$

$$+ \int_{(s,t]} L_n\phi^{(n)}(u)\Pi_{n-1}(C_{n-1}[u,t])dm_{n-1}(u), \quad s,t \in I, \quad s \leq t. \qquad (3)$$

In (3), $m_k(t) := EZ_k(t),\ t \in I$, Π_k is the product measure on I^k given by $\Pi_k := m_0 \times \cdots \times m_{k-1}$ and $C_k(s,t]$ is a simplex-type set in I^k. The operator L_k (resp. its associated process $\mathbb{Z}_k$ as in (1)) is called the derived operator of L_0 (resp. the derived process of $\mathbb{Z}_0$) of order k. Of course, if $Z_0(t) = t,\ t \in I$, (3) gives us the classical Taylor's formula for differentiable functions. On the other hand, applying (3) to the Szász operator S_1 defined in (2), we obtain (cf. Adell and Lekuona [5, Corollary 1 and Proposition 4]) that

$$S_1\phi(t) = E\phi(N(t)) = \sum_{k=0}^{n-1} \frac{(t-s)^k}{k!} E\phi^{(k)}(N(s) + V_k)$$

$$+ \int_s^t E\phi^{(n)}(N(u) + V_n)\frac{(u-s)^{n-1}}{(n-1)!}\,du, \qquad 0 \leq s \leq t, \qquad (4)$$

where $V_k := U_1 + \cdots + U_k$ and $(U_i)_{i \geq 1}$ is a sequence of independent and on the interval [0,1] uniformly distributed random variables. A striking feature is that, in many cases, a Taylor's formula similar to (3) holds for non-smooth functions. This is due to the fact that the operator L_0 takes non-smooth into smooth functions, as it happens when $L_0 = S_1$. Therefore, in this case, we can obtain in a very easy way sharp estimates of the distance between $N(t)$ and $N(s)$ in different metrics by considering suitable sets of test functions ϕ, even if the parameter t is replaced by a random variable T.

The aim of this paper is to apply the differential calculus outlined above in order to give closed form solutions, in an asymptotic sense, to the problem of best Poisson approximation of Poisson mixtures, with respect to the Kolmogorov and the Fortet-Mourier metrics. Such results complement those obtained in a previous paper (cf. Adell and Lekuona [6]).

To this respect, recall that a mixing random variable T is a nonnegative random variable with distribution function F independent of the standard Poisson process $(N(t), t \geq 0)$. The random variable $N(T)$ is called a Poisson mixture with mixing random variable T and its probability law is given by

$$P(N(T) = k) = \int_{[0,\infty)} \frac{e^{-t}t^k}{k!} \, dF(t), \qquad k = 0, 1, \ldots . \tag{5}$$

For instance, if T has the gamma density

$$\rho_{b,s}(\theta) := \frac{b^{bs}}{\Gamma(bs)} \theta^{bs-1} e^{-b\theta}, \qquad \theta > 0 \quad (b > 0, \; s > 0),$$

then $N(T)$ has the negative binomial distribution given by

$$P(N(T) = k) = \binom{bs + k - 1}{k} \left(\frac{1}{b+1}\right)^k \left(\frac{b}{b+1}\right)^{bs}, \qquad k = 0, 1, \ldots .$$

Poisson mixtures are commonly used in modelling different phenomena in applied probability and statistics. A variety of examples coming from biology, physics, reliability or insurance can be found in Johnson et al. [16], Grandell [15] and Denuit and Van Bellegem [14]. As follows from (5), the probability law of a Poisson mixture is, in general, rather involved or quite complex to work with. It seems therefore natural to approximate a Poisson mixture $N(T)$ by a Poisson random variable $N(s)$, provided T be close to s. In this sense, the problem of best Poisson approximation consists of finding the Poisson distribution closest to a Poisson mixture with respect to a given metric. Such a problem was first posed by Serfling [24] in the context of Poisson approximation of Bernoulli convolutions with regard to the total variation distance. In the same context, a deeper approach to the problem, using semigroup techniques, has been developed by Deheuvels and Pfeifer [11, 12] and Deheuvels et al. [13]. Finally, Pfeifer [19] and Adell and Lekuona [6] have considered the problem in the context of Poisson mixtures.

In a strict sense, the problem of best Poisson approximation, with respect to a given probability metric d, consists of finding the parameter μ solving the equation

$$\inf_s d(N(T), N(s)) = d(N(T), N(\mu)).$$

As observed in Deheuvels *et al.* [13, p. 192], there are examples showing that "we cannot hope for an explicit closed solution for the problem ..., not even for commonly used distance measures". The problem, however, may be posed from an asymptotic point of view as follows. Suppose that $(T_n)_{n\geq 0}$ is a sequence of mixing random variables such that $ET_n = s > 0$, $n \geq 0$, and that $\mu_2(n) := E(T_n - s)^2 \to 0$ as $n \to \infty$. If we are able to prove that

$$d\left(N(T_n), N\left(s - \frac{a}{2s}\mu_2(n)\right)\right) = C(s, a)\mu_2(n) + o(\mu_2(n)), \qquad (6)$$

then it is possible to find the parameter $a^* := a^*(s)$ which minimizes the leading constant $C(s, a)$.

Closed form solutions in the aforementioned asymptotic sense are provided in Section 3, when d is the Kolmogorov or the Fortet-Mourier metric. In the first case, $C(s, a^*)$ depends of the integer parts of the roots of a suitable linear combination of the first and second Charlier polynomials, while, in the second, $C(s, a^*)$ depends on the mean absolute deviation with respect to the median of the Poisson distribution with mean s (see Theorems 3.2 and 3.3, which are the main results of this paper). Such results are based on different Taylor's formulae for the standard Poisson process concerning arbitrary exponential bounded functions, as stated in Theorem 2.1 and Corollary 2.3 in Section 2.

2. Differential calculus for the Poisson process

Let us start by giving some notations. We denote by $\mathbb{N}$ the set of nonnegative integers and by $\mathbb{N}^* := \mathbb{N} \setminus \{0\}$. Every function ϕ is a real measurable function defined on $[0, \infty)$ and $\|\phi\|$ is its usual sup-norm. All of the random variables appearing under the same expectation sign E are supposed to be mutually independent. For any $m \in \mathbb{N}^*$, β_m stands for a random variable having the beta density $\rho_m(\theta) := m(1 - \theta)^{m-1}$, $\theta \in [0, 1]$, while $\beta_0 := 1$. For any $\alpha \geq 0$, $\mathcal{E}(\alpha)$ is the set of all functions ϕ such that $|\phi(x)| \leq Ce^{\alpha x}$, $x \geq 0$, for some constant $C \geq 0$. By $\triangle_m\phi$ we denote the mth forward difference of ϕ, i.e.,

$$\triangle_m\phi(x) := \sum_{k=0}^{m}(-1)^{m-k}\binom{m}{k}\phi(x + k), \qquad x \geq 0, \quad m \in \mathbb{N}. \qquad (7)$$

Finally, we denote by $C_m(s; x)$ the mth Charlier polynomial with respect to $N(s)$, that is,

$$C_m(s; x) := \sum_{k=0}^{m}\binom{m}{k}\binom{x}{k}k!(-s)^{-k}, \qquad x \geq 0, \quad s > 0, \quad m \in \mathbb{N}. \qquad (8)$$

Such polynomials satisfy the orthogonality property (cf. Chihara [9, p. 4])

$$EC_k(s; N(s))C_m(s; N(s)) = \frac{m!}{s^m}\delta_{k,m}, \qquad k, m \in \mathbb{N}. \qquad (9)$$

Forward differences and Charlier polynomials are related by the following formula

$$E\phi(N(s))C_m(s; N(s)) = (-1)^k E\triangle_k\phi(N(s))C_{m-k}(s; N(s)), \qquad (10)$$

where $\phi \in \mathcal{E}(\alpha)$, $s > 0$, $m \in \mathbb{N}$ and $k = 0, \ldots, m$. The proof of (10) is based on the three-term recurrence relation satisfied by the Charlier polynomials (see Barbour et al. [8, Lemma 9.1.4] or Roos [21, formula (6)]). An extension of (10) when the parameter s is replaced by a mixing random variable T may be found in Adell and Lekuona [6, Lemma 2.1].

With the preceding notations, we enunciate the following.

Theorem 2.1. *Let $t \geq 0$, $s > 0$ and $m \in \mathbb{N}^*$. Denote by $\gamma_m := \gamma_m(s, t) = s + (t - s)\beta_m$. Then, for any $\phi \in \mathcal{E}(\alpha)$, we have*

$$E\phi(N(t)) = \sum_{k=0}^{m-1} \frac{(t-s)^k}{k!} E\triangle_k\phi(N(s)) + \frac{(t-s)^m}{m!} E\triangle_m\phi(N(\gamma_m))$$

$$= \sum_{k=0}^{m-1} \frac{(-1)^k(t-s)^k}{k!} E\phi(N(s))C_k(s; N(s))$$

$$+ \frac{(-1)^m(t-s)^m}{m!} E\phi(N(\gamma_m))C_m(\gamma_m; N(\gamma_m))$$

$$= \sum_{k=0}^{m-1} \frac{E\phi(N(s))C_k(s; N(s))}{EC_k^2(s; N(s))} EC_k(s; N(t))$$

$$+ \frac{E\phi(N(\gamma_m))C_m(\gamma_m; N(\gamma_m))}{EC_m(s; N(\gamma_m))C_m(\gamma_m; N(\gamma_m))} EC_m(s; N(t)).$$

Proof. The first two equalities in Theorem 2.1 have been shown in Adell and Lekuona [6, Theorem 2.1]. On the other hand, if ϕ is m times differentiable, it can be checked by induction that

$$\triangle_k\phi(x) = E\phi^{(k)}(x + V_k), \quad x \geq 0, \quad k = 0, \ldots, m, \qquad (11)$$

where $V_k := U_1 + \cdots + U_k$ and $(U_i)_{i \geq 1}$ is a sequence of independent and on the interval $[0, 1]$ uniformly distributed random variables.

Let $m \in \mathbb{N}^*$ and $k = 0, \ldots, m$. From (11), se see that $\triangle_l C_k(s; x) = 0$, $x \geq 0$, whenever $l > k$. Therefore, applying the first two equalities in Theorem 2.1 to $\phi(x) = C_k(s; x)$ and using the orthogonality property in (9), we obtain

$$EC_k(s; N(t)) = \frac{(-1)^k(t-s)^k}{k!} EC_k^2(s; N(s)), \quad k = 0, \ldots, m-1, \qquad (12)$$

as well as

$$EC_m(s; N(t)) = \frac{(-1)^m(t-s)^m}{m!} EC_m(s; N(\gamma_m))C_m(\gamma_m; N(\gamma_m)).$$

This shows the third equality and completes the proof of Theorem 2.1. □

Remark 2.2. *Full expansions in Theorem 2.1 are also valid. For instance, we can write*

$$E\phi(N(t)) = \sum_{k=0}^{\infty} \frac{(t-s)^k}{k!} E\triangle_k \phi(N(s)), \qquad \phi \in \mathcal{E}(\alpha),$$

as the series above absolutely converges. This readily follows from (7).

Although the third equality in Theorem 2.1 will not be used in this paper, it may be of interest in the theory of orthogonal polynomials. Actually, L^2-expansions of the form

$$\phi(x) \sim \sum_{k=0}^{\infty} \frac{E\phi(N(s))C_k(s;N(s))}{EC_k^2(s;N(s))} C_k(s;x), \quad \phi \in L^2(\mu_L), \tag{13}$$

where μ_L stands for the Lebesgue measure, are well known (cf. Chihara [9, p. 17]). Remark 2.2 and formulae (10) and (12) imply that if we randomize x by $N(t)$ in (13) and then integrate, we obtain usual convergence rather than L^2-convergence, as we have

$$E\phi(N(t)) = \sum_{k=0}^{\infty} \frac{E\phi(N(s))C_k(s;N(s))}{EC_k^2(s;N(s))} EC_k(s;N(t)), \qquad \phi \in \mathcal{E}(\alpha).$$

On the other hand, setting $\phi = 1_{\{n\}}$, $n \in \mathbb{N}$, in the second equality in Theorem 2.1, we obtain

$$\begin{aligned} P(N(t) = n) &= \sum_{k=0}^{m-1} \frac{(-1)^k (t-s)^k}{k!} \, C_k(s;n) P(N(s) = n) \\ &+ \frac{(-1)^m (t-s)^m}{m!} \, E 1_{\{n\}}(N(\gamma_m)) C_m(\gamma_m; N(\gamma_m)), \end{aligned} \tag{14}$$

where $\gamma_m := s + (t-s)\beta_m$. We point out that (14) coincides, up to changes of notation, with the formula obtained by Roos [22, Lemma 2 and formula (9)] using a different approach. In turn, the formulae in Theorem 2.1 can be derived from (14) by integration. Finally, a multivariate version of (14) is given in Roos [23, Lemma 1].

Let $s > 0$ be fixed and let $(T_n, n \in \mathbb{N})$ be a sequence of mixing random variables converging to s as $n \to \infty$. For any $k, n \in \mathbb{N}$, we denote by $\mu_k(n) := \mu_{k,s}(n) = E(T_n - s)^k$, whenever the expectation exists.

Replacing t by T_n in Theorem 2.1 and applying Fubini's theorem, we obtain the following.

Corollary 2.3. *Let $s > 0$ and $m, n \in \mathbb{N}$. Assume that $E|T_n - s|^m < \infty$. Then, for any $\phi \in \mathcal{E}(\alpha)$ such that $\|\triangle_m \phi\| < \infty$, we have*

$$\begin{aligned} E\phi(N(T_n)) &= \sum_{k=0}^{m} \frac{\mu_k(n)}{k!} E\triangle_k \phi(N(s)) \\ &+ \frac{1}{m!} E(T_n - s)^m (\triangle_m \phi(N(s + (T_n - s)\beta_m)) - \triangle_m \phi(N(s))). \end{aligned} \tag{15}$$

Formula (15) constitutes our main tool to deal with the problem of best Poisson approximation of Poisson mixtures. Finally, since in the following section we shall need to bound the remainder term in (15), we give the following inequality

$$|E\phi(N(t)) - E\phi(N(s))| \le (\|\phi_+\| + \|\phi_-\|)(1 - e^{-|t-s|}), \quad s,t \ge 0, \qquad (16)$$

where $\phi_+ := \max(\phi, 0)$ and $\phi_- := \max(-\phi, 0)$. For a proof of (16), see Adell and Lekuona [6, Lemma 2.2].

3. Best Poisson approximation

When estimating the distance between random variables, the Kolmogorov and the Fortet-Mourier metrics are among the most commonly used (see, for instance, Pfeifer [19], Deheuvels *et al.* [13], Rachev [20] and Roos [21, 22], where other probability metrics are also considered). Recall that given two $\mathbb{N}$-valued random variables X and Y, the Kolmogorov and the Fortet-Mourier distances between X and Y are, respectively, defined by

$$d_0(X, Y) := \sup_{l \in \mathbb{N}} |E1_{[l,\infty)}(X) - E1_{[l,\infty)}(Y)|$$

and

$$d_1(X, Y) := \sum_{l=0}^{\infty} |E1_{[l,\infty)}(X) - E1_{[l,\infty)}(Y)|.$$

From now on, we fix $s > 0$ and assume that the sequence of mixing random variables $(T_n, n \in \mathbb{N})$ satisfies $ET_n = s$ and $\mu_2(n) := E(T_n - s)^2 < \infty$, $n \in \mathbb{N}$. For any real a, let $P_a(s; x)$ be the quadratic polynomial

$$P_a(s; x) := C_2(s; x) - \frac{a}{s} C_1(s; x) = \frac{1}{s^2}(x^2 - (2s + 1 - a)x + s(s - a)), \qquad (17)$$

where the last equality follows from (8). The roots of $P_a(s; x)$ are

$$r_i(a) := s + \frac{1-a}{2} + (-1)^i \sqrt{s + \left(\frac{1-a}{2}\right)^2}, \quad i = 1, 2. \qquad (18)$$

It turns out that the integer parts of these roots are relevant in the problem at hand. The following technical lemma can be easily checked.

Lemma 3.1. *Let $r_i(a)$, $i = 1, 2$, be the functions defined in (18). Then,*

(a) *$r_1(a)$ is a decreasing function, $r_1(a) < 0$ for $a > s$, and*

$$\lim_{a \to -\infty} r_1(a) = s \quad and \quad \lim_{a \to \infty} r_1(a) = -\infty.$$

(b) *$r_2(a)$ is a decreasing function, $r_2(a) \ge s$ for any real a, and*

$$\lim_{a \to -\infty} r_2(a) = \infty \quad and \quad \lim_{a \to \infty} r_2(a) = s.$$

For any real numbers x and y, we denote by $x \vee y := \max(x, y)$ and by $\lfloor x \rfloor$ the integer part of x. Also, we consider the functions

$$f_s(a) := E1_{\{n_1(a)\}}(N(s))\left(C_1(s; N(s)) - \frac{a}{s}\right), \quad a \in \mathbb{R} \tag{19}$$

and

$$g_s(a) := E1_{\{n_2(a)\}}(N(s))\left(\frac{a}{s} - C_1(s; N(s))\right), \quad a \in \mathbb{R}, \tag{20}$$

where $n_i(a) := \lfloor r_i(a) \rfloor$, $i = 1, 2$.

Having in mind the preceding notations, we state the main results of this paper.

Theorem 3.2. *Assume that $\mu_2(n) \to 0$ and that $E|T_n - s|^3 = o(\mu_2(n))$, as $n \to \infty$. Then for any $a \leq (2s^2)/\mu_2(n)$, we have*

$$\left| d_0\left(N(T_n), N\left(s - \frac{a}{2s}\mu_2(n)\right)\right) - \frac{\mu_2(n)}{2}(f_s(a) \vee g_s(a)) \right|$$
$$\leq \frac{2}{3}E|T_n - s|^3 + \left(\frac{a}{2s}\mu_2(n)\right)^2. \tag{21}$$

Moreover, for large enough n, we have

$$\inf\left\{f_s(a) \vee g_s(a) : a \leq \frac{2s^2}{\mu_2(n)}\right\} = f_s(a^*), \tag{22}$$

where a^* is the unique solution to the equation $f_s(a) = g_s(a)$. Equivalently,

$$a^* = \inf\{a \leq s : f_s(a) \leq g_s(a)\}$$
$$= \inf\left\{a \leq s : (s - a - n_1(a))\frac{s^{n_1(a)}}{n_1(a)!} \leq (a + n_2(a) - s)\frac{s^{n_2(a)}}{n_2(a)!}\right\}. \tag{23}$$

Theorem 3.3. *In the setting of Theorem 3.2, we have*

$$\left| d_1\left(N(T_n), N\left(s - \frac{a}{2s}\mu_2(n)\right)\right) - \frac{\mu_2(n)}{2s}E|N(s) - (s - a)| \right|$$
$$\leq \frac{2}{3}E|T_n - s|^3 + \left(\frac{a}{2s}\mu_2(n)\right)^2. \tag{24}$$

Moreover, for large enough n, we have

$$\inf\left\{E|N(s) - (s - a)| : a \leq \frac{2s^2}{\mu_2(n)}\right\} = E|N(s) - Me(s)|, \tag{25}$$

where $Me(s)$ stands for the median of $N(s)$.

Concerning Theorems 3.2 and 3.3, some remarks are in order. First, inequalities (21) and (24) show that the distances d_0 and d_1 fulfill (6). Second, formula (23) allows us to obtain explicit values of a^* for small values of s, as done in Deheuevels and Pfeifer [12, Example 3.1] and Deheuevels *et al.* [13, Lemma 2.3], in the context of Poisson approximation of Bernoulli convolutions. Third, for arbitrary values of s, formula (23) provides an algorithm to evaluate a^* in a finite number of steps. Such an algorithm can be implemented in a system for mathematical computation such as MAPLE. Actually, as follows from (19) and (20), the function f_s is piecewise linear and decreasing, while g_s is piecewise linear and increasing. In addition, it follows from Lemma 3.1 that f_s and g_s have decreasing sequences of knots at $\{r_1^{-1}(i), \ i \le \lfloor s \rfloor - \lfloor \lfloor s \rfloor s^{-1} \rfloor\}$ and $\{r_2^{-1}(i), \ i \ge \lfloor s \rfloor + 1\}$, respectively, where it is understood that i is an arbitrary integer. Therefore, to compute a^*, we give the following algorithm

$$a^* = \frac{(s-k)\,m! - (m-s)\,k!\,s^{m-k}}{m! + k!\,s^{m-k}},$$

where k and m are defined by

$$r_1^{-1}(k) := \min\left\{r_1^{-1}(i): \ f_s(r_1^{-1}(i)) \le g_s(r_1^{-1}(i)), \quad i \le \lfloor s \rfloor - \left\lfloor \frac{\lfloor s \rfloor}{s} \right\rfloor\right\}$$

and

$$r_2^{-1}(m) := \min\left\{r_2^{-1}(i): \ f_s(r_2^{-1}(i)) \le g_s(r_2^{-1}(i)), \quad i \ge \lfloor s \rfloor + 1\right\}.$$

With respect to the Fortet-Mourier metric d_1, we see from Theorem 3.3 that the best Poisson parameter is

$$s + \frac{Me(s) - s}{2s}\,\mu_2(n),$$

for which the leading coefficient of the main term of the approximation is $E|N(s) - Me(s)|/(2s)$. Sharp lower and upper bounds for $Me(s) - s$ and closed form expressions for the mean absolute deviation $E|N(s) - Me(s)|$ are obtained in Adell and Jodrá [4].

Proof of Theorem 3.2. Let $n \in \mathbb{N}$ and let ϕ be a function such that $\|\triangle_2\phi\| < \infty$. Using a Taylor expansion of second order around $N(s)$ as in Corollary 2.3 and Theorem 2.1, we obtain

$$E\phi(N(T_n)) - E\phi\left(N\left(s - \frac{a}{2s}\mu_2(n)\right)\right) = \frac{\mu_2(n)}{2}E\phi(N(s))P_a(s; N(s)) + R_n(\phi), \quad (26)$$

where $P_a(s; x)$ is the polynomial defined in (17) and

$$\begin{aligned}
R_n(\phi) :=& \frac{1}{2}\,E(T_n - s)^2\big(\triangle_2\phi(N(s + (T_n - s)\beta_2)) - \triangle_2\phi(N(s))\big) \\
&+ \frac{a}{2s}\,\mu_2(n)E\left(\triangle_1\phi(N(s - \frac{a}{2s}\mu_2(n)\beta_1)) - \triangle_1\phi(N(s))\right).
\end{aligned} \quad (27)$$

To obtain the main term in (21), denote by $b_l := E1_{[l,\infty)}(N(s))P_a(s; N(s))$, $l \in \mathbb{N}$. Since

$$b_0 = 0, \qquad b_{l+1} - b_l = -E1_{\{l\}}(N(s))P_a(s; N(s)), \quad l \in \mathbb{N},$$

we see that the sequence $(b_l, \ l \in \mathbb{N})$ attains its minimum at $n_1(a) + 1$ (whenever $n_1(a) \ge 0$) and its maximum at $n_2(a) + 1$. Therefore, we have from (10), (19) and

(20) that

$$\sup_{l \in \mathbb{N}} |E1_{[l,\infty)}(N(s))P_a(s; N(s))| = \max_{i=1,2} \{|E1_{[n_i(a)+1,\infty)}(N(s))P_a(s; N(s))|\}$$

$$= f_s(a) \vee g_s(a). \tag{28}$$

With respect to the error bound in (21), we apply (16) and Fubini's theorem in (27) to obtain

$$\sup_{l \in \mathbb{N}} R_n(1_{[l,\infty)}) \le 2E(T_n - s)^2 \left(1 - e^{-|T_n - s|\beta_2}\right)$$

$$+ \frac{|a|}{s} \mu_2(n) E\left(1 - e^{-\frac{|a|}{2s}\mu_2(n)\beta_1}\right) \le \frac{2}{3}E|T_n - s|^3 + \left(\frac{a}{2s}\mu_2(n)\right)^2, \tag{29}$$

where we have used the inequality $1 - e^{-|x|} \le |x|$, $x \ge 0$ and the fact that $E\beta_m = 1/(m+1)$, $m \in \mathbb{N}$. Therefore, (21) follows from (26)–(29).

On the other hand, as follows from (19) and (20), the function $f_s(a)$ is piecewise linear and decreasing, while $g_s(a)$ is piecewise linear and increasing. Therefore, there is a unique solution a^* to the equation $f_s(a) = g_s(a)$, at which the infimum in (22) is attained. Finally, since $n_1(a) < 0$ for $a > s$, as follows from Lemma 3.1(a), we see that $f_s(a) = 0$ for $a > s$. Consequently,

$$a^* = \inf \{a \le s : f_s(a) \le g_s(a)\},$$

which shows (23) and completes the proof of Theorem 3.2.   □

Proof of Theorem 3.3. By (10) and (17), we have

$$\sum_{n=0}^{\infty} |E1_{[n,\infty)}(N(s))P_a(s; N(s))| = \sum_{n=0}^{\infty} \left|E1_{\{n-1\}}(N(s))\left(C_1(s; N(s)) - \frac{a}{s}\right)\right|$$

$$= \frac{1}{s}E|N(s) - (s - a)|. \tag{30}$$

On the other hand, it has been shown in Adell and Lekuona [6, formula (4.8)] that for any $u, v > 0$ and $k \in \mathbb{N}^*$, we have

$$\sum_{l=0}^{\infty} |E\triangle_k 1_{[l,\infty)}(N(u)) - E\triangle_k 1_{[l,\infty)}(N(v))| \le 2^k \left(1 - e^{-|u-v|}\right). \tag{31}$$

As in the proof of (21), inequality (24) follows by applying (26) and (27) to the function $\phi := 1_{[l,\infty)}$, $l \in \mathbb{N}$ and taking into account (30) and (31). Finally, statement (25) follows from the well-known fact that the median of a random variable minimizes its mean absolute deviation with respect to an arbitrary point. The proof of Theorem 3.3 is complete.   □

Acknowledgement

This work was supported by Spanish research project BFM2002-04163-C02-01.

References

[1] Adell, J.A., Badía, F.G. and de la Cal, J.: *Beta-type operators preserve shape properties*. Stoch. Proc. Appl. **48**, 1–8, 1993.

[2] Adell, J.A. and de la Cal, J.: *Using stochastic processes for studying Bernstein-type operators*. Rend. Circ. Mat. Palermo (2) Suppl. **33**, 125–141, 1993.

[3] Adell, J.A. and de la Cal, J.: *Bernstein-type operators diminish the φ-variation*. Constr. Approx. **12**, 489–507, 1996.

[4] Adell, J.A. and Jodrá, P.: *The median of the Poisson distribution*. Preprint, 2003.

[5] Adell, J.A. and Lekuona, A.: *Taylor's formula and preservation of generalized convexity for positive linear operators*. J. Appl. Prob. **37**, 765–777, 2000.

[6] Adell, J.A. and Lekuona, A.: *Sharp estimates in signed Poisson approximation of Poisson mixtures*. Preprint, 2003.

[7] Altomare, F. and Campiti, M.: *Korovkin-type Approximation Theory and Its Applications*. Walter de Gruyter, Berlin, 1994.

[8] Barbour, A.D., Holst, L. and Janson, S.: *Poisson Approximation*. Clarendon Press, Oxford, 1992.

[9] Chihara, T.S.: *An Introduction to Orthogonal Polynomials*. Gordon and Breach, New York, 1978.

[10] Cottin, C. and Gonska, H.H.: *Simultaneous approximation and global smoothness preservation*. Rend. Circ. Mat. Palermo (2) Suppl. **33**, 259–279, 1993.

[11] Deheuvels, P. and Pfeifer, D.: *A semigroup approach to Poisson approximation*. Ann. Probab. **14**, 663–676, 1986a.

[12] Deheuvels, P. and Pfeifer, D.: *Operator semigroups and Poisson convergence in selected metrics*. Semigroup Forum **34**, 203–224, 1986b.

[13] Deheuvels, P., Pfeifer, D. and Puri, M.L.: *A new semigroup technique in Poisson approximation*. Semigroup Forum **38**, 189–201, 1989.

[14] Denuit, M. and Van Bellegem, S.: *On the stop-loss and total variation distances between random sums*. Statist. Prob. Lett. **53**, 153–165, 2001.

[15] Grandell, J.: *Mixed Poisson Processes*. Chapman & Hall, London, 1997.

[16] Johnson, N.L., Kotz, S. and Kemp, A.W.: *Univariate Discrete Distributions, 2nd Edition*. Wiley, New York, 1992.

[17] Khan, M.K. and Peters, M.A.: *Lipschitz constants for some approximation operators of a Lipschitz continuous function*. J. Approx. Theory **59**, 307–315, 1989.

[18] Kratz, W. and Stadtmüller, U.: *On the uniform modulus of continuity of certain discrete approximation operators*. J. Approx. Theory **54**, 326–337, 1988.

[19] Pfeifer, D.: *On the distance between mixed Poisson and Poisson distributions*. Statist. Decisions **5**, 367–379, 1987.

[20] Rachev, S.T.: *Probability Metrics and the Stability of Stochastic Models*. Wiley, Chichester, 1991.

[21] Roos, B.: *Asymptotics and sharp bounds in the Poisson approximation to the Poisson-binomial distribution*. Bernoulli **5**(6), 1021–1034, 1999.

[22] Roos, B.: *Improvements in the Poisson approximation of mixed Poisson distributions*. J. Statist. Plann. Inference **113**(2), 467–483, 2003a.

[23] Roos, B.: *Poisson approximation of multivariate Poisson mixtures.* J. Appl. Probab. **40**, 376–390, 2003b.

[24] Serfling, R. J.: *Some elementary results on Poisson approximation in a sequence of Bernoulli trials.* SIAM Rev. **20**, 567–579, 1978.

[25] Zhou, D.X. *On a problem of Gonska.* Results Math. **28**, 169–183, 1995.

José Antonio Adell and Alberto Lekuona
Universidad de Zaragoza
Departamento de Métodos Estadísticos
Facultad de Ciencias
50009 Zaragoza, Spain
e-mail: {adell,lekuona}@unizar.es

Trends and Applications in Constructive Approximation
(Eds.) M.G. de Bruin, D.H. Mache & J. Szabados
International Series of Numerical Mathematics, Vol. 151, 13–26

On Some Classes of Diffusion Equations and Related Approximation Problems

Francesco Altomare and Ioan Rasa

Abstract. Of concern is a class of second-order differential operators on the unit interval. The C_0-semigroup generated by them is approximated by iterates of positive linear operators that are introduced here as a modification of Bernstein operators. Finally, the corresponding stochastic differential equations are also investigated, leading, in particular to the evaluation of the asymptotic behaviour of the semigroup.

1. Introduction

In this paper we study the positive semigroups $(T(t))_{t>0}$ generated by the differential operators

$$Wu(x) = x(1-x)u''(x) + (a+1 - (a+b+2)x)u'(x)$$

defined on suitable domains of $C[0,1]$ which incorporate several boundary conditions.

In the spirit of a general approach introduced by the first author ([1], see also, e.g., [4], [5], [7], [14], [23]–[25] and the references given there) we show that the semigroups can be approximated by iterates of some positive linear operators that are introduced here, perhaps, for the first time.

These operators are a simple modification of the classical Bernstein operators and, as them, are of interpolatory type.

We also investigate some shape preserving properties of these operators that, in turn, imply similar ones for the semigroup $(T(t))_{t\geq 0}$.

By following a recent approach due to the second author ([24]–[25]) we also consider the solutions $(Y_t)_{t\geq 0}$ of the stochastic equations associated with W and which are formally related to the semigroups by the formula

$$T(t)f(x) = E^x f(Y_t) \quad (0 \leq x \leq 1,\ t \geq 0).$$

For suitable values of a and b we get information about $(T(t))_{t\geq 0}$ and $(Y_t)_{t\geq 0}$, in particular about their asymptotic behaviour. Analogous results have been obtained in [25] considering the operators described in [16].

We finally point out that the generation properties of the operators W have been also investigated in $L^1[0,1]$ (see [7]). In the spirit of [24], Section 7, W may be viewed also as generator of a C_0-semigroup on $L^2[0,1]$; details will appear elsewhere.

2. The semigroup

Let $C[0,1]$ be the space of all real-valued continuous functions, endowed with the supremum norm and the usual order.

For $a, b \in \mathbb{R}$ consider the differential operator

$$Wu(x) = x(1-x)u''(x) + (a+1-(a+b+2)x)u'(x), \ 0 < x < 1, \ u \in C^2(0,1).$$

The corresponding diffusion equation, i.e.,

$$u_t(t,x) = x(1-x)u_{xx}(t,x) + (a+1-(a+b+2)x)u_x(t,x)$$
$$(0 < x < 1, \ t \geq 0) \tag{1}$$

occurs in some stochastic model from genetics discussed in [13] (see also [9] and [27]).

Usually the above diffusion equation is coupled with some initial boundary conditions.

Let

$$D_V(W) = \{u \in C[0,1] \cap C^2(0,1) \ : \lim_{x\to 0,1} Wu(x) = 0\},$$

$$D_M(W) = \{u \in C[0,1] \cap C^2(0,1) \ : Wu \in C[0,1]\},$$

$$D_{VM}(W) = \{u \in C[0,1] \cap C^2(0,1) \ : \lim_{x\to 0} Wu(x) = 0, \lim_{x\to 1} Wu(x) \in \mathbb{R}\},$$

$$D_{MV}(W) = \{u \in C[0,1] \cap C^2(0,1) \ : \lim_{x\to 1} Wu(x) = 0, \lim_{x\to 0} Wu(x) \in \mathbb{R}\}$$

and set

$$D(W) = \begin{cases} D_V(W), & \text{if } a,b < 0; \\ D_M(W), & \text{if } a,b \geq 0; \\ D_{VM}(W), & \text{if } a < 0, b \geq 0; \\ D_{MV}(W), & \text{if } a \geq 0, b < 0. \end{cases}$$

For every $u \in D(W), Wu$ can be continuously extended to $[0,1]$.

We shall continue to denote by Wu this extension and so we obtain a linear operator $W: D(W) \longrightarrow C[0,1]$.

As a particular case of the results of [7], we have:

Theorem 2.1. *In each of the following cases*

    1) $a \geq 0$,   2) $b \geq 0$,   3) $a, b \leq -1$,   4) $-1 < a, b < 0$,

$(W, D(W))$ *is the infinitesimal generator of a strongly continuous positive semi-group* $(T(t))_{t\geq 0}$ *on* $C[0,1]$. *Moreover,* $T(t)1 = 1, t \geq 0$, *i.e.,* $(T(t))$ *is a contraction semigroup.*

For $u \in C[0,1] \cap C^2(0,1)$ define the boundary conditions:

$$N_a \begin{cases} u \in C^1[0,\frac{1}{2}], \ u'(0) = 0, \lim_{x\to 0} xu''(x) = 0 & \text{if } a < -1; \\ \lim_{x\to 0} xu''(x) = 0 & \text{if } a = -1; \\ u \in C^1[0,\frac{1}{2}], \lim_{x\to 0} xu''(x) = 0 & \text{if } a \geq 0. \end{cases}$$

$$N_b \begin{cases} u \in C^1[\frac{1}{2},1], u'(1) = 0, \lim_{x\to 1}(1-x)u''(x) = 0 & \text{if } b < -1; \\ \lim_{x\to 1}(1-x)u''(x) = 0 & \text{if } b = -1; \\ u \in C^1[\frac{1}{2},1], \lim_{x\to 1}(1-x)u''(x) = 0 & \text{if } b \geq 0. \end{cases}$$

As a particular case of [7], Theorem 2.3, we get

Theorem 2.2. *Let* $a, b \in (-\infty, -1] \cup [0, +\infty)$ *and* $u \in C[0,1] \cap C^2(0,1)$. *Then*

(i) $u \in D(W)$ *if and only if* u *satisfies* N_a *and* N_b.

(ii) $C^2[0,1] \cap D(W)$ *is a core of* $(W, D(W))$.

3. Approximation of the semigroup by modified Bernstein operators

In this section we shall introduce a modification of Bernstein operators and we shall approximate the semigroup considered in the previous section by suitable iterates of these modified operators.

From now on we shall assume that $a, b \in [-1, +\infty)$. For every $n \geq M_0 := \max\{a+1, b+1\}$ we shall consider the following positive linear operator $L_n : C[0,1] \longrightarrow C[0,1]$ defined by

$$L_n f(x) := \sum_{h=0}^{n} \binom{n}{h} x^h (1-x)^{n-h} f\left(\left(1 - \frac{a+b+2}{2n}\right)\frac{h}{n} + \frac{a+1}{2n}\right) \tag{2}$$

for every $f \in C[0,1]$ and $x \in [0,1]$.

Note that if we consider the auxiliary function

$$v(x) = a + 1 - (a+b+2)x \qquad (0 \leq x \leq 1) \tag{3}$$

then

$$L_n f = B_n\left(f \circ \left(e_1 + \frac{v}{2n}\right)\right) \qquad (f \in C[0,1]) \tag{4}$$

where B_n denotes the nth Bernstein operator and $e_1(x) := x$ $(0 \leq x < 1)$.

From formula (4) and from well-known properties of Bernstein operators it is possible to obtain the approximation properties of the sequence $(L_n)_{n\geq M_0}$.

As usual we set $e_j(x) := x^j$ $(0 \leq x \leq 1)$, $j = 0, 1, \ldots$.

Then for every $n \geq M_0$,

$$L_n e_0 = e_0, \tag{5}$$

$$L_n e_1 = e_1 + \frac{v}{2n}, \tag{6}$$

$$L_n e_2 = \left(1 - \frac{a+b+2}{2n}\right)^2 \left(e_2 + \frac{e_1 - e_2}{n}\right)$$
$$+ \frac{a+1}{n}\left(1 - \frac{a+b+2}{2n}\right)e_1 + \frac{(a+1)^2}{4n^2}e_0. \tag{7}$$

We are now in the position to state the following result.

Theorem 3.1. *For every $f \in C[0,1]$,*

$$\lim_{n \to \infty} L_n f = f \quad \text{uniformly on} \quad [0,1].$$

Moreover, for every $x \in [0,1]$ and $n \geq M_0$,

$$|L_n f(x) - f(x)| \leq \omega_1\left(f, \frac{|v(x)|}{2n}\right) + M\omega_2\left(f, \left(1 - \frac{a+b+2}{2n}\right)\sqrt{\frac{x(1-x)}{n}}\right),$$

where ω_1 and ω_2 denote the ordinary first and second moduli of smoothness and M is a suitable constant independent of f, n and x.

Proof. The first statement follows from formulae (5)–(7) and the Korovkin theorem. As regards the subsequent estimate, taking formula (5.2.43) of [4] into account, we get

$$|L_n f(x) - f(x)|$$

$$\leq \left|B_n\left(f \circ \left(e_1 + \frac{v}{2n}\right)\right)(x) - f\left(x + \frac{v(x)}{2n}\right)\right| + \left|f\left(x + \frac{v(x)}{2n}\right) - f(x)\right|$$

$$\leq M\omega_2\left(f \circ \left(e_1 + \frac{v}{2n}\right), \sqrt{\frac{x(1-x)}{n}}\right) + \omega_1\left(f, \frac{|v(x)|}{2n}\right)$$

$$\leq M\omega_2\left(f, \left(1 - \frac{a+b+2}{2n}\right)\sqrt{\frac{x(1-x)}{n}}\right) + \omega_1\left(f, \frac{|v(x)|}{2n}\right). \qquad \square$$

We are now going to show some shape preserving properties of the operators L_n. As usual, for $0 < \alpha \leq 1$ and $M \geq 0$ we set

$$\mathrm{Lip}(\alpha, M) := \{f \in C[0,1] : |f(x) - f(y)| \leq M|x - y|^\alpha \quad \text{for every} \quad x, y \in [0,1]\}.$$

Proposition 3.2. *The following statements hold true:*

(i) *Each operator L_n maps increasing continuous functions into increasing continuous functions and convex continuous functions into convex continuous functions.*

(ii) *For every $n \geq M_0$, $0 < \alpha \leq 1$, $M \geq 0$,*

$$L_n(\mathrm{Lip}(\alpha, M)) \subset \mathrm{Lip}\left(\alpha, M\left(1 - \frac{a+b+2}{2n}\right)^\alpha\right).$$

(iii) *For every $n \geq M_0$, $f \in C[0,1]$ and $0 \leq \varepsilon \leq 1/2$,*

$$\omega_1(L_n f, \varepsilon) \leq 2\omega_1\left(f, \left(1 - \frac{a+b+2}{2n}\right)\varepsilon\right)$$

and

$$\omega_2(L_n f, \varepsilon) \leq 3\omega_2\left(f, \left(1 - \frac{a+b+2}{2n}\right)\varepsilon\right).$$

Proof. Since the function v is affine and increasing statement (i) can be easily proved by using (4) together with the property of Bernstein operators of leaving invariant the cone of continuous increasing functions as well as the cone of convex continuous functions.

As regards statement (ii), if $f \in \text{Lip}(\alpha, M)$ then

$$f \circ \left(e_1 + \frac{v}{2n}\right) \in \text{Lip}\left(\alpha, M\left(1 - \frac{a+b+2}{2n}\right)^\alpha\right)$$

and then the result follows because $\text{Lip}\left(\alpha, M\left(1 - \frac{a+b+2}{2n}\right)^\alpha\right)$ is invariant under the Bernstein operator B_n. Finally, the inequalities

$$\omega_1(B_n\varphi, \varepsilon) \leq 2\omega_1(\varphi, \varepsilon) \quad \text{and} \quad \omega_2(B_n\varphi, \varepsilon) \leq 3\omega_2(\varphi, \varepsilon) \quad (\varphi \in C[0,1], \varepsilon > 0)$$

obtained, respectively, in [6] and [22], imply statement (iii) because

$$\omega_i\left(f \circ \left(e_1 + \frac{v}{2n}\right), \varepsilon\right) \leq \omega_i\left(f, \left(1 - \frac{a+b+2}{n}\right)\varepsilon\right), \quad i = 1, 2. \qquad \square$$

The sequence $(L_n)_{n \geq M_0}$ verifies the following asymptotic formula.

Theorem 3.3. *For every $f \in C^2[0,1]$,*

$$\lim_{n \to \infty} n(L_n f(x) - f(x)) = \frac{x(1-x)}{2} f''(x) + \frac{v(x)}{2} f'(x)$$

uniformly with respect to $x \in [0,1]$.

Proof. Since

$$L_n f(x) - f(x) = \left(B_n\left(f \circ \left(e_1 + \frac{v}{2n}\right)\right)(x) - B_n f(x)\right) + (B_n f(x) - f(x)),$$

we have to determine only the limit of the expression

$$n\left(B_n\left(f \circ \left(e_1 + \frac{v}{2n}\right)\right)(x) - B_n f(x)\right),$$

because, as it is well known, $n(B_n f(x) - f(x)) \longrightarrow \frac{x(1-x)}{2} f''(x)$ uniformly with respect to $x \in [0,1]$.

Let $p_{nk}(x) = \binom{n}{k}x^k(1-x)^{n-k}$. The above-mentioned expression becomes successively

$$n\sum_{k=0}^{n} p_{nk}(x)\left(f\left(\frac{k}{n}+\frac{1}{2n}v\left(\frac{k}{n}\right)\right)-f\left(\frac{k}{n}\right)\right)$$

$$= n\sum_{k=0}^{n} p_{nk}(x)\frac{1}{2n}v\left(\frac{k}{n}\right)f'(c_{nk})$$

$$= \frac{1}{2}\sum_{k=0}^{n} p_{nk}(x)v\left(\frac{k}{n}\right)f'\left(\frac{k}{n}\right)+\frac{1}{2}\sum_{k=0}^{n} p_{nk}(x)v\left(\frac{k}{n}\right)\left(f'(c_{nk})-f'\left(\frac{k}{n}\right)\right)$$

$$= \frac{1}{2}\sum_{k=0}^{n} p_{nk}(x)v\left(\frac{k}{n}\right)f'\left(\frac{k}{n}\right)+\frac{1}{2}\sum_{k=0}^{n} p_{nk}(x)v\left(\frac{k}{n}\right)f''(d_{nk})\left(c_{nk}-\frac{k}{n}\right),$$

where c_{nk} and d_{nk} are between $\frac{k}{n}$ and $\frac{k}{n}+\frac{1}{2n}v\left(\frac{k}{n}\right)$.

As $n\to\infty$, the uniform limit of the first sum is $\frac{1}{2}v(x)f'(x)$, while the uniform limit of the second sum is zero. So the theorem is proved. $\square$

By using Theorem 3.3 we can quickly proceed to give a representation of the semigroup studied in Section 2 in terms of iterates of the operators L_n. However this representation will be proved only for the particular cases $a \geq 0$, $b \geq 0$ or $a = -1$ and $b \geq 0$ or $a \geq 0$ and $b = -1$. In the remaining cases, the problem of constructing a suitable approximation process whose iterates approximate the semigroup remains open.

We finally point out that, if $a = b = -1$, then $L_n = B_n$ and the corresponding result about the representation of the semigroup is well known (see, e.g., [4, Ch. VI]).

When $a \geq 0$ and $b \geq 0$, the semigroup can be also represented by iterates of the operators introduced in [7, Theorem 4.6] or in [17] (see [23]).

According to the previous section we set

$$D(W) = \begin{cases} D_M(W) & \text{if } a \geq 0, b \geq 0, \\ D_{VM}(W) & \text{if } a = -1, b \geq 0, \\ D_{MV}(W) & \text{if } a \geq 0, b = -1 \end{cases}$$

and denote by $(T(t))_{t\geq 0}$ the semigroup generated by $(W, D(W))$.

Theorem 3.4. *In each of the above-mentioned cases, for every $f \in C[0,1]$ and $t \geq 0$,*

$$T(t)f = \lim_{n\to\infty} L_n^{k(n)}f \quad \text{uniformly on} \quad [0,1]$$

where $(k(n))_{n\geq 1}$ is an arbitrary sequence of positive integers such that $\frac{k(n)}{n} \to 2t$ and $L_n^{k(n)}$ denotes the iterate of L_n of order $k(n)$.

Proof. From Theorem 3.3 it follows that

$$\lim_{n\to\infty} n(L_n u - u) = \frac{1}{2} W u \quad \text{in} \quad C[0,1]$$

for every $u \in C^2[0,1] \subset D(W)$.

Since $C^2[0,1]$ is a core for $(W, D(W))$ (see Theorem 2.2) and $\|L_n^p\| \leq 1$ for every $n \geq M_0$ and $p \geq 1$, then, by a result of Trotter [27], denoting by $(S(t))_{t\geq 0}$ the semigroup generated by $(\frac{1}{2}W, D(W))$, for every $f \in C[0,1]$ and $t \geq 0$ and for every sequence $(k(n))_{n\geq 1}$ of positive integers such that $\dfrac{k(n)}{n} \to t$ we have

$$S(t)f = \lim_{n\to\infty} L_n^{k(n)} f \quad \text{in} \quad C[0,1].$$

Now the result obviously follows since $T(t) = S(2t)$ for every $t \geq 0$.   □

Taking Theorem 3.4 and Proposition 3.2 into account, we can easily derive the following properties of the semigroup $(T(t))_{t\geq 0}$. These properties can be immediately translated into the corresponding ones for the solutions $u(t,x)$ of the diffusion equation (1) coupled with an initial condition $u_0 \in D(W)$ and which is given by

$$u(t,x) = T(t)u_0(x) \quad (0 \leq x \leq 1, \ t \geq 0).$$

Corollary 3.5. *In each of the three cases considered in Theorem 3.4, the following statements hold true:*

(i) *Each operator $T(t)$ $(t \geq 0)$ maps increasing continuous functions into increasing continuous functions as well as convex continuous functions into convex continuous functions.*

(ii) *For every $0 < \alpha \leq 1$, $M \geq 0$ and $t \geq 0$*

$$T(t)(\mathrm{Lip}(\alpha, M)) \subset \mathrm{Lip}(\alpha, M \exp(-(a+b+2)\alpha t)).$$

Proof. We need only to prove (ii). Let $f \in \mathrm{Lip}(\alpha, M)$ and $t \geq 0$. Consider a sequence $(k(n))_{n\geq 1}$ of positive integers such that $\frac{k(n)}{n} \to 2t$. Replacing, if necessary, f by f/M, we can always assume that $M = 1$. Then for every $n \geq 1$

$$L_n^{k(n)} f \in \mathrm{Lip}\left(\alpha, \left(1 - \frac{a+b+2}{2n}\right)^{\alpha k(n)}\right).$$

Passing to the limit as $n \to \infty$ we get that $T(t)f \in \mathrm{Lip}(\alpha, \exp(-(a+b+2)\alpha t))$ by virtue of Theorem 3.4.   □

The limit behaviour of the semigroup, i.e., the limit $\lim_{t\to\infty} T(t)$, will be studied in the next section.

4. The stochastic equation

Consider the stochastic equation associated to W:

$$dY_t = \sqrt{2Y_t(1 - Y_t)}dB_t + (a + 1 - (a + b + 2)Y_t)dt \;, \; t \geq 0,$$

with $Y_0 = x \in (0, 1)$. (see, e.g., [10], [11], [12], [15], [21], [26]).

Feller's test is applicable; we omit the proof and give only the final result:

Theorem 4.1. *Let ζ be the lifetime of the solution $(Y_t)_{t \geq 0}$.*

(i) *If $a \geq 0$ and $b \geq 0$, then $P(\zeta = \infty) = 1$ and*
$$P(\inf_{0 \leq t < \infty} Y_t = 0) = P(\sup_{0 \leq t < \infty} Y_t = 1) = 1.$$

(ii) *If $a \geq 0$ and $b < 0$, we have $P(\zeta < \infty) = 1$ and*
$$P(\inf_{0 \leq t < \zeta} Y_t > 0) = P(\lim_{t \to \zeta} Y_t = 1) = 1.$$

(iii) *For $a < 0$ and $b \geq 0$, $P(\zeta < \infty) = 1$ and*
$$P(\lim_{t \to \zeta} Y_t = 0) = P(\sup_{0 \leq t < \zeta} Y_t < 1) = 1.$$

(iv) *For $a < 0$ and $b < 0$, $E\zeta < \infty$ and*
$$P(\lim_{t \to \zeta} Y_t = 1) = 1 - P(\lim_{t \to \zeta} Y_t = 0) = \varphi(x), \text{ where}$$

$$\varphi(x) = \frac{\int_0^x u^{-a-1}(1 - u)^{-b-1}du}{\int_0^1 u^{-a-1}(1 - u)^{-b-1}du}.$$

For $a \geq 0$ and $b \geq 0$ it is possible to apply results from [18], [19] in order to prove that the semigroup $(T(t))$ is compact and $||T(t) - T|| \longrightarrow 0$ exponentially, as $t \to \infty$, where

$$Tf(x) = \frac{\int_0^1 u^a(1 - u)^b f(u)du}{\int_0^1 u^a(1 - u)^b du} \;, \; f \in C[0, 1], \; x \in [0, 1].$$

(See also [24], (10.8).)

The kernel of the integral representation of $T(t)$ for $a \geq 0$ and $b \geq 0$ (see [14], [8], and [18], Theorem 4.4) is

$$p(t, x, y) = \sum_{n=0}^{\infty} e^{-n(n+a+b+1)t} J_n^{(a,b)}(x) J_n^{(a,b)}(y) y^a (1 - y)^b,$$

where $J_n^{(a,b)}(x)$ are the Jacobi polynomials orthonormal on the interval $[0, 1]$ with weight $x^a(1 - x)^b$.

The asymptotic behavior of the semigroup $(T(t))$ in the remaining cases is suggested by Theorem 4.1 (see also ([24], (10.10)) and described in

Theorem 4.2. *For all $f \in C[0,1]$,*

1. $\lim\limits_{t \to \infty} T(t)f = f(0)$ *if* $a < 0$, $b \geq 0$;

2. $\lim\limits_{t \to \infty} T(t)f = f(1)$ *if* $a \geq 0$, $b < 0$;

3. $\lim\limits_{t \to \infty} T(t)f = f(0)(1 - \varphi) + f(1)\varphi$ *if* $a, b \leq -1$ *or* $-1 < a, b < 0$, *where the function* φ *is defined in Theorem 4.1.*

Proof. 1. Let $v(x) = x^{-a}$, $x \in [0,1]$. Then $v \in C[0,1] \cap C^2(0,1)$ and $Wv = a(b+1)v$. We deduce that $v \in D_{VM}(W) = D(W)$ and

$$T(t)v = e^{a(b+1)t}v, \ t \geq 0.$$

Now let $T : C[0,1] \to C[0,1]$, $Tf(x) = f(0)$, $f \in C[0,1]$, $x \in [0,1]$.
Then $\lim\limits_{t \to \infty} T(t)1 = T1$ and $\lim\limits_{t \to \infty} T(t)v = Tv$.
An application of Theorem 3.4.3 [4] shows that

$$\lim\limits_{t \to \infty} T(t)f = Tf, \ f \in C[0,1],$$

which is the first statement of the theorem.

2. The proof of the second statement is similar.

3. Let $w(x) = x^{-a}(1 - x)^{-b}$, $x \in [0,1]$. Then $w \in C[0,1] \cap C^2(0,1)$ and $Ww = (a+b)w$. It follows that $w \in D_V(W) = D(W)$ and

$$T(t)w = e^{(a+b)t}w, \ t \geq 0.$$

On the other hand, $\varphi \in C[0,1] \cap C^2(0,1)$ and $W\varphi = 0$, which means that $T(t)\varphi = \varphi$, $t \geq 0$.

Let $T : C[0,1] \to C[0,1]$, $Tf = f(0)(1 - \varphi) + f(1)\varphi$, $f \in C[0,1]$. For $t \to \infty$ we have $T(t)1 \to T1$, $T(t)\varphi \to T\varphi$, $T(t)w \to Tw$. To conclude the proof it suffices to apply Theorem 3.4.3 [4]. □

From the above proof we can deduce also quantitative versions of Theorem 4.2:

1. Let $a < 0$, $b \geq 0$. If $f \in C[0,1]$ and

$$|f(x) - f(0)| \leq C_f x^{-a}, \ x \in [0,1] \tag{8}$$

for some constant C_f, then

$$|T(t)f(x) - f(0)| \leq C_f e^{a(b+1)t}x^{-a}, \ x \in [0,1], \ t \geq 0.$$

2. Let $a \geq 0$, $b < 0$, and $f \in C[0,1]$ with

$$|f(x) - f(1)| \leq K_f(1 - x)^{-b}, \ x \in [0,1]. \tag{9}$$

Then we have

$$|T(t)f(x) - f(1)| \leq K_f e^{(a+1)bt}(1 - x)^{-b}, \ x \in [0,1], \ t \geq 0.$$

3. Let $a, b \leq -1$ or $-1 < a$, $b < 0$. Then φ satisfies both (8) and (9). If a function $f \in C[0, 1]$ also satisfies both (8) and (9), then for $x \in [0, 1]$, $t \geq 0$,

$$|T(t)f(x) - f(0)(1 - \varphi(x)) - f(1)\varphi(x)| \tag{10}$$
$$\leq (C_f K_\varphi + C_\varphi' K_f)e^{(a+b)t}x^{-a}(1 - x)^{-b}.$$

4. Let $a = b = -1$; then $\varphi(x) = x$. Let $(T(t))$ be the corresponding semigroup. Consider also the semigroup$(S(t))$ associated with the classical Bernstein operators (see [4], Theorem 6.3.5). The generator of $(S(t))$ is $\frac{1}{2}W$, which means that $S(t) = T(t/2)$. From (10) we get

$$|S(t)f(x) - f(0)(1 - x) - f(1)x| \leq (C_f + K_f)e^{-t}x(1 - x). \tag{11}$$

Another proof of (11) can be obtained using the approximation of $S(t)$ by iterates of classical Bernstein operators.

5. Explicit solutions

We shall use Lamperti's method [12, pp. 294–295] and the method of Doss-Sussmann [12, pp. 295–296], [26, pp. 382–383] in order to get information about the solution Y_t.

By using Lamperti's method in our setting, take $U = \sqrt{2}arcsin\sqrt{Y}$. Itô's formula yields

$$dU = dB + \frac{a - b + (a + b + 1)\cos(\sqrt{2}U)}{\sqrt{2}\sin(\sqrt{2}U)}dt,$$

with $U_0 = \sqrt{2}\arcsin\sqrt{x}$.

So the generator of the process $(U_t)_{t \geq 0}$ is

$$\frac{1}{2}\frac{d^2}{du^2} + \frac{\sqrt{2}}{2}\left(\frac{a - b}{\sin(\sqrt{2}u)} + (a + b + 1)\cot(\sqrt{2}u)\right)\frac{d}{du}.$$

In particular, for $b = a$ the generator of U becomes

$$\frac{1}{2}\frac{d^2}{du^2} + \frac{\sqrt{2}}{2}(2a + 1)\cot(\sqrt{2}u)\frac{d}{du},$$

which means that U is a *Legendre process* [26, p. 357].

So we have

Theorem 5.1. $Y_t = \sin^2\frac{U_t}{\sqrt{2}}$. *For $b = a$, U is a Legendre process.*

The Doss-Sussmann method yields

Theorem 5.2. $Y_t = \sin^2\left(\dfrac{B_t}{\sqrt{2}} + \arcsin\sqrt{X_t}\right),$ *where* X_t *is the solution of the equation*

$$\frac{dX}{dt} = \frac{\sqrt{X(1-X)}(a-b+(a+b+1)\cos(B\sqrt{2}+2\arcsin\sqrt{X}))}{\sin(B\sqrt{2}+2\arcsin\sqrt{X})},$$

with $X_0 = x$.

When $a = b = -\frac{1}{2}$, the following explicit solution can be obtained by using either Theorem 5.1 or Theorem 5.2:

Corollary 5.3. *For* $a = b = -\dfrac{1}{2}$ *we have*

$$Y_t = \sin^2\left(\frac{B_t}{\sqrt{2}} + \arcsin\sqrt{x}\right), \ 0 \le t < \zeta.$$

Similar solutions for corresponding problems are given in [15], Section 4.4, and [21], Chapter 5.

We conclude with two examples.

I. Let $a = b = -\frac{1}{2}$, i.e.,

$$Wu(x) = x(1-x)u''(x) + (\frac{1}{2} - x)u'(x),$$

with $D(W) = D_V(W)$ described in Section 2.

By using the expression of Y_t given in Corollary 5.3 it can be seen that $Wu = \lim_{n\to\infty} n(M_n u - u)$ for every $u \in C^2[0,1]$, where

$$M_n f(x) := E f(Y_{1/n}) - (W_{1/4n}(f \circ g))(g^{-1}(x)),$$

$g(x) = \sin^2(x)$ and $(W_t)_{t>0}$ are the classical Gauss-Weierstrass convolution operators (see [4], (5.2.78)).
Now let $f_n(x) = (\arcsin\sqrt{x})^n$, $0 \le x \le 1$, $n \ge 1$.

According to Corollary 5.3,

$$u_n(t,x) := E f_n(Y_t) = E\left(\left(\frac{B_t}{\sqrt{2}} + \arcsin\sqrt{x}\right)^n\right)$$

$$= \sum_{j=0}^{[n/2]} \binom{n}{2j} \frac{(2j)!}{j!} \left(\frac{t}{4}\right)^j (\arcsin\sqrt{x})^{n-2j}.$$

Consequently, the function $u_n(t,x)$ satisfies

$$\begin{cases} \dfrac{\partial}{\partial t} u_n(t,x) = W u_n(t,x) \\ u_n(0,x) = f_n(x) \end{cases}$$

for $t \ge 0$, $0 \le x \le 1$.

Since f_1 and $u_1(t,\cdot)$ are in $D(W)$, we get

$$T(t)f_1(x) = u_1(t,x) = f_1(x).$$

Thus $T(t)f_1 = f_1$; this is also a consequence of the fact that $Wf_1 = 0$.

II. Let now $a \geq 0$, $b \geq 0$. Then

$$W J_n^{(a,b)} = -n(n+a+b+1)J_n^{(a,b)} , \quad n \geq 0,$$

and

$$T(t)J_n^{(a,b)} = e^{-n(n+a+b+1)t}J_n^{(a,b)} , \quad n \geq 0, \ t \geq 0.$$

Since 0 and 1 are entrance boundaries, the diffusion (Y_t) extends to a continuous Feller process on $[0,1]$ (see [11], Theorem 23.13). We have also

$$E J_n^{(a,b)}(Y_t) = e^{-n(n+a+b+1)t}J_n^{(a,b)}(x).$$

Now it is possible to compute the moments of Y_t. For example,

$$EY_t = \frac{a+1}{a+b+2} + \frac{1}{a+b+2}e^{-(a+b+2)t}j_1(x)$$

and

$$EY_t^2 = \frac{(a+1)(a+2)}{(a+b+2)(a+b+3)}$$
$$+ \frac{2(a+2)}{(a+b+2)(a+b+4)}e^{-(a+b+2)t}j_1(x)$$
$$+ \frac{1}{(a+b+3)(a+b+4)}e^{-2(a+b+3)t}j_2(x),$$

where

$$j_1(x) = (a+b+2)x - (a+1)$$

and

$$j_2(x) = (a+b+3)(a+b+4)x^2 - 2(a+2)(a+b+3)x + (a+1)(a+2)$$

differ from $J_1^{(a,b)}(x)$, respectively $J_2^{(a,b)}(x)$, only by some constant factors.

Taking into account the asymptotic behavior of the semigroup $(T(t))$ we get also

$$\lim_{t\to\infty} E(Y_t^n) = \frac{(a+1)(a+2)\ldots(a+n)}{(a+b+2)(a+b+3)\ldots(a+b+n+1)}.$$

The same results can be achieved by using the fact that the probability density of Y_t is the function $p(t,x,\cdot)$ from Section 4.

Moreover,

$$EY_t^{-a} = e^{a(b+1)t}x^{-a},$$
$$E(1-Y_t)^{-b} = e^{(a+1)bt}(1-x)^{-b},$$
$$EY_t^{-a}(1-Y_t)^{-b} = e^{(a+b)t}x^{-a}(1-x)^{-b}.$$

Acknowledgement

Part of this work has been developed during a stay at both Bommerholz-Witten and Oberwolfach in the framework of the RiP programme of the Mathematisches Forschungsinstitut.

We would like to express our thanks to Professor D.Mache and to MFO for the stimulating atmosphere and the warm hospitality.

References

[1] Altomare, F., *Approximation theory methods for the study of diffusion equations*, in: Approximation Theory, Proc. IDOMAT'95, M.W. Müller, M. Felten, D.H. Mache(Eds), Mathematical Research, vol.86, pp. 9–26, Akademie Verlag, Berlin, 1995.

[2] Altomare, F. and Attalienti, A., *Degenerate evolution equations in weighted continuous function spaces, Markov processes and the Black-Scholes equation – Part I*, Result. Math. 42 (2002), 193–211.

[3] Altomare, F. and Attalienti, A., *Degenerate evolution equations in weighted continuous function spaces, Markov processes and the Black-Scholes equation – Part II*, Result. Math. 42(2002), 212–228.

[4] Altomare, F. and Campiti, M., *Korovkin-type Approximation Theory and its Applications*, W. de Gruyter, Berlin-New York, 1994.

[5] Altomare, F. and Rasa, I., *Feller semigroups, Bernstein type operators and generalized convexity associated with positive projections*, in: New Developments in Approximation Theory, (Eds.) M.D. Buhmann, M. Felten, D.H. Mache and M.W. Müller, Int. Series of Numerical Mathematics Vol. 132, Birkhäuser Verlag, Basel (1999), 9–32.

[6] Anastassiou, G.A., Cottin, C. and Gonska, H.H., *Global smoothness of approximating functions*, Analysis 11 (1991), 43–57.

[7] Attalienti, A. and Campiti, M., *Degenerate evolution problems and Beta-type operators*, Studia Math. 140 (2000), 117–139.

[8] Da Silva, M.R., *Nonnegative order iterates of Bernstein polynomials and their limiting semigroup*, Portugal Math. 42 (1984), 225–248.

[9] Feller, W., *Diffusion processes in genetics*, Proceedings of the Second Berkeley Symposium on Mathematical Statistics and Probability, 1951, 227–246.

[10] Hackenbroch, W. and Thalmaier, A., *Stochastische Analysis*, B.G. Teubner, Stuttgart (1994).

[11] Kallenberg, O., *Foundations of Modern Probability*, Springer-Verlag, New York (2002).

[12] Karatzas, I. and Shreve, S.E., *Brownian Motion and Stochastic Calculus*, Springer-Verlag, New York (2000).

[13] Karlin, S. and McGregor J., *On a genetics model of Moran*, Proceedings Cambridge Ph. Soc., Math. and Physical Sciences, 58 (1962), 299–311.

[14] Karlin, S. and Ziegler, Z., *Iteration of positive approximation operators*, J. Approx. Theory 3 (1970), 310–339.

[15] Kloeden, P.E. and Platen, E., *Numerical Solution of Stochastic Differential Equations,* Springer-Verlag, Berlin (1999).

[16] López-Moreno, A.-J. and Muñoz-Delgado, F.-J., *Asymptotic behavior of Kantorovich type modifications of the Meyer-König and Zeller operators, in:* Proceedings of the "Tiberiu Popoviciu" Itinerant Seminar, (Ed.) Elena Popoviciu, Editura SRIMA, Cluj-Napoca (2002), 119–129.

[17] Mache, D.H., *A link between Bernstein polynomials and Durrmeyer polynomials with Jacobi weights,* in: Approximation Theory VIII, Vol. 1, Approximation and Interpolation, C.K. Chui and L.L. Schumaker (Eds.), World Scientific Publ. Co., 1995, 403–410.

[18] Metafune, G., Pallara, D. and Wacker, M., *Feller semigroups on $\mathbb{R}^N$,* Semigroup Forum 65 (2002), 159–205.

[19] Metafune, G., Pallara, D. and Wacker, M., *Compactness properties of Feller semigroups* Studia Math. 153 (2002), 179–206.

[20] Nagel, R. (Ed.), *One-Parameter Semigroups of Positive Operators,* Springer-Verlag, Berlin (1986).

[21] Øksendal, B., *Stochastic Differential Equations,* Springer-Verlag, Berlin (2000).

[22] Păltănea, R., *On the transformation of the second order modulus by Bernstein operators,* Anal. Numér. Théor. Approx. 27 (1998), 309–313.

[23] Rasa, I., *Semigroups associated to Mache operators, in:* Advanced Problems in Constructive Approximation, (Eds.) M.D. Buhmann and D.H. Mache, Int. Series of Numerical Mathematics Vol. 142, Birkhäuser Verlag, Basel (2002), 143–152.

[24] Rasa, I., *Positive operators, Feller semigroups and diffusion equations associated with Altomare projections,* Conf. Sem. Mat. Univ. Bari 284 (2002), 1–26.

[25] Rasa, I., *One-dimensional diffusions and approximation,* IV International Meeting on Approximation Theory, Jaén University, Úbeda (Spain), June 2003 (to appear).

[26] Revuz, D. and Yor, M., *Continuous Martingales and Brownian Motion,* Springer-Verlag, Berlin (1999).

[27] Trotter, H.F., *Approximation of semigroups of operators,* Pacific J. Math. 8 (1958), 887–919.

Francesco Altomare
University of Bari
Department of Mathematics
Via E.Orabona, 4
I-70125 Bari (Italy)
e-mail: altomare@dm.uniba.it

Ioan Rasa
Technical University of Cluj-Napoca
Department of Mathematics
Str.C. Daicoviciu, 15
R-3400 Cluj-Napoca (Romania)
e-mail: Ioan.Rasa@math.utcluj.ro

Trends and Applications in Constructive Approximation
(Eds.) M.G. de Bruin, D.H. Mache & J. Szabados
International Series of Numerical Mathematics, Vol. 151, 27 51
© 2005 Birkhäuser Verlag Basel/Switzerland

Recent Developments in Barycentric Rational Interpolation

Jean-Paul Berrut, Richard Baltensperger and Hans D. Mittelmann

Abstract. In 1945, W. Taylor discovered the barycentric formula for evaluating the interpolating polynomial. In 1984, W. Werner has given first consequences of the fact that the formula usually is a rational interpolant. We review some advances since the latter paper in the use of the formula for rational interpolation.

Mathematics Subject Classification (2000). 65D05, 65D15, 41A20.

Keywords. interpolation, rational interpolation, optimal interpolation.

1. Introduction: polynomial interpolation in barycentric form

Interpolation is one of the central tools of (numerical) mathematics: analysis problems are being solved by first choosing from the infinite complexity of an arbitrary function a few of its (possibly unknown) values and replacing it with an interpolant of these values. The solution of the original problem is then approximated with the solution of the corresponding problem for the interpolant. Examples abound: differentiation, integration, multistep methods for ordinary differential equations, collocation methods for partial differential equations and other functional equations, etc. Interpolation is therefore a core subject of any course in numerical analysis and most books devote a chapter to it. They usually start with Lagrange interpolation, before enumerating its supposed practical drawbacks [Ber-Tre] and going over to presumably better methods.

This view of Lagrange interpolation contrasts with its ubiquity in practice and research: almost every volume of a numerical analysis journal contains some application of Lagrange cardinal functions. The present paper reviews some of the recent advances in the *practical* use of interpolation in the form of the so-called *barycentric formula*.

This work has been supported by the Swiss National Science Foundation, grant Nr. 20–66754.01.

We will focus on the simplest case, namely interpolation in one dimension between distinct points in an interval I. Many extensions, e.g., interpolation using information on derivatives at some or all interpolation points [Bul-Rut, Hen1] as well as interpolation in several dimensions [Bul-Rut, Bal-Ber2, She] can be handled in the barycentric setting as well.

Let thus $N+1$ distinct interpolation points (nodes) x_j, $j = 0(1)N$, be given, together with corresponding numbers f_j, $j = 0(1)N$, which can be values of a function f $\left(f_j := f(x_j)\right)$ or not. Our aim is to find an infinitely differentiable interpolant of f between the x_j. We will denote by $\mathcal{P}_m$ the linear space of all polynomials of degree at most m. The simplest problem is that of finding $p \equiv p_N \in \mathcal{P}_N$ for which

$$p(x_j) = f_j, \qquad j = 0(1)N$$

and its solution is given by the *Lagrange interpolation formula*

$$p(x) = \sum_{j=0}^{N} f_j \ell_j(x), \qquad \ell_j(x) := \frac{\prod\limits_{k \neq j}(x - x_k)}{\prod\limits_{k \neq j}(x_j - x_k)}. \tag{1.1}$$

As mentioned already, many books list drawbacks of this representation of the (unique) interpolating polynomial, in particular the fact that each evaluation of p at some x requires $\mathcal{O}(N^2)$ floating point operations (flops). However, if one has first computed the denominators of the Lagrange fundamental polynomials ℓ_j as

$$w_j := \frac{1}{\prod\limits_{k \neq j}(x_j - x_k)}, \qquad j = 0(1)N, \tag{1.2}$$

in $\mathcal{O}(N^2)$ flops, then every evaluation of p written as

$$p(x) = \ell(x) \sum_{j=0}^{N} \frac{w_j}{x - x_j} f_j, \tag{1.3}$$

where

$$\ell(x) := (x - x_0)(x - x_1) \cdots (x - x_N),$$

needs only $\mathcal{O}(N)$ flops [Hen1, Ber-Tre]. ℓ is often denoted by ω or Ω in the literature. w_j is called the *weight* corresponding to the point x_j. For particular sets of points, such as equidistant or Chebyshev ones, the weights can be computed analytically [Hen1]. Nodes like Chebyshev's, distributed like $1/\sqrt{1 - x^2}$, are especially important since they lead to exponential convergence for holomorphic functions (see [Ber-Tre] for citations). Exponential convergence is a special case of *spectral convergence*, i.e., faster than polynomial.

N. Higham [Hig] has just shown that (1.3) is the formula of choice if one wishes to evaluate p as stably as possible (even when p is a very bad approximant of f). 20 years ago already, W. Werner [WeW] has given an algorithm for updating the weights w_j in $\mathcal{O}(N)$ operations when a point is added. Werner's algorithm contains an unstable expression for the weight corresponding to the new point. In [Ber-Tre], the first author[1] and N. Trefethen have suggested to use (1.2) for this point also, that way maintaining the stability of the formula. Their algorithm used with formula (1.3) sweeps up the last arguments against scalar Lagrange interpolation: (1.3) is *the* formula to use if one wishes the result to be as close as possible to the value of the interpolating polynomial.

Higham has also proven the stability of the so-called *barycentric formula* [Hen1, Ber-Tre]

$$p(x) = \frac{\displaystyle\sum_{j=0}^{N} \frac{w_j}{x - x_j} f_j}{\displaystyle\sum_{j=0}^{N} \frac{w_j}{x - x_j}} \tag{1.4}$$

when the set of nodes is such that interpolating constant functions is as well-conditioned a problem as the original one of interpolating f. (1.4) is obtained from (1.3) by dividing by the corresponding interpolant for the function 1 and simplifying $\ell(x)$. Since interpolating with nodes leading to ill-conditioned interpolation of constant functions is seldom indicated (see however [Gau] for an example), the barycentric formula is usually the method of choice for evaluating p.

(1.4) indeed has several advantages over (1.3) [Ber-Tre], some of which we will encounter on our way. First of all, the weights now arise in the denominator as well as in the numerator, so that any common factor independent of j may be ignored, leading to so-called *simplified weights* w_j^* [Hen1]. Most important in practice (see [Mul-Hua-Slo, Tre, Bat-Tre] and below) are the so-called *Chebyshev points of the second kind*

$$x_j = \cos j \frac{\pi}{N}, \qquad j = 0(1)N, \tag{1.5}$$

for which one simply has [Sal]

$$w_j^* = w_j^{(2)} := (-1)^j \delta_j, \qquad \delta_j = \begin{cases} 1/2, & j = 0 \text{ or } j = N, \\ 1, & \text{otherwise,} \end{cases} \tag{1.6}$$

so that the barycentric formula reads

$$p(x) = \frac{\displaystyle\sum_{j=0}^{N}{}'' \frac{(-1)^j}{x - x_j} f_j}{\displaystyle\sum_{j=0}^{N}{}'' \frac{(-1)^j}{x - x_j}}, \tag{1.7}$$

---

[1]When authors are not specified we mean "of the present work".

where the double prime means that the first and last terms of the sum should be halved. These factors $1/2$ at the extremities of the interval come from the fact that Chebyshev interpolation is in fact even trigonometric interpolation between equidistant points, and that on the circle the interior points arise twice, the extremal points a single time [Ber1]. Polynomials interpolating between Chebyshev points may also be evaluated rapidly and stably by means of the fast Fourier transform, albeit less transparently and more expensively for a small number of evaluation points x. Since Chebyshev nodes are located in the interval $[-1, 1]$, the interpolation problem is often considered to have been moved there, something we will also assume in most of what follows.

Formula (1.7) is extremely stable: it has been used with N up to tens of thousands [Ber2, Bat-Tre], whereas MATLAB's `polyfit` does not even handle $N = 50$ with the same points.

Another sign of the importance of Chebyshev points is the fact that they are clustered at the extremities of the interval of interpolation in just the right way for the corresponding linear projection to have a small norm – see [Ber-Tre] for a short description of such norms, called Lebesgue constants, and books on interpolation [Dav, Sza-Vér, Phi] for a more thorough treatment. Chapter 6 of [Mas-Han] contains further results on interpolation between Chebyshev nodes.

In this context, we draw the reader's attention to the new MATLAB software for functions, called `@chebfun` and due to Battles and Trefethen [Bat-Tre], in which an object is not a vector of the standard MATLAB, but a vector of values of a function f at enough Chebyshev points of the second kind for f to be approximated with machine precision by the corresponding interpolating polynomial. The subroutines then perform the classical operations of calculus such as differentiation, integration, etc., by Chebyshev methods.

2. From polynomial to rational interpolation

What about the barycentric formula (1.4) with points for which interpolating constant functions is ill-conditioned? The formula will likely be unstable, the computed $\tilde{p}$ far away from p. Is that bad? Not necessarily! Indeed, the condition of p itself will likely be poor and p should not be used for solving problems. This is the case when the weights vary enormously, i.e., when the quotient $\frac{\max |w_j|}{\min |w_j|}$ is large [Ber-Tre]. In particular, the computed denominator (multiplied back by ℓ) will not be the function 1, but another polynomial in $\mathcal{P}_N$. The right-hand side of (1.4) becomes a rational function $r \in \mathcal{R}_{N,N}$, where $\mathcal{R}_{m,n}$ denotes the set of all rationals with numerator in $\mathcal{P}_m$ and denominator in $\mathcal{P}_n$. By the following lemma, r interpolates the value f_j at x_j whenever $w_j \neq 0$.

Lemma 2.1. *Let* $\{(x_j, f_j)\}$, $j = 0(1)N$, *be* $N+1$ *pairs of real numbers with* $x_j \neq x_k$, $j \neq k$, *and let* $\{u_j\}$ *be* $N + 1$ *real numbers. Then*

a) *if $u_k \neq 0$, the rational function*

$$r(x) = \frac{\displaystyle\sum_{j=0}^{N} \frac{u_j}{x - x_j} f_j}{\displaystyle\sum_{j=0}^{N} \frac{u_j}{x - x_j}} \in \mathcal{R}_{N,N} \qquad (2.1)$$

interpolates f_k at x_k: $\lim_{x \to x_k} r(x) = f_k$;

b) *conversely, every rational interpolant $r \in \mathcal{R}_{N,N}$ of the f_j may be written as in (2.1) for some u_j.*

Proof. Statement a) is just an immediate calculation. For b), let q denote the denominator of r and let $q_j := q(x_j)$, $j = 0(1)N$. Lagrange formula (1.3) then permits to write q as $q(x) = \ell(x) \sum_{j=0}^{N} \frac{w_j q_j}{x - x_j}$. Moreover, the q_j determine the numerator as the polynomial $p \in \mathcal{P}_N$ with $p(x_j) = q_j f(x_j)$. (2.1) with $u_j := w_j q_j$, all j, then is a representation of r. $\qquad \square$

a) expresses a second advantage of the barycentric formula: interpolation is warranted even when the w_j are computed with errors. This opens a wide field of research: whereas polynomial interpolation merely permits the choice of the nodes x_j, rational interpolation allows for that of the u_j also. The proof further shows that the rational interpolation problem is completely solved once the values of q at the nodes are known.

Why use rational interpolation? For two reasons, at least:

- when one cannot choose the nodes, polynomial interpolation may diverge even for well-behaved functions. An especially important case is when the f_j are the result of sampling at equidistant points. Then the sequence of p for increasing N will not converge if the f_j are the values of a function with singularities not too far from the interval of interpolation, as in Runge's example [Epp].

- even when one can choose nodes with a good distribution on the interval, such as those of Chebyshev and Legendre, polynomial interpolation may not be the answer for it may converge much too slowly for practical purposes. Indeed, Markov's inequality implies that a polynomial of moderate degree cannot have a large slope on an interval where it does not take large values. On the interval $[-1, 1]$ the inequality says that any polynomial q of degree at most n satisfies

$$\|q'\|_\infty \leq n^2 \|q\|_\infty,$$

where $\| \cdot \|_\infty$ denotes the maximum or L_∞-norm. It follows that, for a polynomial p of degree n to be a good approximation of f with p' a good approximation of f', one must have

$$n \geq \sqrt{\|f'\|_\infty / \|f\|_\infty}.$$

Approximating a function that behaves like erf(δx), where erf denotes the error function, therefore requires too large an N with very large δ (see Figure 3 in [Ber-Mit2]).

The classical answer to these difficulties is to use piecewise functions as approximants. Our interest here, however, is to stay with analytic functions. They are simpler to program and often converge faster, i.e., exponentially or spectrally [Ber-Tre]. Rational interpolation is the next such possibility, and the function $1/(1 + ax^2)$ for large a shows that no limitation such as Markov's inequality is in the way there.

3. Classical rational interpolation

The problem here is to find $r \in \mathcal{R}_{m,n}$ that interpolates the f_j, i.e., $p \in \mathcal{P}_m$ and $q \in \mathcal{P}_n$ such that

$$r(x_j) = \frac{p(x_j)}{q(x_j)} = f_j, \qquad j = 0(1)N. \tag{3.1}$$

In the canonical representation, p and q together have $m + n + 2$ coefficients, of which one may be set to 1 by dividing both polynomials by it. The $N + 1$ interpolation conditions (3.1) thus are equally numerous as the coefficients when

$$N = m + n. \tag{3.2}$$

This condition characterizes the classical rational interpolation problem, see [Sto] (or [Gut] for a more general treatment). Grosse's catalogue [Gro] contains a long list of papers on rational interpolation.

The problem need not have a solution (see the examples in [Sto, p. 50] or [Ber-Mit1, p. 367]), but it usually does and, if so, then the solution is unique [Sto, p. 51] and may be written in barycentric form (Lemma 2.1).

3.1. Classical rational interpolation in barycentric form

Schneider and Werner [Sch-Wer] have been the first to determine barycentric representations of rational interpolants. Their method uses a classical way of determining the Newton form of the interpolant before applying an algorithm of Werner to pass from the Newton to the barycentric form. In [Ber-Mit1], the first and last authors have given a method for directly finding the corresponding weights u_j when $n \leq m$.

Theorem 3.1. *If a solution r of the classical rational interpolation problem (3.1)–(3.2) with $n \leq m$ exists, then $\mathbf{u} = [u_0, u_1, \ldots, u_N]$ is a vector of weights in one of* *its barycentric representations* (2.1) *iff* **u** *belongs to the kernel of the* $N \times (N+1)$-*matrix*

$$\mathbf{A} := \begin{bmatrix} 1 & 1 & 1 & \cdots & 1 \\ x_0 & x_1 & x_2 & \cdots & x_N \\ x_0^2 & x_1^2 & x_2^2 & \cdots & x_N^2 \\ \vdots & \vdots & \vdots & & \vdots \\ x_0^{m-1} & x_1^{m-1} & x_2^{m-1} & \cdots & x_N^{m-1} \\ f_0 & f_1 & f_2 & \cdots & f_N \\ f_0 x_0 & f_1 x_1 & f_2 x_2 & \cdots & f_N x_N \\ f_0 x_0^2 & f_1 x_1^2 & f_2 x_2^2 & \cdots & f_N x_N^2 \\ \vdots & \vdots & \vdots & & \vdots \\ f_0 x_0^{n-1} & f_1 x_1^{n-1} & f_2 x_2^{n-1} & \cdots & f_N x_N^{n-1} \end{bmatrix} . \tag{3.3}$$

In order to save space, we introduce some notation: $\mathbf{V}_{P,Q}$, $P \leq Q$, will be the matrix made up of the first P rows of the transposed Vandermonde matrix corresponding to the $Q+1$ values $x_0, \ldots, x_Q$, and $\mathbf{F}_Q = \mathrm{diag}(f_0, \ldots, f_Q) \in \mathbb{R}^{Q+1,Q+1}$ will be the diagonal matrix of values $f_0, \ldots, f_Q$. Then $\mathbf{A}$ in (3.3) may be written $\mathbf{A} = [\mathbf{V}_{m,N}^T, \mathbf{F}_N^T \mathbf{V}_{n,N}^T]^T$.

The proof of Theorem 3.1 consists in showing that the first m equations express that the degree of the denominator written in the form $\ell(x) \sum_{j=0}^{N} \frac{u_j}{x - x_j}$ in (2.1) is at most n, the last n equations that the numerator degree is at most m. Lemma 2.1 a) implies that r in (2.1) interpolates in x_j if $u_j \neq 0$; the latter is not necessary, though.

As a corollary, Theorem 3.1 delivers the kernel of the matrix $\mathbf{V}_{N-1,N}$: it is just the space spanned by the vector of the polynomial barycentric weights w_j.

The customary way of coping with the case $n > m$ is to determine the reciprocal of r by interpolating the values $1/f_j$. This requires a special treatment in case some of the f_j vanish. Brimmeyer [Bri] has discovered how to stay within the barycentric context by considering the corresponding x_j as poles and using the method given in [Ber5] and described below to preassign such poles in the determination of the weights of $1/r$.

In our experience, the algorithm given in [Ber-Mit1] for computing the kernel of $\mathbf{A}$ in (3.3) is much more efficient than computing the singular value decomposition of $\mathbf{A}$ with MATLAB's routine svd. It consists in triangulating (3.3) in two steps: one analytical, which leads to divided differences, the other numerical, through Gaussian elimination with column pivoting. Though, in contrast with p, r is a good approximation even with equidistant interpolation points for N large enough, Chebyshev nodes again lead to much better conditioned problems. Note, however, that these nodes must be reordered for a stable computation of the divided differences. The algorithm is then extremely stable, see the examples in [Ber-Mit1].

As with the interpolating polynomial, the degrees of p and q may be smaller than m and n. This manifests itself in the kernel of $\mathbf{A}$ having dimension larger than 1. A way of coping with this, suggested in [Ber-Mit1], is to decrease n by 1, restart the computation, and repeat until a kernel of dimension 1 is obtained for some $n^* < n$. $\mathbf{u}$ then yields the barycentric representation (2.1) of a *reduced* r, i.e., one in which the linear factors corresponding to common zeros of p and q have been simplified. We will denote it by r^*. If $u_k = 0$ for some k and $r^*(x_k) \neq f_k$, then x_k is an *unattainable point* [Sto] and the problem does not have a solution. This is the first part of the following theorem, due to Schneider and Werner [Sch-Wer, Ber-Mit1], which allows an easy detection of two drawbacks of classical rational interpolation directly from the weights.

Theorem 3.2. *Let u be barycentric weights of a reduced rational interpolant $r^* = p^*/q^*$. Then*

a) *a point x_k is unattainable iff $u_k = 0$;*

b) *if $u_k \neq 0$ for all k, if the interpolation points have been ordered as $x_0 < x_1 < \cdots < x_N$ and if $\mathrm{sign}\, u_j = \mathrm{sign}\, u_{j+1}$, then r^* has an odd number of poles between x_j and x_{j+1}.*

Proof. a) follows from the above discussion and Lemma 2.1a, b) from noticing that q^* changes sign, thus has a zero, between x_j and x_{j+1}, and p^* does not, for otherwise r^* would not be reduced. □

$u_k = 0$ in (2.1) simply means that the node x_k is ignored. One may then eliminate the pair (x_k, f_k) and solve the problem in $\mathcal{R}_{m,n}$, $m+n = N-1$, $n = n^*-1$ [Ber5]. (Such elimination of data might be problematic if several u_j vanish.)

Zhu and Zhu [Zhu-Zhu] have recently suggested an equivalent, and maybe even more elegant way of finding r by directly determining the values $\mathbf{q} := [q_0, q_1, \ldots, q_N]^T$ of its denominator. Let $\mathbf{W} := \mathrm{diag}(w_0, w_1, \ldots, w_N)$ be the diagonal matrix of the polynomial weights to $x_0, x_1, \ldots, x_N$. Then the proof of Lemma 2.1 shows that one has the bijective map $\mathbf{u} = \mathbf{W}\mathbf{q}$; the kernel equation $\mathbf{A}\mathbf{u} = \mathbf{0}$ may thus be written $\mathbf{A}\mathbf{W}\mathbf{q} = \mathbf{0}$.

Corollary 3.3. *If a solution r of the classical rational interpolation problem (3.1)– (3.2) with $n \leq m$ exists, then the vector $\mathbf{q}$ of its denominator values belongs to the kernel of the $N \times (N+1)$-matrix $\mathbf{A}\mathbf{W}$.*

The elements of $\mathbf{A}\mathbf{W}$ are given in [Zhu-Zhu].

3.2. Reduced complexity barycentric rational interpolation

Besides its many advantages, the barycentric representation of $r \in \mathcal{R}_{m,n}$ also has some disadvantages in comparison with the canonical $r(x) = (a_m x^m + \cdots + a_1 x + a_0)/(b_n x^n + \cdots + b_1 x + b_0)$. One of them is the fact that its evaluation at a particular point x requires about twice as many flops.

In [Ber6], the first author has suggested a method for improving upon this. Since the numerator and the denominator have degrees at most $L := \max\{m, n\}$, they may both be written as interpolating polynomials between any number of points larger than L. Let therefore M with $L \leq M \leq N$ be given. By Lemma 2.1b, the solution of problem (3.1) may be written as

$$r(x) = \sum_{j=0}^{M} \frac{u_j}{z - z_j} f_j \Big/ \sum_{j=0}^{M} \frac{u_j}{z - z_j}. \qquad (3.4)$$

(3.4) has been called a *reduced complexity barycentric representation of r* in [Ber6]. With it, only $M + 1$ unknown u_j, $j = 0(1)M$, are to be determined, but interpolation is guaranteed by Lemma 2.1b only at the corresponding x_j, called *primary nodes* in [Ber6]. The degree conditions that make up the matrix $\mathbf{A}$ in Theorem 3.1 are also less numerous, since the denominator, respectively numerator degree must be decreased merely by $M - n$, resp. $M - m$ units. In total, $N - M$ less weights are to be determined, but $2(N - M)$ less degree diminishing conditions hold. The remaining $N - M$ equations are the interpolation conditions at the *secondary nodes* $x_{M+1}, x_{M+2}, \ldots, x_N$. Interpolation is warranted at any of these x_k by [Ber6]

$$\sum_{j=0}^{M} f[x_j, x_k] u_j = 0, \qquad (3.5)$$

where as usual $f[x_{i_0}, \ldots, x_{i_s}]$ denotes the divided difference of order s for the points $x_{i_0}, \ldots, x_{i_s}$.

The matrix $\mathbf{B}$ with elements $b_{ij} := f[x_i, x_j]$, $i = 0(1)M$, $j = M + 1(1)N$, of the divided differences in (3.5) is the *Löwner matrix* corresponding to the sets of points $\{x_0, \ldots, x_M\}$ and $\{x_{M+1}, \ldots, x_N\}$.

Theorem 3.4. *If a solution r of the classical rational interpolation problem (3.1)– (3.2) with $n \leq m$ exists, and if $x_0, \ldots, x_M$ are the primary nodes, then any vector $\mathbf{u} = [u_0, \ldots, u_M]$ of its weights in a reduced complexity barycentric representation (3.4) belongs to the kernel of the $M \times (M + 1)$-matrix*

$$\mathbf{A} = \begin{bmatrix} \mathbf{V}_{M-n,M} \\ \mathbf{V}_{M-m,M} \mathbf{F}_M \\ \mathbf{B} \end{bmatrix},$$

where $\mathbf{B}$ is the Löwner matrix given above.

The partitioning of the nodes in primary and secondary ones is important. Numerical experiments demonstrate in particular that the extreme points should be primary nodes, for otherwise the interpolation at the secondary ones, enforced by the equations $\mathbf{Bu} = \mathbf{0}$, may not be accurate.

The decisive advantages of the reduced complexity representation are the smaller number of points in the (often ill-conditioned) Vandermonde matrices and the fact that the divided differences arising in the computation of the kernel of $\mathbf{A}$ have only $M - n + 2$ arguments, as opposed to $N - n + 1$ when working with the representation (2.1) and the matrix (3.3). The improvement in equidistant interpolation may be spectacular, see Table 3 in [Ber6].

Steffen [Ste] has adapted the above reduced complexity rational interpolation to the method of Zhu and Zhu, thereby computing only the $M + 1$ values of the denominator at the primary nodes. Her numerical results are very close to those of [Ber6].

3.3. Monitoring the poles

The poles are at the same time the curse and the blessing of classical rational interpolation. Their bright side, mentioned above, is their capability of accomodating large gradients of f. Their dark side is the fact that, for N small, they may show up about everywhere in the plane, thus also in the immediate vicinity of, or even on, the interval of interpolation when interpolating a perfectly innocuous function (see Cordellier's example in [WeH] or [Ber-Mit2]; one may also read p. 357 of [Ber-Mit1]). When more is known about f than just the interpolated points in the plane, one should therefore try to incorporate the extra information into the interpolant to monitor the poles.

Sometimes the location of some or all of the poles is known a priori (see §8 below for an example). The above method for determining the barycentric weights may be modified to determine the rational interpolant in this case [Ber5]. Assume that P poles of the denominator are prescribed; denote them by z_k, $k = 1, \ldots, P$ and their multiplicity by ν_k with $\nu := \sum_{k=1}^{P} \nu_k$; assume further that $z_k \neq x_j$ for all j and all k. The problem is now to compute rational interpolants with prescribed poles, i.e., to find

$$r = p/q \in \mathcal{R}_{m,n+\nu}, \qquad m + n = N, \qquad n + \nu \leq N,$$

such that (3.1) is satisfied and r has the ν preassigned poles. If the interpolant exists, its denominator will contain as a factor the polynomial

$$d(z) := a \prod_{k=1}^{P} (z - z_k), \quad a \neq 0 \in \mathbb{C} \text{ arbitrary.} \tag{3.6}$$

Let

$$d_j := d(x_j), \qquad j = 0(1)N, \tag{3.7}$$

be the values of d at the nodes. Then the part of r remaining to be determined,

$$r^* = p^*/q^* := r \cdot d \in \mathcal{R}_{m,n}, \qquad m \geq n,$$

must take the values

$$r^*(x_j) = g_j, \qquad g_j := f_j \cdot d_j, \qquad j = 0(1)N, \tag{3.8}$$

at the nodes. Accounting for the difficulties of unattainable points and/or multiple solutions mentioned above, we are left with the following problem:

(R) *Find the largest possible $n^* \leq n$ and the corresponding unique $r^* \in \mathcal{R}_{m^*,n^*}$ with $m^* + n^* = N$, $n^* \leq m^*$, that satisfies the interpolation conditions (3.8).*

Theorem 3.1 implies as a corollary that, if such a r^* exists, its barycentric weights $\mathbf{b} = [b_0, b_1, \ldots, b_N]^T$ make up the one-dimensional kernel of the matrix

$$\mathbf{A} = \begin{bmatrix} \mathbf{V}_{m^*,N} \\ \mathbf{V}_{n,N}\mathbf{G}_N \end{bmatrix},$$

where $\mathbf{G}_N = \mathrm{diag}(g_0, g_1, \ldots, g_N)$. Once this kernel has been determined, e.g., by the algorithm of [Ber-Mit1], barycentric weights of r in (3.1) are given by [Ber5]

$$u_j = d_j \cdot b_j,$$

a result extended in the following theorem proven in [Ber5].

Theorem 3.5. *If some $r \in \mathcal{R}_{m,n+\nu}$, $n \le m \le N$, $n+\nu \le N$, exist with $r(x_j) = f_j$, $j = 0(1)N$, and with poles of order ν_k at the points z_k, $k = 1(1)P$, then one of them is given by (3.1) with $u_j = d_j b_j$. On the other hand, if*

$$\sum_{j=0}^{N} \prod_{k \ne j} (x_j - z_k) f_j \ne 0, \tag{3.9}$$

r as in (2.1) with $u_j = d_j b_j$ has a pole at z_k.

The proof implies that, if the numerator has degree $\le N - \nu$, the conditions (3.9) are not satisfied, meaning that r cannot be guaranteed to display the prescribed poles. This may reflect the fact that there is no $r \in \mathcal{R}_{N,N}$ interpolating the values f_j and possessing the desired poles. For instance, for $c \in \mathbb{R}$ constant there is no $r \not\equiv c$ interpolating the values $f_j \equiv c \forall j$, since every r as in (3.1) with constant f_j is this same constant for all choices of u_j. This is reasonable from an approximation point of view. Whether (3.9) is necessary for the presence of a pole at z_k is an open question.

When all the poles are prescribed, one has $n = 0$, r^* is the interpolating polynomial and one gets the following corollary, which will be important in the subsequent sections.

Corollary 3.6. *If $r \in \mathcal{R}_{N,\nu}$, $\nu \le N$, exists with $r(x_j) = f_j$, $j = 0(1)N$, and with poles of order ν_k at z_k, $k = 1(1)P$, $\nu := \sum_{k=1}^{P} \nu_k$, then its barycentric weights are given up to a constant by $u_j = w_j b_j$, $j = 0(1)N$.*

r is then the quotient of the polynomial of degree at most N interpolating $f \cdot d$ between the x_j and the polynomial d [Ber-Mit2]. It cannot have any other pole than the z_k, which eliminates the potentially harmful free poles of classical rational interpolation. The rational interpolants arising in the optimal approximation of functionals in the Hardy space H^2 make an interesting example [Ber3].

4. Optimal attachment of poles to the interpolating polynomial

Corollary 3.6 shows that, when one can give all of the poles of the interpolant, rational interpolation is as simple as polynomial interpolation. Though in some cases the poles can be obtained a priori (see below), this is usually not the case.

In [Ber-Mit2], the first and last authors have suggested a way of determining an optimal position of the poles *when the function f taking the values $f(x_i) = f_i$ is known everywhere on the interval I* (as in the application of §9).

Consider the (potential) poles z_k, $k = 1(1)P$, in a rational interpolant

$$r(x) = \frac{\displaystyle\sum_{j=0}^{N} \frac{w_j d_j}{x - x_j} f_j}{\displaystyle\sum_{j=0}^{N} \frac{w_j d_j}{x - x_j}} = \frac{\displaystyle\sum_{j=0}^{N} \frac{w_j \prod_{k=1}^{P}\left(1 - \frac{x_j}{z_k}\right)}{x - x_j} f_j}{\displaystyle\sum_{j=0}^{N} \frac{w_j \prod_{k=1}^{P}\left(1 - \frac{x_j}{z_k}\right)}{x - x_j}} \tag{4.1}$$

with denominator d as in (3.6) as variables (the last equality of (4.1) only holds when $z_k \neq 0 \forall k$ and is obtained by dividing by $\prod_{k=1}^{P}(-z_k)$ – to simplify the notation, we consider from here on a pole of multiplicity ν_k as ν_k separate poles). The goal is to choose the z_k in such a way that the interpolant r in (4.1) is as good an approximation of f as possible. It has been suggested in [Ber-Mit2] to consider functions continuous on I and to minimize the infinity norm of $r - f$, i.e., to solve the following min-max, or nonlinear Chebyshev, approximation problem:

(A) *Minimize* $\|r - f\|_\infty := \max_{x \in I} |r(x) - f(x)|$, *with r as in (4.1), with respect to the z_k, $k = 1(1)P$.*

Theorem 4.1 (Ber-Mit2). *Problem* (A) *always has a solution.*

As mentioned in [Ber-Mit2], the solution is usually not unique (constant functions are counterexamples). The more interesting question of the unicity of the optimal interpolant r is open.

These interpolants have very nice properties. There can be no unattainable point nor unwanted pole. r is always at least as good as the interpolating polynomial, for the latter is the case when all z_k are at infinity (see the right-hand side of (4.1)). Moreover, attachment of another pole can never result in a worsening of the approximation, since the already optimized poles constitute a feasible point for the optimization.

The optimization problem has been numerically solved with success in [Ber-Mit2] with standard modern optimization algorithms. The nice properties just mentioned occur in practice, and the numerical results, e.g., with Cordellier's example, for which classical rational interpolation with small N is useless, are quite impressive.

We mention that the authors of [Ber-Mit2] originally intended to solve the more ambitious problem of minimizing $\|r - f\|_\infty$ with respect to all u_j in a representation (2.1) of r, but that they encountered difficulties, both theoretical ones (existence of an optimum, of an alternating sequence, etc.) and practical ones (too many parameters to optimize).

5. Differentiation of rational interpolants

Already in their 1986 paper [Sch-Wer] Schneider and Werner have given very elegant general formulae for the derivatives of rational interpolants in the form (2.1). These formulae, one for differentiation between the nodes and one at the nodes, cover arbitrary orders of differentiation. Outside the nodes they are just barycentric formulae with the values f_i at the nodes replaced by divided differences of the corresponding order. For simplicity, we give them here only for the first and second order derivatives that will be needed in an application below:

$$r'(x) = \begin{cases} \displaystyle\sum_{j=0}^{N} \frac{u_j}{x - x_j} r[x, x_j] \Big/ \sum_{j=0}^{N} \frac{u_j}{x - x_j}, & x \neq x_i, \ i = 0(1)N, \\ \displaystyle -\Big(\sum_{\substack{j=0 \\ j \neq i}}^{N} u_j r[x_i, x_j] \Big) \Big/ u_i, & x = x_i \end{cases}$$

(5.1a)

and

$$r''(x) = \begin{cases} \displaystyle 2\sum_{j=0}^{N} \frac{u_j}{x - x_j} r[x, x, x_j] \Big/ \sum_{j=0}^{N} \frac{u_j}{x - x_j}, & x \neq x_i, \ i = 0(1)N, \\ \displaystyle -2\Big(\sum_{\substack{j=0 \\ j \neq i}}^{N} u_j r[x_i, x_i, x_j] \Big) \Big/ u_i, & x = x_i, \end{cases}$$

(5.1b)

with $r[z, z, x_j] = \frac{r'(z) - r[z, x_j]}{z - x_j}$.

6. From nonlinear to linear rational interpolation

With all approximants given in §§3 and 4 except that for which all the poles are prescribed (Corollary 3.6), the barycentric weights u_j, and therefore the denominator, depend on the interpolated function: the approximation operator is nonlinear. In practice, however, many problems are addressed by means of linear approximants: also nonlinear ones are often solved with a sequence of linear approximations.

To every fixed set of nodes x_j and every set of corresponding fixed weights u_j there corresponds a linear interpolant: for example, with $u_j = w_j \forall j$ this is just polynomial interpolation. For Chebyshev points such as (1.5), the polynomial weights (1.2) lead to a very well-conditioned polynomial interpolant, whereas this notoriously is not the case with equidistant nodes. *It seems therefore natural to try finding good weights corresponding to a given set of nodes* $\{x_j\}$.

For a given vector of nodes, the vector of weights $\mathbf{b} = [b_0, b_1, \dots, b_N]^T$ defines the linear vector space $\mathcal{R}_N^{(b)}$ of all rational interpolants (1.4) with these weights

[Bal-Ber2]. The set of rational functions

$$L_j^{(b)}(x) := \frac{b_j}{x - x_j} \bigg/ \sum_{k=0}^{N} \frac{b_k}{x - x_k}, \quad j = 0, 1, \ldots, N, \tag{6.1}$$

constitutes a basis of $\mathcal{R}_N^{(b)}$.

The first author has suggested in [Ber2] to use the weights (1.6), i.e., the interpolant (1.7), for every set of nodes. This linear rational interpolant never has poles in the interval of interpolation and is extremely well conditioned in practice, even with random points. Unfortunately, its convergence is not fast enough for approximating higher order derivatives for many sets of nodes, in particular for the very important equidistant grid (the first author will give a description of this convergence in a forthcoming paper). There are, however, some sets of nodes for which it converges about as fast as the best polynomial interpolation: conformal maps of Chebyshev nodes!

7. Conformal point shifts

Polynomial interpolation between Chebyshev points is trigonometric interpolation of even functions between equidistant points [Ber1, Tre]. It has very nice properties: fast convergence for very smooth functions, small operator norm (Lebesgue constant), very stable barycentric formula. It also has some drawbacks, in particular the $\mathcal{O}(N^{-2})$-concentration of the nodes at the extremities of the interval of interpolation, which results in (at least) three difficulties:

  a) ill-conditioning of the derivative near the extremities;

  b) bad distribution of the information over the interval;

  c) mediocre approximation of functions with shocks close to the center, where points are scarcer.

Several experts in the solution of differential equations by means of Chebyshev interpolants [Bay-Tur, Boy, Kos-Tal] have suggested to address these difficulties by a conformal shift of the nodes toward the equidistant position. This does not change the order of convergence of the interpolant.

To be more precise, consider, beside the x-space in which f is to be approximated, another space, with variable y, and the $N + 1$ Chebyshev points of the second kind $y_j = \cos j\pi/N$, $j = 0(1)N$, on the interval $J := [-1, 1]$ in this y-space. Let further g be a conformal map from a domain $\mathcal{D}_1$ containing J (in the y-space) to a domain $\mathcal{D}_2$ containing I (in the x-space). This defines new interpolation points on I, $x_j := g(y_j)$, and the conformal transplantation [Hen2] $F(y) := f(x)$ of any function in the x-space back into y-space. (The transplantation of an x-space-function will be denoted by the corresponding upper case letter.) Then, one may consider at least two approximants of a function f:

– the transplantation a_N of the polynomial A_N interpolating F between the y_j:

$$A_N(y) := \sum_{j=0}^{N} F(y_j)L_j(y) = \sum_{j=0}^{N} f(x_j)L_j\left(g^{-1}(x)\right) =: a_N(x); \qquad (7.1)$$

– the rational interpolant mentioned at the end of § 6:

$$r_N(x) = \sum_{j=0}^{N} \frac{w_j^{(2)}}{x - x_j} f_j \Bigg/ \sum_{j=0}^{N} \frac{w_j^{(2)}}{x - x_j}, \qquad w_j^{(2)} \text{ from } (1.6). \qquad (7.2)$$

The most favorable case is that in which the function f to be interpolated is analytic in a domain containing I. It follows from a classical result that, if $f : \mathcal{D}_2 \mapsto \mathbb{C}$ is such that the composition $f \circ g : \mathcal{D}_1 \mapsto \mathbb{C}$ is analytic inside and on an ellipse E_ρ with foci at ± 1 and sum of its axes equal to 2ρ, $\rho > 1$, then [For, p. 28]

$$|a_N(x) - f(x)| = |A_N(y) - F(y)| = \mathcal{O}(\rho^{-N}) \text{ for every } x \in [-1, 1].$$

The corresponding result for r_N in (7.2) has been proven in [Bal-Ber-Noë]:

Theorem 7.1. *Let $\mathcal{D}_1$, $\mathcal{D}_2$ be two domains of $\mathbb{C}$ containing $J = [-1, 1]$, respectively I ($\in \mathbb{R}$), let g be a conformal map $\mathcal{D}_1 \to \mathcal{D}_2$ such that $g(J) = I$, and f be a function $\mathcal{D}_2 \to \mathbb{C}$ such that the composition $f \circ g : \mathcal{D}_1 \to \mathbb{C}$ is analytic inside and on an ellipse C_ρ ($\subset \mathcal{D}_1$), $\rho > 1$, with foci at ± 1 and with the sum of its major and minor axes equal to 2ρ. Let r_N be the rational function (7.2) interpolating f between the transformed Chebyshev points $x_k := g(y_k)$. Then, for every $x \in [-1, 1]$,*

$$|r_N(x) - f(x)| = \mathcal{O}(\rho^{-N}).$$

Conformal point shifts thus preserve exponential convergence. They may markedly lessen (though not eliminate) the difficulties a)–c) enumerated at the section's onset. The ill-conditioning of the derivatives near the extremities with Chebyshev points is due to the accumulation of points there, which for large N are so close that a small change in a f_j has a strong impact on the derivative around x_j. (This may be quantitatively studied with the pseudospectrum, see [Tre, p. 108].) Following Kosloff and Tal-Ezer, one may improve upon this by moving the points closer to equidistant. With the approximation (7.1), the derivatives are given by

$$a_N'(x) = A_N'(y) \cdot \left[g^{-1}(x)\right]' = \frac{A_N'(y)}{g'(y)} \qquad (7.3a)$$

and

$$a_N''(x) = \frac{1}{[g'(y)]^2} A_N''(y) - \frac{g''(y)}{[g'(y)]^3} A_N'(y), \qquad (7.3b)$$

in which $A_N'(y)$ and $A_N''(y)$ may be computed by (5.1). With the approximation (7.2), the derivatives are simply given by the formulae (5.1) with $u_j = w_j^{(2)} \forall j$.

Kosloff and Tal-Ezer have suggested the map

$$g(y) = \frac{\arcsin(\alpha y)}{\arcsin \alpha}, \qquad 0 < \alpha < 1. \tag{7.4}$$

In the limiting cases, $\alpha \to 0$ keeps the points at their Chebyshev position, whereas $\alpha \to 1$ renders them equidistant. The derivatives of g to be used in (7.3) are given by

$$g'(y) = \frac{\alpha}{\arcsin \alpha} \frac{1}{\sqrt{1 - (\alpha y)^2}}, \qquad g''(y) = \frac{\alpha^3}{\arcsin \alpha} \frac{y}{\sqrt{\left(1 - (\alpha y)^2\right)^3}},$$

so that in (7.3b)

$$\frac{g''(y)}{[g'(y)]^3} = (\arcsin^2 \alpha) y.$$

The effect of this map upon the derivatives has been extensively studied, see, e.g., [Red-Wei-Nor, Mea-Ren, Abr-Gar]. It is indeed significant. And the map is even more successful in alleviating the drawbacks b) and c) mentioned above: through the refurbishment of the center with nodes it approximates functions with steep gradients or oscillations there much better than the simple polynomial interpolating between Chebyshev points, see the examples in [Ber-Mit4].

One may combine such changes of variable with the optimal attachment of poles of §4. In [Ber-Mit4], the first and last authors have done this by attaching poles v_k to A_N in y-space, which corresponds to attaching poles $z_k := g(v_k)$ to a_N in x-space. For a function with a steep gradient in the center of the interval, such as Hemker's example

$$f(x) = \cos \pi x + \frac{\operatorname{erf}(\delta x)}{\operatorname{erf}(\delta)}, \qquad \delta = \sqrt{250}, \qquad -1 \le x \le 1,$$

the improvement obtained by means of the poles is still more pronounced than that coming from the change of variable. However, this is no longer the case with a function oscillating in the center: the optimization problem with a number of poles large enough to cope with the oscillations seems beyond the capability of existing optimization software.

The second author has just noticed that the pole attachment may usually also be performed with r_N from (7.2) by multiplying every weight $w_j^{(2)}$ by $\prod_k (x_j - z_k)$. The resulting rational interpolant is the quotient of two rational functions with the same denominator as r_N, one interpolating $f \cdot d$, the other interpolating d from (3.6). Contrary to the attachment via (7.1), further poles than the z_k may arise in $\mathbb{C}$, but they move infinitely far as $N \to \infty$. One advantage is the more elegant formula (5.1) for the derivatives which avoids the chain rule. Exponential convergence is maintained, as will be shown in further work. The numerical examples in [Bal-Ber2] suggest that this interpolant should be just a little less good (no more than one digit) than a_N with the same attached poles.

8. An application: the linear rational pseudospectral method for boundary value problems

In order to demonstrate the use of linear rational interpolation in practical problems we shall now solve the following simple boundary value problem: find u on the interval $[-1, 1]$ that satisfies the differential equation

$$u''(x) + p(x)u'(x) + q(x)u(x) = h(x) \qquad (8.1a)$$

at every $x \in (-1, 1)$ and takes the values

$$u(-1) = u_\ell, \qquad u(1) = u_r \qquad (8.1b)$$

at the boundary points. We assume that the functions p, q, and h are such that the problem is well posed.

Suppose first that one knows a good location for poles z_k of an approximation of u, e.g., because the equation is fuchsian or from an application of the WKB-method [Wei, Bal-Ber-Dub]. To solve (8.1), one may then try substituting for u a linear rational interpolant $\widetilde{u}$ written in the basis (6.1),

$$\widetilde{u}(x) = \sum_{j=0}^{N} u_j L_j^{(b)}(x), \qquad (8.2)$$

to obtain

$$\sum_j u_j L_j^{(b)''}(x) + p(x) \sum_j u_j L_j^{(b)'}(x) + q(x) \sum_j u_j L_j^{(b)}(x) = h(x). \qquad (8.3)$$

Since the boundary values should enter the solution of the problem, one takes here a set of nodes containing -1 and 1, e.g., Chebyshev points of the second kind. Then (8.3) yields an equation for the unknown values $u_1, \ldots, u_{N-1}$ of $\widetilde{u}$ at $x_1, \ldots, x_{N-1}$ (u_0 and u_N being known from the boundary conditions). If the exact solution does not miraculously happen to belong to $\mathcal{R}_N^{(b)}$, (8.3) does not have a solution (8.2) and one may collocate, i.e., merely require that the two sides of (8.3) agree in as many values x_i of x as there are unknowns, here $N - 1$:

$$\sum_j u_j L_j^{(b)''}(x_i) + p(x_i) \sum_j u_j L_j^{(b)'}(x_i) + q(x_i) \sum_j u_j L_j^{(b)}(x_i) = h(x_i), \ i = 1(1)N - 1.$$
$$(8.4)$$

This is a system of linear equations for the unknown values $u_1, \ldots, u_{N-1}$, which may be written as

$$\mathbf{A}\mathbf{u} = \mathbf{h} \qquad (8.5)$$

with $\mathbf{A} := \mathbf{D}^{(2)} + \mathbf{P}\mathbf{D}^{(1)} + \mathbf{Q}$ and

$$\mathbf{u} := [u_1, u_2, \ldots, u_{N-1}]^T,$$

$$\mathbf{D}^{(1)} = \left(D_{ij}^{(1)}\right), \quad D_{ij}^{(1)} := L_j^{(b)'}(x_i),$$

$$\mathbf{D}^{(2)} = \left(D_{ij}^{(2)}\right), \quad D_{ij}^{(2)} := L_j^{(b)''}(x_i),$$

$$\mathbf{P} := \text{diag}\left(p(x_i)\right), \quad \mathbf{Q} := \text{diag}\left(q(x_i)\right),$$

$$\mathbf{h} := [h(x_i) - u_r\left(L_0^{(b)''}(x_i) + p(x_i)L_0^{(b)'}(x_i)\right) - u_\ell\left(L_N^{(b)''}(x_i) + p(x_i)L_N^{(b)'}(x_i)\right)]^T,$$

$$i, j = 1, \ldots, N - 1.$$

An advantage of the barycentric representation of $\tilde{u}$ is the simplicity of the formulae for the elements of $\mathbf{D}^{(1)}$ and $\mathbf{D}^{(2)}$, which may be given as

$$D_{ij}^{(1)} = \begin{cases} \dfrac{b_j/b_i}{x_i - x_j}, & i \neq j, \\ -\sum_{k \neq i} D_{ik}^{(1)}, & i = j, \end{cases}$$

$$D_{ij}^{(2)} = \begin{cases} 2D_{ij}^{(1)}\left(D_{ii}^{(1)} - \dfrac{1}{x_i - x_j}\right), & i \neq j, \\ -\sum_{k \neq i} D_{ik}^{(2)}, & i = j. \end{cases}$$

(8.6)

The weights b_j enter explicitly only through the remarkably simple formula for the non-diagonal elements of $\mathbf{D}^{(1)}$! The use of these formulae was first advocated in [Bal-Ber1] for the polynomial case, in [Bal-Ber2] and [Ber-Bal] for the rational case. They may be obtained from (5.1); a simple direct proof is given in [Ber-Tre].

The matrix $\mathbf{A}$ in (8.5) is full, as opposed to those of the finite difference and the finite element methods. The reason for the efficiency of the method already in the polynomial case lies in the spectral convergence of $\tilde{u}$ toward u – for good nodes – when all functions arising in the problem are analytic within ellipses containing the interval $[-1, 1]$ in their interior, see the examples in [Tre]. This remains true when one takes $b_j = w_j d_j$ as in (4.1) to solve a problem whose solution has the same poles z_k:

Theorem 8.1 (Bal-Ber-Dub). *Let the solution u of (8.1) be meromorphic with poles at $z_1, \ldots, z_P$. Then the linear rational collocation method with trial function (8.2) for Chebyshev points of the second kind converges exponentially toward u, and at least as fast as the corresponding polynomial collocation solution of an associated boundary value problem for $u \cdot d$.*

The associated boundary value problem is explictly known [Pér-Cas-Hay, Ber-Bal-Dub]. Of particular importance is the fact that prescribing the poles results in a better solution for small N. Indeed, when using spectral elements in several dimensions, one cannot increase N at will in each element since the system of equations would become too large. The figure displays the solution of Example 2 of [Bal-Ber-Dub] with $N = 17$ together with the polynomial solution (left) and

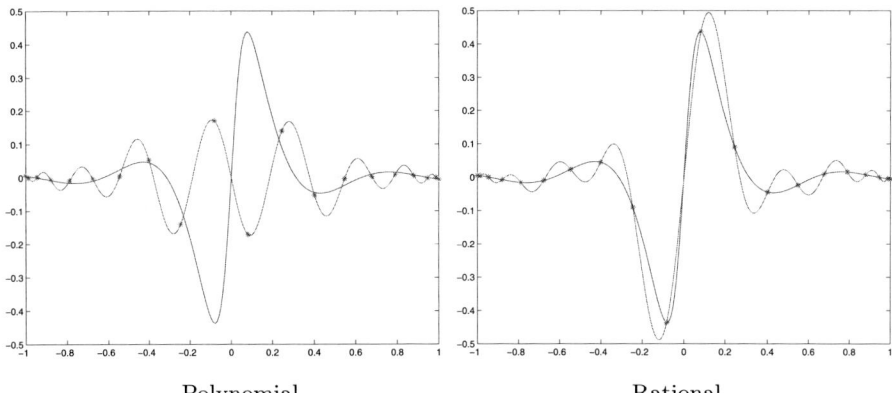

Polynomial   Rational

Comparison of two pseudospectral solutions of a boundary value problem (8.1)

the rational solution with the correct poles (right). Note in particular that the u_j (big dots) are very good approximations to $u(x_j)$ in the second case: they could be interpolated with splines to avoid oscillations for N small. Weideman has proposed in [Wei] a method that is mathematically equivalent to ours when all the poles are fixed a priori but that does not employ the barycentric representation.

Every set of points x_j and every set of b_j determine another linear rational collocation method. A natural choice are shifted points x_j and the weights $b_j = w_j^{(2)} \forall j$. $\mathcal{R}_N^{(b)}$ then is the set of all interpolants (7.2) and takes advantage of the improved condition of the derivatives and/or of the better approximation in the center of I, see § 7.

The better condition of the derivatives is not so important when solving (8.5) directly, that is by Gaussian elimination, for this amounts to applying $\mathbf{A}^{-1}$, i.e., to integration [Ber4, Tan-Tru]. However, it might be important when solving (8.5) for more complicated problems, and is definitely an advantage when one applies $\mathbf{A}$ itself, e.g.:

- when solving time evolution partial differential equations with the pseudospectral *method of lines* (pseudospectral discretization in space followed by a time-stepping algorithm such as extrapolation or Runge-Kutta for solving the resulting system of ordinary differential equations in time). This is the application that led Kosloff and Tal-Ezer to advocate their shift (7.4). The use of the corresponding points x_j in the just mentioned $\mathcal{R}_N^{(b)}$ has been studied in [Bal-Ber2]. It decreases the number of time steps by a factor of two to three without any noticeable change in the computing effort for spectral discretization.

- when solving spatial systems of equations such as (8.5) by iteration [Ber-Bal]: a good choice of the parameter α in (7.4) (about $\alpha = .99$) may decrease the number of iterations by a third, and this again with no change in the computer code but the command computing the x_j.

9. Back to nonlinear: adaptive point shifts and poles

As already mentioned in § 7, a side effect of Kosloff and Tal-Ezer's point shift is the improvement in the approximation of functions with steep gradients in the center of I by polynomials interpolating between Chebyshev points; in many examples it becomes the most desirable effect. It is too blunt, however: it materializes only when there is only one front and when the latter lies precisely in the center. This is by no means the usual case: fronts may arise everywhere in I and there may be several of them.

To cope with the first difficulty, Bayliss and Turkel [Bay-Tur] have studied some two-parameter shifts, with one parameter for the location of the front, the other for its intensity. The following observation then led the first and last authors to a solution of the several fronts problem: with one steep front, the inverse change of variable is itself steep at the front, but has about zero slope away from it, so that adding another inverse shift preserves both slopes. This naturally leads to the inverse shift [Ber-Mit5]

$$y(x) = g^{[-1]}(x) = \mu + \frac{1}{\lambda} \sum_{q=1}^{Q} \arctan[\alpha_q(x - \beta_q)], \qquad (9.1)$$

where the parameters β_q and α_q determine location and intensity of the qth front, and where λ and μ ensure that $g^{[-1]}(-1) = -1$ and $g^{[-1]}(1) = 1$. Whenever needed, the shift $g(y)$ itself is obtained by inverting $g^{[-1]}$, a simple task in the two-front case [Ber-Mit5]. In many instances, as in the solution of differential equations below, only $g^{[-1]}$ (or, rather, its derivatives) arises in the solution of the collocation problem; g is just needed for evaluating the final solution between the nodes.

For approximation, if one has some (possibly vague) information about localization and intensity of the front, one may try adjusting the parameters α_q and β_q by trial and error; this may be quite effective in practice. In [Ber-Mit5], the parameters were optimized in a more sophisticated way by minimizing

$$\|R - F\|_\infty := \max_{y \in [-1,1]} |R(y) - F(y)|,$$

where R denotes the rational approximation in y-space, with modern simulated annealing software. The effect of the shift is quite impressive for interior fronts, much more pronounced than that of the optimized poles.

The latter are still useful, though. Firstly, they may bring a noticeable bit of extra precision at the fronts. Secondly, and more importantly, they are necessary for accomodating boundary layers. Indeed, by moving the points toward the interior fronts, the shifts have the negative effect of depleting the extremities, thus worsening the approximation of incidental boundary layers. Poles, which may be located everywhere and in particular have real part outside I, remain efficient in that case, as the results of [Ber-Mit5] demonstrate. We just reproduce here Table 1, whose captions should be self explanatory and may be found in [Ber-Mit5].

β	α	Poles	$\|r - f\|$	$\|r' - f'\|$	$\|r'' - f''\|$
$*$	$*$		$1.867e - 1$	$2.361e + 1$	$3.454e + 3$
$*$	$*$	$(-.4902, \pm 2.011e - 2)$ $(-.5056, \pm 2.228e - 2)$ $(-.5178, \pm 5.613e - 2)$	$5.224e - 4$	$2.158e - 1$	$8.241e + 1$
$-.5185$	7.408		$9.447e - 9$	$5.012e - 6$	$1.138e - 2$
$-.4976$	8.273	$(-1.027, \pm 3.147e - 3)$	$1.279e - 11$	$6.270e - 9$	$1.654e - 5$
$-.4981$	8.519	$(-1.030, \pm 3.574e - 3)$ $(1.062, \pm 5.346e - 3)$	$2.495e - 12$	$1.754e - 9$	$5.659e - 6$

Table 1. Effect of an optimized Bayliss-Turkel point shift on rational approximation with and without optimized poles in an example with $N = 100$

The advances in approximation brought by optimized poles and point shifts may be used for improving solutions of problems. As an example we take again the boundary value problem (8.1). By means of the chain rule one sees that the point shift transplants (8.1a) into the equation

$$[y'(x)]^2 U''(y) + [y''(x) + P(y)y'(x)]U'(y) + Q(y)U(y) = H(y) \qquad (9.2)$$

in y-space, which will be solved by collocation at Chebyshev points of the second kind y_j.

The first and last authors have suggested in [Ber-Mit3] and [Ber-Mit6] the following two-step recursive procedure for solving (9.2):

Step 1. Compute the approximate solution $\mathbf{U}^{(k)} = [U_1^{(k)}, \ldots, U_{N-1}^{(k)}]^T$ of (9.2) – with the boundary conditions (8.1b) – by the linear rational collocation method with $b_j = w_j d_j$, $d_j = \prod_{k=1}^{P}(y_j - v_k)$, with the poles $v_1, \ldots, v_P$ in y-space ($d_j \equiv 1$ for $k - 1$) and the inverse point shift (9.1). This amounts to solving a system (8.5) with matrix

$$\mathbf{A} := \mathbf{G}_1^2 \mathbf{D}^{(2)} + (\mathbf{G}_2 + \mathbf{G}_1 \mathbf{P})\mathbf{D}^{(1)} + \mathbf{Q},$$

where $\mathbf{D}_1$ and $\mathbf{D}_2$ are the Chebyshev differentiation matrices (8.6) and where $\mathbf{G}_1$ and $\mathbf{G}_2$ denote the diagonal matrices of the derivatives of $g^{[-1]}$ at the nodes x_i,

$$\mathbf{G}_1 = \text{diag}\big(y'(x_1), \ldots, y'(x_{N-1})\big), \qquad \mathbf{G}_2 = \text{diag}\big(y''(x_1), \ldots, y''(x_{N-1})\big),$$

while $\mathbf{P}$ and $\mathbf{Q}$ contain the values $P(y_i) = p(x_i)$, resp. $Q(y_i) = q(x_i)$.

Step 2. Minimize the residual norm

$$\|[y']^2 R'' + [y'' + Py']R' + QR - H\|_\infty$$

of the differential equation for the rational interpolant

$$R(y) := \sum_{j=0}^{N} \frac{w_j^{(2)} \prod_{\ell=1}^{P}\left(1 - \frac{y_j}{v_\ell}\right)}{y - y_j} U_j^{(k)} \Bigg/ \sum_{j=0}^{N} \frac{w_j^{(2)} \prod_{\ell=1}^{P}\left(1 - \frac{y_j}{v_\ell}\right)}{y - y_j}$$

with respect to the poles v_k in y-space, $k = 1, \ldots, P$, and the shift parameters α_q, β_q, $q = 1, \ldots, Q$. This changes the b_j to yield a new interpolant to the $U_j^{(k)}$.

The derivatives $y'(x)$ and $y''(x)$ needed in the computation of $\mathbf{G}_1$ and $\mathbf{G}_2$ are given by the simple formulae

$$y'(x) = \frac{1}{\lambda} \sum_{q=1}^{Q} \frac{\alpha_q}{1 + s_q^2},$$

$$y''(x) = -\frac{2}{\lambda} \sum_{q=1}^{Q} \frac{\alpha_q^2 s_q}{(1 + s_q^2)^2}, \quad s_q := \alpha_q(x - \beta_q).$$

In all the examples we tried [Ber-Mit6] the method approximates the solution with an error merely about ten times as large as the direct approximation of the exact solution in [Ber-Mit5] (one example of which is given in the table above), a splendid performance. The convergence of the method has not been proven yet; however, an L_2-Galerkin version has been shown to reduce the energy norm of the error at each step of the algorithm.

10. Conclusion

We hope that the present article has convinced the reader that applications of the barycentric representation of rational interpolants brings interesting advances in infinitely smooth practical approximation. Its use in classical rational interpolation yields a very stable way of computing the interpolant and allows for a relatively simple detection of unattainable points and poles. The latter may also be easily monitored in the complex plane and their location optimized to yield new rational interpolants which approximate a given function with an error that diminishes with the number of the poles. In view of the globality of the interpolants, fronts are handled with conformal shifts of variables which may be optimized as well. Though expensive to determine, the resulting approximants display an impressive accuracy. They may be used in the solution of problems such as differential and integral equations.

Acknowledgement

This work has been supported by the Swiss National Science Foundation, grant Nr. 20-66754.01.

References

[Abr-Gar] Abril-Raymundo M.R., García-Archilla B., Approximation properties of a mapped Chebyshev method, *Appl. Numer. Math.* **32** (2000) 119–136.

[Bal-Ber1] Baltensperger R., Berrut J.-P., The errors in calculating the pseudospectral differentiation matrices for Čebyšev-Gauss-Lobatto points, *Comput. Math. Appl.* **37** (1999) 41–48. Errata **38** (1999) 119.

[Bal-Ber2] Baltensperger R., Berrut J.-P., The linear rational collocation method, *J. Comput. Appl. Math.* **134** (2001) 243–258.

[Bal-Ber-Dub] Baltensperger R., Berrut J.-P., Dubey Y., The linear rational pseudospectral method with preassigned poles, *Numer. Algorithms* **33** (2003) 53–63.

[Bal-Ber-Noël] Baltensperger R., Berrut J.-P., Noël B., Exponential convergence of a linear rational interpolant between transformed Chebyshev points, *Math. Comp.* **68** (1999) 1109–1120.

[Bat-Tre] Battles Z., Trefethen L.N., An extension of MATLAB to continuous functions and operators, *SIAM J. Sci. Comput* **25** (2004) 1743–1770.

[Bay-Tur] Bayliss A., Turkel E., Mappings and accuracy for Chebyshev pseudo-spectral approximations, *J. Comput. Phys.* **101** (1992) 349–359.

[Ber1] Berrut J.-P., Baryzentrische Formeln zur trigonometrischen Interpolation (I), *Z. angew. Math. Phys. (ZAMP)* **35** (1984) 91–105.

[Ber2] Berrut J.-P., Rational functions for guaranteed and experimentally well-conditioned global interpolation, *Comput. Math. Appl.* **15** (1988) 1–16.

[Ber3] Berrut J.-P., Barycentric formulae for some optimal rational approximants involving Blaschke products, *Computing* **44** (1990) 69–82.

[Ber4] Berrut J.-P., *A pseudospectral Čebyšev method with preliminary transform to the circle: ordinary differential equations*, Report No. 252, Mathematisches Institut, Technische Universität München, 1990; revised Université de Fribourg, 1995.

[Ber5] Berrut J.-P., The barycentric weights of rational interpolation with prescribed poles, *J. Comput. Appl. Math.* **86** (1997) 45–52.

[Ber6] Berrut J.-P., A matrix for determining lower complexity barycentric representations of rational interpolants, *Numer. Algorithms* **24** (2000) 17–29.

[Ber-Bal] Berrut J.-P., Baltensperger R., The linear rational pseudospectral method for boundary value problems, *BIT* **41** (2001) 868–879.

[Ber-Mit1] Berrut J.-P., Mittelmann H., Matrices for the direct determination of the barycentric weights of rational interpolation, *J. Comput. Appl. Math.* **78** (1997) 355–370.

[Ber-Mit2] Berrut J.-P., Mittelmann H., Rational interpolation through the optimal attachment of poles to the interpolating polynomial, *Numer. Algorithms* **23** (2000) 315–328.

[Ber-Mit3] Berrut J.-P., Mittelmann H.D., The linear rational pseudospectral method with iteratively optimized poles for two-point boundary value problems, *SIAM J. Sci. Comput.* **23** (2001) 961–975.

[Ber-Mit4] Berrut J.-P., Mittelmann H.D., Point shifts in rational interpolation with optimized denominator, in: J. Levesley, I.J. Anderson and J.C. Mason, eds., *Algorithms for Approximation IV, Proceedings of the 2001 International Symposium* (The University of Huddersfield, 2002) 420–427.

[Ber-Mit5] Berrut J.-P., Mittelmann H.D., Adaptive point shifts in rational approximation with optimized denominator, *J. Comput. Appl. Math.* **164–165** (2004) 81–92.

[Ber-Mit6] Berrut J.-P., Mittelmann H.D., Optimized point shifts and poles in the linear rational pseudospectral method for boundary value problems, to appear in *J. Comput. Phys.*

[Ber-Tre] Berrut J.-P., Trefethen L.N., Barycentric Lagrange interpolation, *SIAM Rev.* **46** (2004) 501–517.

[Boy] Boyd J.P., The arctan/tan and Kepler-Burgers mappings for periodic solutions with a shock, front, or internal boundary layer, *J. Comput. Phys.* **98** (1992) 181–193.

[Bri] Brimmeyer, M., *Forme barycentrique de complexité réduite pour interpolants rationnels avec pôles prescrits* (Master's thesis, University of Fribourg (Switzerland), 2001).

[Bul-Rut] Bulirsch R., Rutishauser H., Interpolation und genäherte Quadratur, in: Sauer R., Szabó I., Hsg., *Mathematische Hilfsmittel des Ingenieurs* (Grundlehren der math. Wissenschaften Bd. 141, Springer, Berlin-Heidelberg, 1968) 232–319.

[Dav] Davis P.J., *Interpolation and Approximation* (Dover, New York, 1975).

[Epp] Epperson J.F., On the Runge example, *Amer. Math. Monthly* **94** (1987) 329–341.

[For] Fornberg B., *A Practical Guide to Pseudospectral Methods* (Cambridge Univ. Press, Cambridge 1996).

[Gau] Gautschi W., Moments in quadrature problems, *Comput. Math. Appl.* **33** (1997) 105–118.

[Gro] Grosse E., A catalogue of algorithms for approximation, in: Mason J.C., Cox M.G. (eds.), *Algorithms for Approximation II* (Chapman and Hall, London, 1990) 479–514.

[Gut] Gutknecht M.H., The rational interpolation problem revisited, *Rocky Mt. J. Math.* **21** (1991) 263–280.

[Hen1] Henrici P., *Essentials of Numerical Analysis with Pocket Calculator Demonstrations* (Wiley, New York, 1982).

[Hen2] Henrici P., *Applied and Computational Complex Analysis, Vol. 3* (Wiley, New York, 1986).

[Hig] Higham N., The numerical stability of barycentric Lagrange interpolation, *IMA J. Numer. Anal.* **24** (2004) 547–556.

[Kos-Tal] Kosloff D., Tal-Ezer H., A modified Chebyshev pseudospectral method with an $\mathcal{O}(N^{-1})$ time step restriction, *J. Comput. Phys.* **104** (1993) 457–469.

[Mas-Han] Mason J.C., Handscomb D.C., *Chebyshev Polynomials* (Chapman & Hall, Boca Raton, 2003).

[Mea-Ren] Mead J.L., Renaut R.A., Accuracy, resolution, and stability properties of a modified Chebyshev method, *SIAM J. Sci. Comput.* **24** (2002) 143–160.

[Mul-Hua-Slo] Mulholland L.S., Huang W.Z., Sloan D.M., Pseudospectral solution of near-singular problems using numerical coordinate transformations based on adaptivity, *SIAM J. Sci. Comput.* **19** (1998) 1261–1289.

[Pér-Cas-Hay] Pérez-Acosta F., Casasús L., Hayek N., Rational collocation for linear boundary value problems, *J. Comput. Appl. Math.* **33** (1990) 297–305.

[Phi]   Phillips G.M., *Interpolation and Approximation by Polynomials* (Springer, New York, 2003).

[Red-Wei-Nor] Reddy S.C., Weideman J.A.C., Norris G.F., *On a modified Chebyshev pseudospectral method*, Report, Oregon State University, 1999.

[Sal]   Salzer H.E., Lagrangian interpolation at the Chebyshev points $x_{n,\nu} = \cos(\nu\pi/n)$, $\nu = 0(1)n$; some unnoted advantages, *The Computer J.* **15** (1972) 156–159.

[Sch-Wer] Schneider C., Werner W., Some new aspects of rational interpolation, *Math. Comp.* **47** (1986) 285–299.

[She]   Shepard D., A two-dimensional interpolation function for irregularly-spaced data, *Proc. 23rd Nat. Conf. ACM* (1968) 517–524.

[Ste]   Steffen S., *Eine Methode zur direkten Bestimmung der Nennerwerte rationaler Interpolierender mit reduzierter Komplexität* (Master's thesis, University of Fribourg (Switzerland), 2002).

[Sto]   Stoer J., *Einführung in die Numerische Mathematik I* (4. Aufl., Springer, Berlin-Heidelberg-New York, 1983).

[Sza-Vér] Szabados J., Vértesi P., *Interpolation of Functions* (World Scientific, Singapore, 1990).

[Tan-Tru] Tang T., Trummer M.R., Boundary layer resolving pseudospectral methods for singular perturbation problems, *SIAM J. Sci. Comput.* **17** (1996) 430–438.

[Tre]   Trefethen L.N., *Spectral Methods in MATLAB* (SIAM, Philadelphia, 2000).

[Wei]   Weideman J.C., Spectral methods based on nonclassical orthogonal polynomials, in: Gautschi W. et al., eds., *Applications and computation of orthogonal polynomials* (Birkhäuser, ISNM 131, 1999) 239–252.

[WeH]   Werner H., Algorithm 51: A reliable and numerically stable program for rational interpolation of Lagrange data, *Computing* **31** (1983) 269–286.

[WeW]   Werner W., Polynomial interpolation: Lagrange versus Newton, *Math. Comp.* **43** (1984) 205–217.

[Zhu-Zhu] Zhu X., Zhu G., A method for directly finding the denominator values of rational interpolants, *J. Comput. Appl. Math.* **148** (2002) 341–348.

Jean-Paul Berrut and Richard Baltensperger
Département de Mathématiques
Université de Fribourg
CH–1700 Fribourg/Pérolles
Switzerland
e-mail: Jean-Paul.Berrut@unifr.ch.
e-mail: Richard.Baltensperger@unifr.ch.

Hans D. Mittelmann
Department of Mathematics and Statistics
Arizona State University
Tempe, Arizona 85287–1804, USA
e-mail: Mittelmann@asu.edu.

D.H. Mache, D. Braess and M. de Bruin

J. Prestin, D. Braess and D.H. Mache

Trends and Applications in Constructive Approximation
(Eds.) M.G. de Bruin, D.H. Mache & J. Szabados
International Series of Numerical Mathematics, Vol. 151, 53–60
© 2005 Birkhäuser Verlag Basel/Switzerland

Approximation on Simplices and Orthogonal Polynomials

Dietrich Braess

Abstract. Inequalities of Jackson and Bernstein type are derived for polynomial approximation on simplices with respect to Sobolev norms. Although we do not find simple bases when looking at 120 years of research of orthogonal polynomials on triangles, sharp estimates are obtained from a decomposition into orthogonal subspaces. The formulas reflect the symmetries of simplices, but analogous estimates on rectangles show that we cannot expect rotational invariance of the terms with derivatives. An essential tool are selfadjoint differential operators that have already been used by other authors for the study of various approximation properties.

1. Introduction

The approximation of functions by polynomials with respect to a weighted L_2-norm is strongly related to orthogonal polynomials. This is well known for functions on the real interval $[-1, +1]$. The orthogonal polynomials for constant weights are the Legendre polynomials P_n which satisfy

$$\int_{-1}^{+1} P_n P_m dx = \frac{2}{2n+1} \delta_{nm}.$$

The Legendre polynomials are eigenfunctions of the singular Legendre differential operator,

$$\mathcal{L} P_n = \mu_n P_n, \quad \mu_n = n(n+1)$$

where $\mathcal{L}$ is given by $(\mathcal{L}v)(x) := -((1-x^2)v')'$. We therefore have also orthogonality of the derivatives with respect to a weight function which vanishes at the boundaries

$$\int_{-1}^{+1} (1-x^2) P_n' P_m' dx = \mu_n \frac{2}{2n+1} \delta_{nm}.$$

If we expand an L_2-function with respect to the Legendre polynomials for the natural normalization $v = \sum_{k=0}^{\infty} b_k \left(k + \frac{1}{2}\right)^{1/2} P_k$, then we have obviously,

$$\|v\|_0^2 := \int_{-1}^{+1} v^2 dx = \sum_{k=0}^{\infty} |b_k|^2,$$

$$|v|_{1,w}^2 := \int_{-1}^{+1} (1 - x^2)(v')^2 dx = \sum_{k=1}^{\infty} \mu_k |b_k|^2 \tag{1}$$

and more generally, for any $\ell \in \mathbb{N}_0$,

$$|v|_{\ell,w}^2 := (-1)^\ell \int_{-1}^{+1} v \mathcal{L}^\ell v dx = \sum_{k=1}^{\infty} (\mu_k)^\ell |b_k|^2$$

which is to be understood in the sense that the series converge if and only if $|v|_{\ell,w}$ is finite. We obtain from the definitions for v, with $|v|_{m,w} < \infty$, $\ell, m \in \mathbb{N}_0$, $m \geq \ell$, the *approximation property* (direct estimate)

$$\inf_{p \in \mathcal{P}_n} |v - p|_{\ell,w} \leq (\mu_{n+1})^{-(m-\ell)/2} |v|_{m,w} \tag{2}$$

and the *inverse estimate*

$$|p|_{m,w} \leq (\mu_n)^{(m-\ell)/2} |p|_{\ell,w} \quad \text{for } p \in \mathcal{P}_n. \tag{3}$$

Remark 1.1. *This fits into the following general framework. Let X be a Banach space which is compactly imbedded into Y. Therefore $\|\cdot\|_X$ is a finer norm than $\|\cdot\|_Y$. Moreover, let V_m, $m \in \mathbb{N}$ be a family of finite-dimensional subspaces of X. The pair X, Y is appropriate for the family (V_m) if there are parameters c_m and a constant C such that the direct approximation property*

$$\inf_{p \in V_m} \|v - p\|_Y \leq c_m \|v\|_X \quad \forall v \in X \tag{4}$$

and the inverse estimate

$$\|p\|_X \leq C c_m^{-1} \|p\|_Y \quad \forall p \in V_m$$

hold. [We note that we cannot have $\|p\|_X \leq o(c_m^{-1})\|p\|_Y \;\forall p \in V_m$ together with (4).] – Classical pairs of spaces that fit in this sense are given by C^0 and C^m due to Jackson's and Bernstein's theorems. Finite element spaces are another example; see, e.g., [4, p. 85] for h-FEM, i. e. when convergence is achieved by refinements of the meshes. Recently the p-FEM has attracted much interest, i. e. the approximation is improved by increasing the degree of the polynomials [21]. Here the theory is less complete.

Direct and inverse estimates for the rectangle are easily obtained from these results by tensor product arguments [5]. Those results show already that we cannot expect rotational invariance of the inequalities.

The situation on triangles and more generally on simplices in $\mathbb{R}^d$ is more involved. There are two approaches in the literature for orthogonal polynomials on triangles/simplices, but none of them can be used directly for our purpose. We will demonstrate that by an algebraic counterpart. The remedy is that we are content with a decomposition into *orthogonal subspaces*. In particular, we will use some selfadjoint differential operators that have been discovered independently by several authors for different purposes.

2. Orthogonal polynomials on triangles

The one-dimensional example in the introduction showed already the relation between orthogonal polynomials and the approximation problem under consideration. There are two different approaches to orthogonal polynomials on triangles.

In 1881 Appell [1] introduced polynomials F_{mn} which give rise to a biorthogonal system F_{mn} and E_{mn} on triangles. The polynomials (and some generalizations)

$$F_{mn}(x,y) := \frac{\partial^{m+n}}{\partial x^m \partial y^n} \left[x^m y^n (1-x-y)^{m+n} \right]$$

are now called Appell's polynomials. Obviously F_{mn} is a polynomial of degree $m+n$.

Let

$$\mathcal{P}_N := \text{span}\{x^m y^n;\ m+n \leq N\} \quad and \quad \mathcal{Q}_N := \mathcal{P}_N \cap \mathcal{P}_{N-1}^\perp.$$

Then F_{mn} is orthogonal to $\mathcal{P}_{m+n-1}$.

We provide the (simple) proof since the technique (from 1881) is typical also for recent constructions. It is sufficient to verify the orthogonality for monomials $x^k y^l$ with $k+l \leq m+n$. Without loss of generality we assume that $k < m$. By partial integration we obtain

$$\int_0^{1-y} x^k y^l \frac{\partial^{m+n}}{\partial x^m \partial y^n} \left[x^{m+\alpha} y^{n+\beta} (1-x-y)^{m+n+\gamma} \right] dx$$

$$= (-1)^m \int_0^{1-y} \left(\frac{\partial^m}{\partial x^m} x^k y^l \right) \frac{\partial^n}{\partial y^n} \left[x^{m+\alpha} y^{n+\beta} (1-x-y)^{m+n+\gamma} \right] dx$$

$$= 0$$

for $0 < y < 1$. This is a standard argument with Rodriguez' formula. After integrating over y we have the orthogonality. $\square$

Although Appell's polynomials F_{mn}, $m+n \leq N$, span $\mathcal{Q}_N$, the polynomials F_{mn} and F_{kl} with $k+l = m+n$ and $(k,l) \neq (m,n)$ are unfortunately not orthogonal. It is difficult to provide an orthogonal basis without destroying the symmetry of the triangle.

[The situation is comparable to that of the eigenvalue problem with the matrix

$$\begin{pmatrix} -2 & 1 & 1 \\ 1 & -2 & 1 \\ 1 & 1 & -2 \end{pmatrix}.$$

The matrix is invariant under permutations of the coordinates. There is an eigenvector $(1, 1, 1)$ with the eigenvalue 0. The orthogonal subspace consists of eigenvectors, but we cannot provide a basis without destroying the symmetry.] – We will refer to invariant subspaces due to this feature.

Investigations of orthogonal polynomials based on Appell's polynomials were done e. g. by Appell and Kampé de Fériet (1926), Gröbner (1948), Erdélyi, Magnus, Oberhettinger, and Tricomi (1953), Fackerell and Littler (1974), Derriennic (1985).

Another approach to orthogonal polynomials is obtained from a transformation of the triangle to the square; see Proriol (1957), Karlin and McGregor (1964), Szegő (1974), Koornwinder (1975), Mysovski (1981), Dunkl (1984), Suetin (1988), Dubiner (1991), Xu (1998). Consider the product

$$p_m\left(\frac{x}{1-y}\right)(1-y)^m q_{n,m}(y) \tag{5}$$

where p_m is the mth orthogonal polynomial for the weight 1 and $q_{n,m}$ is the nth orthogonal polynomial for the weight $(1-y)^m$. Obviously the products provide orthogonal polynomial for the triangle and can be expressed in terms of Jacobi polynomials. Unfortunately these polynomials are less suited for our intention since the transformation makes that the derivatives of the fractions in (5) give rise to expressions that are more involved.

For completeness we also refer to [17].

3. Estimates on the simplex in $\mathbb{R}^d$

Now we are prepared to consider the original approximation problem on a d-simplex S^d. The simplex is the convex hull of its $d+1$ vertices $a_0, a_1, \ldots, a_d \in \mathbb{R}^d$ which do not lie on a $(d-1)$-dimensional hyperplane. In order to keep the symmetry we refer to the barycentric coordinates $\lambda_0, \lambda_1, \ldots, \lambda_d$ of the points $x = \sum_j \lambda_j a_j \in S^d$. Specifically we have

$$\lambda_j \geq 0, \ j = 0, 1, \ldots, d, \quad \sum_j \lambda_j = 1,$$

We will make use of multiindex notation, in particular

$$\lambda^m := \lambda_0^{m_0} \lambda_1^{m_1} \ldots \lambda_d^{m_d}, \qquad \lambda^\alpha = \lambda_0^{\alpha_0} \lambda_1^{\alpha_1} \ldots \lambda_d^{\alpha_d},$$

and $|m| = \sum_j m_j$, $|\alpha| = \sum_j \alpha_j$. We assume that $\alpha_j > -1$ for all j. Hence, $w_\alpha := \lambda^\alpha$ is a weight function for which the inner product

$$(f, g) = \int_{S^d} f g w_\alpha \tag{6}$$

and the weighted L_2-norm $\|f\|_{0,w}^2 := (f, f)$ is well defined. As before, we set

$$\mathcal{P}_n := \operatorname{span}\{\lambda^m; \ |m| \leq n\} \quad \text{and} \quad \mathcal{Q}_n := \mathcal{P}_n \cap \mathcal{P}_{n-1}^\perp.$$

Due to the condition $\sum \lambda_j = 1$, the representation of a function given in terms of barycentric coordinates is not unique. Nevertheless we can write the directional derivative for the direction from a_k to a_j in the form

$$\frac{\partial}{\partial \lambda_j} - \frac{\partial}{\partial \lambda_k} \quad \text{or for short} \quad \partial_j - \partial_k.$$

Lemma 3.1. *Let $j \neq k$. Then the differential operator of second order*

$$\mathcal{L}_0 := -\lambda^{-\alpha}(\partial_j - \partial_k)\,\lambda_j \lambda_k \lambda^\alpha\,(\partial_j - \partial_k) \tag{7}$$

is selfadjoint with respect to the inner product $(\cdot, \cdot)$. It maps $\mathcal{P}_n$ into $\mathcal{P}_n$ and $\mathcal{Q}_n$ into $\mathcal{Q}_n$.

Sketch of proof. Consider a segment on a line parallel to the direction from a_k to a_j. The product $\lambda_j \lambda_k$ vanishes at the two points at which the line intersects the boundary of S^d. No boundary terms occur when performing partial integration. Therefore $\mathcal{L}_0$ is selfadjoint.

The degree of a polynomial is not augmented by the application of $\mathcal{L}_0$ since the multiplication by the quadratic polynomials is compensated by two differentiations. The arguments of Appell show that also the orthogonal complement $\mathcal{Q}_n$ is mapped into itself. □

Now combinations of the differential operators of the form (7) have been used for several purposes [3, 5, 6, 8, 7, 14, 20, 24]. In particular,

$$\mathcal{L}_w := -\lambda^{-\alpha} \sum_{j<k} (\partial_j - \partial_k)\lambda_j \lambda_k \lambda^\alpha (\partial_j - \partial_k) \tag{8}$$

can be regarded as a Laplacian for the simplex due to its symmetry. Special cases of the eigenvalue problem (9) have already been stated by Appell and Kampé de Fériet [?] in terms of Appell's polynomials. Proofs can be found in the literature cited above. A simple proof in [5] makes use of the fact that $\mathcal{L}_w$ maps $\mathcal{Q}_n$ into itself and that it is sufficient to determine the image $\mathcal{L}_w p$ merely modulo $\mathcal{P}_{n-1}$.

Theorem 3.2. *The operator $\mathcal{L}_w$ is selfadjoint and*

$$\mathcal{L}_w p = \mu_n p \quad \text{for all } p \in \mathcal{Q}_n. \tag{9}$$

with the eigenvalues μ_n explicitly given by

$$\mu_n = \mu_n(d, \alpha) := n(n + d + |\alpha|), \quad n = 1, 2, \ldots . \tag{10}$$

In accordance with (1) we now define a weighted H^1-seminorm which will form an appropriate pair together with $\| \cdot \|_{0,w}$

$$|f|^2_{1,w} := \sum_{j<k} \int_{S^d} |(\partial_j - \partial_k)f|^2 \lambda_j \lambda_k w_\alpha.$$

We obtain our essential tool from the fact that $\mathcal{L}_w$ is selfadjoint

$$|f|^2_{1,w} = \int_{S^d} f(\mathcal{L}_w f)w_\alpha. \tag{11}$$

In particular assume that f is expanded into polynomials from the orthogonal subspaces

$$f = \sum_{k=0}^{\infty} p_k \quad \text{with } p_k \in \mathcal{Q}_k.$$

From the orthogonality of $\mathcal{Q}_k$ and $\mathcal{Q}_l$, $k \neq l$, and Theorem 3.2 we conclude that

$$\|f\|_{0,w}^2 = \sum_{k=0}^{\infty} \|p_k\|_{0,w}^2,$$

$$|f|_{1,w}^2 = \sum_{k=0}^{\infty} \int_{S^d} p_k(\mathcal{L}_w p_k) w_\alpha = \sum_{k=0}^{\infty} \mu_k \|p_k\|_{0,w}^2,$$

and, more generally, for any $\ell \in \mathbb{N}_0$,

$$|f|_{\ell,w}^2 := \sum_{k=0}^{\infty} \int_{S^d} p_k(\mathcal{L}_w^\ell p_k) w_\alpha = \sum_{k=0}^{\infty} (\mu_n)^\ell \|p_k\|_{0,w}^2.$$

The last equality is understood in the sense that the infinite series converges if and only if $|f|_{\ell,w}$ is finite. Similar to $|f|_{1,w}$, the seminorm $|f|_{\ell,w}$ admits the following representation in terms of f and its derivatives:

$$|f|_{\ell,w}^2 = \begin{cases} \int_{S^d} (\mathcal{L}_w^m f)^2 w_\alpha & \text{if } \ell = 2m, \\ \int_{S^d} (\mathcal{L}_w^m f)\mathcal{L}_w(\mathcal{L}_w^m f) w_\alpha & \text{if } \ell = 2m+1. \end{cases} \tag{12}$$

Accordingly, for $m \in \mathbb{N}_0$, we define the weighted spaces

$$V_w^m(S^d) := \left\{ v \in L^2(S^d); \ |f|_{\ell,w} < \infty \text{ for } \ell = 0,1,\ldots,m \right\}.$$

In the literature cited above there are several results on the approximation by polynomials on simplices. The following theorem from [5] fits into the framework of Remark 1.1 and admits a formulation such that there is no gap between the direct and the inverse estimate.

Theorem 3.3. *Let ℓ, m be nonnegative integers and $m \geq \ell$ and denote by $\mu_n = n(n+d+|\alpha|)$ the eigenvalues of $\mathcal{L}_w$. Then, for any $v \in V_w^m(S^d)$, the approximation property*

$$\inf_{p \in \mathcal{P}_n} |v - p|_{\ell,w} \leq (\mu_{n+1})^{-(m-\ell)/2} |v|_{m,w} \quad n = 0,1,2,\ldots$$

holds, and for any $p \in \mathcal{P}_n$ we have the inverse estimate

$$|p|_{m,w} \leq (\mu_n)^{(m-\ell)/2} |p|_{\ell,w}.$$

Both inequalities are sharp.

The operator $\mathcal{L}_w$ annihilates constants, but its kth power does not annihilate $\mathcal{P}_{k-1}$. Recently, Jetter and Stöckler [14] have constructed symmetric differential operators of higher order which do not have this defect. Their operators can be used for improving the results of Theorem 3.3.

References

[1] P. Appell (1881), Sur des polynômes de deux variables analogues aux polynômes de Jacobi. Arch. Math. Phys. 66, 238–245

[2] P. Appell and J. Kampé de Fériet (1926), Fonctions hypergéométriques et hyper-sphériques. Polynômes d'Hermite, Gauthier-Villars, Paris

[3] H. Berens, H.J. Schmid, and Yuan Xu (1992), Bernstein-Durrmeyer polynomials on a simplex. J. Approximation Theory 68, 247–261

[4] D. Braess (2001), Finite Elements. Cambridge University Press

[5] D. Braess and C. Schwab (2000), Approximation on simplices with respect to weighted Sobolev norms. J. Approximation Theory 103, 329–337

[6] W. Chen and Z. Ditzian (1993), A note on Bernstein-Durrmeyer operators in $L_2(S)$. J. Approximation Theory 72, 234–236

[7] M.-M. Derriennic (1985), Polynômes orthogonaux de type Jacobi sur un triangle. C. R. Acad. Sci., Paris, Ser. I 300, 471–474

[8] Z. Ditzian (1995), Multidimensional Jacobi-type Bernstein-Durrmeyer operators. Acta Sci. Math. 60, No.1-2, 225–243 (1995)

[9] M. Dubiner (1991), Spectral methods on triangles and other domains. J. Sci. Comput. 6, No.4, 345–390

[10] C.F. Dunkl (1984), Orthogonal polynomials with symmetry of order three. Can. J. Math. 36, 685–717

[11] A. Erdélyi, W. Magnus, F. Oberhettinger, and F.G. Tricomi (eds.) (1953) Higher transcendental functions. Vol. II. pp 269–273 (Bateman Manuscript Project.) New York-Toronto-London: McGraw-Hill

[12] E.D. Fackerell and R.A. Littler (1974), Polynomials biorthogonal to Appell's polynomials. Bull. Austral. Math. Soc. 11, 181–195

[13] W. Gröbner (1948), Über die Konstruktion von Systemen orthogonaler Polynome in ein- und zweidimensionalen Bereichen. (German). Monatsh. Math. 52, 38 54

[14] K. Jetter and Stöckler (2003), New polynomial preserving operators on simplices. Preprint 242, University Dortmund

[15] S. Karlin and J. McGregor (1964), On some stochastic models in genetics. Stochastic models in medicine and biology, 245–279 (Proc. Sympos. University of Wisconsin, Madison, June 1963. University of Wisconsin Press, Madison)

[16] T. Koornwinder (1975), Two-variable analogues of the classical orthogonal polynomials. In "Theory and Application of Special Functions", Proc. adv. Semin., (R.A. Askey, ed.) Madison 1975, pp 435–495

[17] H.L. Krall and I.M. Sheffer (1967), Orthogonal polynomials in two variables. Ann. Mat. Pura Appl., IV. Ser. 76, 325–376

[18] I.P. Mysovski (1981), Interpolatory Cubature Formulas [in Russian]. Nauka, Moscow

[19] J. Proriol (1957), Sur une famille de polynômes a deux variables orthogonaux dans un triangle. C. R. Acad. Sci., Paris 245, 2459–2461

[20] T. Sauer (1994), The genuine Bernstein-Durrmeyer operator on a simplex. Results in Math. 26, 99–130

[21] Ch. Schwab (1998), p- and hp-Finite Element Methods. Clarendon, Oxford

[22] P.K. Suetin (1988), Orthogonal Polynomials in Two Variables [in Russian]. Nauka, Moscow

[23] G. Szegő (1974), Orthogonal Polynomials. (Third edition). Amer. Math. Soc., Providence, RI

[24] Y. Xu (1998), Orthogonal polynomials and cubature formulae on spheres and on balls. SIAM J. Math. Anal. 29, 779–793

[25] Y. Xu (1998), Summability of Fourier orthogonal series for Jacobi weight functions on the simplex in $\mathbb{R}^d$. Proc. AMS 126, 3027–3036

Dietrich Braess
Faculty of Mathematics
Ruhr-University
D-44780 Bochum, Germany
e-mail: `braess@num.ruhr-uni-bochum.de`

Trends and Applications in Constructive Approximation
(Eds.) M.G. de Bruin, D.H. Mache & J. Szabados
International Series of Numerical Mathematics, Vol. 151, 61–70
© 2005 Birkhäuser Verlag Basel/Switzerland

(0, 2) Pál-type Interpolation:
A General Method for Regularity

Marcel G. de Bruin and Detlef H. Mache

Abstract. The methods of proof of regularity for interpolation problems often are dependent on the problem at hand. In case of given pairs of node generating polynomials the method of deriving an ordinary differential equation for the interpolating polynomial or that of exploiting the specific form of the node generator have mainly been used up to now.

Recently another method was used in the case of Pál-type interpolation where 'only' one of the node generators is fixed in advance: a 'general' method of deriving a companion generator that leads to a regular interpolation problem. Using $(0, 2)$ Pál-type interpolation, it is shown that each of the methods has its merits and for sake of simplicity we will restrict ourselves to the case that the nodes are the zeros of pairs of polynomials of the following form: $\{p(z)q(z), p(z)\}$ with p, q co-prime and both having simple zeros.

Mathematics Subject Classification (2000). 41A05.

Keywords. Pál type interpolation, regularity.

1. Introduction

The study of Hermite-Birkhoff interpolation is a well-known subject (cf. the excellent book [2]). Recently the regularity of some interpolation problems on non-uniformly distributed nodes on the unit circle has been studied.

Along with the continuing interest in interpolation in general, a number of papers on Pál-type interpolation have appeared, cf. [3], [4], [6].

In this paper the attention will be focused on so-called $(0, 2)$ Pál-type interpolation problems on the pair of node generators $\{p(z)q(z), p(z)\}$:

- given two co-prime polynomials $p(z)$ resp. $q(z)$, with simple zeros $\{z_i\}_{i=1}^{n} \in \mathbb{C}$ resp. $\{w_j\}_{j=1}^{m} \in \mathbb{C}$ (*nodes generators*),
- given *data* $\{c_i\}_{i=1}^{n+m}$, $\{d_j\}_{j=1}^{n} \in \mathbb{C}$,

find $P_k \in \Pi_k$, $k = m + 2n - 1$ with $P_k(z_i) = c_i$ $(1 \leq i \leq n)$, $P_k(w_j) = c_{n+j}$ $(1 \leq j \leq m)$ and $P_k''(z_i) = d_i$ $(1 \leq i \leq n)$.

Here Π_k is the set of polynomials of degree at most k with complex coefficients. This type of interpolation problems started with the paper [1] by L.G. Pál in 1975.

Although very often the method of proof of regularity depends on the problem at hand, one can, nevertheless, distinguish two main tools as indicated in [5]:

A. Prove that the square system of homogeneous linear equations for the unknown coefficients of the polynomial P_k has a non-vanishing determinant.

B. Find a differential equation for P_k (or for a factor of P_k) and show that if this equation has a polynomial solution, the solution must be the trivial one.

Recently a new method has been introduced by the authors in [7] for $(0,1)$ Pál-type interpolation:

C. Given $p(z)$ 'only', apply a 'reduction method' and determine 'companion polynomial(s)' $q(z)$ that make the problem regular.

The layout of the paper is as follows: in section 2 general results for method **C** will be given, followed in section 3 by new results on $(0,2)$ Pál-type interpolation. In section 4 the general theorems from section 2 will be proved and in section 5 the proofs for the new examples will be given, using each of the methods **A**, **B** and **C**, along with a discussion of the relative merits of the three methods. Finally some references will be given.

2. General results for method C

Consider the node-generating polynomials

$$p(z) = \prod_{i=1}^{n}(z - z_i) \tag{1}$$

and

$$q(z) = \prod_{j=1}^{m}(z - w_j), \tag{2}$$

co-prime and each having simple zeros.

Remark. It is *not* allowed that p and q have (a) common zero(es).

We then have the following result

Theorem 2.1. *If there exist polynomials* $g(z)$, $r_1(z)$, $r_2(z)$ *such that*

$$2p'(z)q(z) = (\alpha_0 + \alpha_1 z)g(z) + r_1(z)p(z), \tag{3a}$$

$$p''(z)q(z) + 2p'(z)q'(z) = \beta_0 g(z) + r_2(z)p(z), \tag{3b}$$

satisfying the condition

$$\alpha_0 + \alpha_1 z \not\equiv 0, \ g(z_i) \neq 0, \ 1 \leq i \leq n, \tag{4}$$

then $(0,2)$ *Pál-type interpolation on the zeros of* $\{p(z)q(z), p(z)\}$ *is regular*

1. *for* $\alpha_1 = 0$ *if and only if* $\beta_0 \neq 0$,
2. *for* $\alpha_1 \neq 0$ *if and only if* $-\beta_0/\alpha_1 \notin \{0, 1, 2, \ldots, n-1\}$.

Remark. The case $\alpha_0 = \alpha_1 = 0$ leads to a contradiction with (3a) as the zeros of p are simple and the polynomials p, q are co-prime.

More general, using simple conditions on the factors of $g(z)$ from (5a), (5b):

Theorem 2.2. *If there exist polynomials* $g(z)$, $r_1(z)$, $r_2(z)$ *such that*

$$2p'(z)q(z) = (\alpha_0 + \alpha_1 z + \alpha_2 z^2)g(z) + r_1(z)p(z), \tag{5a}$$

with $\alpha_0 + \alpha_1 z + \alpha_2 z^2$ *having two different (complex) roots* σ_1, σ_2, *and*

$$p''(z)q(z) + 2p'(z)q'(z) = (\beta_0 + \beta_1 z)g(z) + r_2(z)p(z), \tag{5b}$$

satisfying the conditions

$$g(z_i) \neq 0, \ 1 \leq i \leq n, \tag{6}$$

and

$$A := \frac{\beta_0 + \beta_1 \sigma_1}{\sigma_1 - \sigma_2} > 0, \ B := \frac{\beta_0 + \beta_1 \sigma_2}{\sigma_2 - \sigma_1} > 0, \tag{7}$$

$$\int_{\sigma_1}^{\sigma_2} (\zeta - \sigma_1)^{A-1}(\zeta - \sigma_2)^{B-1} p(\zeta) d\zeta \neq 0, \tag{8}$$

then the (0, 2) *Pál-type interpolation problem on the zeros of* $\{p(z)q(z), p(z)\}$ *is regular.*

3. New regular problems

In this section some new results on regularity are given.

Theorem 3.1. *The* (0, 2) *Pál-type interpolation problem on the zeros of the pair* $\{p(z)q(z), p(z)\}$, *with* p, q *co-prime and having simple zeros, is regular for the following choices of the node generators:*

1. $p(z) = z^n - \alpha^n$, $\alpha \neq 0$; $q(z) = z$, $n \geq 1$.
2. $p(z) = z^n - \alpha^n$, $\alpha \neq 0$; $q(z) = z^n - \beta^n$, $\beta \neq 0$ *and*

$$\begin{cases} \alpha^n \neq \beta^n & \text{for } n = 1, \\ \alpha^n \neq \beta^n, \ (3n + 2k - 1)\alpha^n \neq (n + 2k - 1)\beta^n & \text{for } n \geq 2. \end{cases}$$

3. $p(z) = z^n - \alpha^n$, $q(z) = (z^{kn} - \alpha^{kn})/(z^n - \alpha^n)$, $\alpha \neq 0$, $k \geq 2$.
4. $p(z) = z^n - \alpha^n$, $q(z) = z(z - z_0)(z^n - \frac{3n+1}{n+1}\alpha^n)$ *with*

$$\alpha, z_0 \neq 0; \ z_0 \neq \alpha \exp\left(\frac{2\pi i k}{n}\right), \ k = 0, 1, \ldots, n - 1; \ z_0^n \neq \frac{3n+1}{n+1}\alpha^n.$$

5. $p(z) = z^n - \alpha^n$, $q(z) = z(z^2 - \xi^2)(z^n - \frac{3n+1}{n+1}\alpha^n)$ *with*

$$\alpha, \xi \neq 0; \ \xi^2 \neq \alpha^2; \ \begin{cases} \xi^n \neq \pm\alpha^n, \ \pm\frac{3n+1}{n+1}\alpha^n & \text{for } n \text{ odd}, \\ \xi^n \neq \alpha^n, \ \frac{3n+1}{n+1}\alpha^n, \ (n+1)\alpha^n & \text{for } n \text{ even}. \end{cases}$$

4. Proofs for method C

The interpolation problem has been formulated in the introduction as:

- given polynomials $p(z)$ and $q(z)$ of degrees n and m,
- with simple zeros z_i, w_j, respectively, all different,
- find a polynomial $P(z)$ of degree at most $n + m - 1$ with

$$P(z_i) = P(w_j) = 0, \ P''(z_i) = 0. \tag{9}$$

Because of the first two sets of conditions in (9), we can write

$$P(z) = p(z)q(z)Q(z), \ \text{degree}\, Q(z) \leq n - 1. \tag{10}$$

The final conditions of (9) then lead to

$$2p'(z_i)q(z_i)Q'(z_i) + \{p''(z_i)q(z_i) + 2p'(z_i)q'(z_i)\}Q(z_i) = 0, \tag{11}$$

with z_i the n zeros of $p(z)$.

Proof of Theorem 2.1. Inserting (3) into (11) and using (4) we find

$$(\alpha_0 + \alpha_1 z_i)Q'(z_i) + \beta_0 Q(z_i) = 0, \ 1 \leq i \leq n. \tag{12}$$

Because of the degree restriction on Q, at most $n - 1$, this immediately implies

$$(\alpha_0 + \alpha_1 z)Q'(z) + \beta_0 Q(z) = 0. \tag{13}$$

Solving this linear first order ordinary differential equation for the cases $\alpha_1 = 0$ (distinguishing $\alpha_0 = 0$ or $\alpha_0 \neq 0$) and $\alpha_1 \neq 0$, we find that $Q(z)$ has to be identically zero under the condition stated in the theorem ($\alpha_1 \neq 0$ was the only case that (13) really had a non-trivial polynomial solution of degree at most $n - 1$; that is where $-\beta_0/\alpha_1 \notin \{1, 2, \ldots, n - 1\}$ comes in). $\square$

Proof of Theorem 2.2. Proceeding as in the previous proof, but now the degree of the polynomial on the left-hand side of the equation could be equal to the degree of $p(z)$, we arrive at the differential equation

$$(\alpha_0 + \alpha_1 z + \alpha_2 z^2)Q'(z) + (\beta_0 + \beta_1 z)Q(z) = Cp(z) \tag{14}$$

for the polynomial Q of degree at most $n - 1$. The equation (14) can be solved with an integrating factor $\mu(z)$ following from

$$\frac{\mu'(z)}{\mu(z)} = \frac{\beta_0 + \beta_1 z}{\alpha_0 + \alpha_1 z + \alpha_2 z^2}$$

and we find

$$(\alpha_0 + \alpha_1 z + \alpha_2 z^2)Q(z) = C \int_{\sigma_1}^{z} (\zeta - \sigma_1)^{A-1}(\zeta - \sigma_2)^{B-1}p(\zeta)d\zeta + D. \tag{15}$$

Now the left-hand side has a zero for $z = \sigma_1$ and $z = \sigma_2$; the first gives $D = 0$ and the second, in view of the condition stated in (8), that $C = 0$. Thus $Q \equiv 0$, implying $P \equiv 0$. $\square$

5. Proofs for the new regular problems

The five cases of theorem 3.1 will each be proved using each of the methods **A, B** and **C**.

5.1. Method A

Proof of 1. The interpolating polynomial has degree at most $2n$; write

$$P(z) = \sum_{k=0}^{n-1} a_k z^k + \sum_{k=0}^{n-1} b_k z^{n+k} + c z^{2n}.$$

This has to vanish at the n zeros z_j of $z^n - \alpha^n$:

$$\sum_{k=0}^{n-1} (a_k + \alpha^n b_k) z^k + c\alpha^{2n} = 0,$$

leading to a polynomial of degree at most $n - 1$ having n zeros, thus:

$$a_0 + \alpha^n b_0 + \alpha^{2n} c_0 = 0, \tag{16a}$$

$$a_k + \alpha^n b_k = 0, 1 \le k \le n - 1. \tag{16b}$$

The condition $P(0) = 0$ implies:

$$a_0 = 0. \tag{17}$$

As the derivative only has to be looked at in the points $z_j \ne 0$, we can as well put $z_j^2 P''(z_j) = 0$:

$$\sum_{k=0}^{n-1} \{k(k-1)a_k + (n+k)(n+k-1)\alpha^n b_k\} z^k + 2n(2n-1)\alpha^{2n} c = 0,$$

which implies

$$n(n-1)\alpha^n b_0 + 2n(2n-1)\alpha^n c_0 = 0, \tag{18a}$$

$$k(k-1)a_k + (n+k)(n+k-1)\alpha^n b_k = 0, 1 \le k \le n - 1. \tag{18b}$$

The equations (16b) and (18b) immediately imply $a_k = b_k = 0$ for $1 \le k \le n-1$ (the determinant of the matrix for the 2×2 system for fixed k is $n(n+2k-1)\alpha^n \ne 0$).

Inserting (17) in (16a) and (18a) gives a 2×2 system for b_0, c_0 with determinant $n(3n-1)\alpha^{3n} \ne 0$, leading to $b_0 = c_0 = 0$ and thus $P \equiv 0$. □

Proof of 2. This time the interpolating polynomial has degree at most $3n - 1$ and we write

$$P(z) = \sum_{k=0}^{n-1} \left(a_k + b_k z^n + c_k z^{2n} \right) z^k$$

This has to vanish at the zeros of both $z^n - \alpha^n$ and $z^n - \beta^n$; as in the previous proof, we find

$$a_k + \alpha^n b_k + \alpha^{2n} c_k = 0, \ 0 \le k \le n - 1, \tag{19a}$$

$$a_k + \beta^n b_k + \beta^{2n} c_k = 0, \ 0 \le k \le n - 1. \tag{19b}$$

Moreover, the second derivative of P has to vanish in the zeros of $z^n - \alpha^n$ and, looking at $z^2 P''(z)$ as before, we find

$$k(k-1)a_k + (n+k)(n+k-1)\alpha^n b_k + (2n+k)(2n+k-1)\alpha^{2n} c_k = 0, \ 0 \le k \le n-1. \tag{20}$$

Combining (19) and (20), we see that the triple $\{a_k, b_k, c_k\}$ satisfies, for each fixed k from $\{0, 1, \ldots, n-1\}$ the linear system

$$\mathcal{A}_k \begin{pmatrix} a_k \\ b_k \\ c_k \end{pmatrix} = \begin{pmatrix} 0 \\ 0 \\ 0 \end{pmatrix}$$

with

$$\mathcal{A}_k = \begin{pmatrix} 1 & \alpha^n & \alpha^{2n} \\ 1 & \beta^n & \beta^{2n} \\ k(k-1) & (n+k)(n+k-1)\alpha^n & (2n+k)(2n+k-1)\alpha^{2n} \end{pmatrix}.$$

As

$$\det \mathcal{A}_k = n\alpha^n(\beta^n - \alpha^n)\left[(3n+2k-1)\alpha^n - (n+2k-1)\beta^n\right] \ne 0,$$

we conclude that, given the conditions, the system has the trivial solution only. $\quad\square$

Proof of 3. In this case we incorporate the node generator $z^{kn} - \alpha^{kn}$ in the definition of the interpolating polynomial:

$$P(z) = \sum_{j=0}^{kn-1} \left(a_j + b_j z^{kn}\right) z^j$$

with $b_j = 0$ for $n \le j \le kn - 1$. Using the method as in the previous cases, we find

$$a_j + \alpha^{kn} b_j = 0, \ 0 \le j \le kn - 1.$$

Combining this with the known values for the b_j, this leads to $a_j = 0$ for $n \le j \le kn - 1$; thus the interpolating polynomial reduces to

$$P(z) = \sum_{j=0}^{n-1} \left(a_j + b_j z^{kn}\right) z^j$$

with

$$a_j + \alpha^{kn} b_j = 0, \ 0 \le j \le n - 1. \tag{21}$$

Just as in the previous cases, the conditions $z_j^2 P''(z_j) = 0$ give

$$j(j-1)a_j + (nk+j)(nk+j-1)\alpha^{kn} b_j = 0, \ 0 \le j \le n - 1. \tag{22}$$

Combination of (21) and (22) leads, for each fixed j in the range, to a system with non-vanishing determinant [here $kn\alpha^{kn}(kn + 2j - 1)$]. $\quad\square$

Proof of 4. and 5. It is quite obvious that the method used in the previous proofs does not lead to a 'linear algebra' problem that can be managed so easily. $\quad\square$

5.2. Method B

Proof of 1. Put

$$P(z) = z(z^n - \alpha^n)Q(z)$$

with $\deg Q \le n - 1$. Inserting the zeros z_j of $z^n - \alpha^n$ in the second derivative, we find

$$0 = P''(z_j) = 2nz_j^n Q'(z_j) + n(n+1)z_j^{n-1}Q(z_j).$$

Dividing by $z_j^{n-1} \ne 0$, we see that the polynomial

$$zQ'(z) + \frac{n+1}{2}Q(z),$$

which is of degree at most $n - 1$, has n zeros. Therefore it satisfies

$$zQ'(z) + \frac{n+1}{2}Q(z) = 0.$$

This differential equation has the solution $Cz^{-(n-1)/2}$: for $n \ge 2$ the only polynomial solution is the trivial one. Moreover, from the method of proof it is clear that in the case $n = 1$ we do not find a differential equation, but just $Q(\alpha) = 0$ with Q a constant: $Q \equiv 0$. $\qquad\square$

Proof of 2. This time we put

$$P(z) = (z^n - \alpha^n)(z - \beta^n)Q(z) \text{ with } \deg Q \le n - 1.$$

Inserting the zeros z_j of $z^n - \alpha^n$ in the second derivative, we find

$$0 = P''(z_j) = 2nz_j^{n-1}Q'(z_j) + \{n(n-1)z_j^{n-2}(\alpha^n - \beta^n) + 2n^2 z_j^{n-2}\}Q(z_j),$$

and after simplification

$$z_j Q'(z_j) + \frac{(3n-1)\alpha^n - (n-1)\beta^n}{2(\alpha^n - \beta^n)}Q(z_j) = 0.$$

Dropping the index on z we again arrive at a polynomial having more zeros than its degree, showing that Q satisfies a simple linear, homogenous differential equation of order 1. The solution can be written as

$$Q(z) = Ce^{-\xi z}, \ C \in \mathbb{C}, \ \xi = \frac{(3n-1)\alpha^n - (n-1)\beta^n}{2(\alpha^n - \beta^n)}.$$

Under the conditions on α, β given in the theorem, Q never reduces to a polynomial of degree at most $n - 1$: the problem is regular. $\qquad\square$

Proof of 3. Now the interpolating polynomial is determined by

$$P(z) = (z^{kn} - \alpha^{kn})Q(z) \text{ with } \deg Q \le n - 1.$$

As in the previous proofs, the interpolation conditions for the second derivative lead to an ordinary differential equations for Q:

$$zQ'(z) + \frac{kn - 1}{2}Q(z) = 0.$$

Just as in the proof for case **1.** this leads to $Q \equiv 0$. $\qquad\square$

Proof of 4. Put

$$P(z) = z(z - z_0)(z^n - \alpha^n)(z^n - \frac{3n+1}{n+1}\alpha^n)Q(z)$$

with $\deg Q \le n - 1$. Proceeding as before, we find after simplification

$$(z_j - z_0)Q'(z_j) + Q(z_j) = 0, \ 1 \le j \le n.$$

Again the number of zeros implies a differential equation

$$(z - z_0)Q'(z) + Q(z) = 0$$

which can be integrated immediately: $(z - z_0)Q(z) = C$, showing that $Q \equiv 0$.
The conditions on α, z_0 ensure that $p(z)$ and $q(z)$ have simple zeros only. □

Proof of 5. Now

$$P(z) = z(z^2 - \xi^2)(z^n - \alpha^n)(z^n - \frac{3n+1}{n+1}\alpha^n)Q(z)$$

with $\deg Q \le n - 1$. Proceeding as in **4.** the differential equation for Q turns out to be

$$(z^2 - \xi^2)Q'(z) + 2zQ(z) = C(z^n - \alpha^n)$$

as the degree on the right-hand side is equal to the number of zero conditions.
Solving this equation we find

$$(z^2 - \xi^2)Q(z) = C \int_0^z (t(n - \alpha^n)dt + D = Cz\left(\frac{z^n}{n+1} - \alpha^n\right) + D.$$

As the left-hand side has a zero for $z = \pm\xi$, this leads to a 2×2 system for the unknown constants C, D with determinant

$$\xi\left[\{1 + (-1)^n\}\frac{\xi^n}{n+1} - 2\alpha^n\right]$$

and the conditions on α, ξ ensure that this system has the trivial solution only. □

5.3. Method C

Proof of 1. The choices

$$\alpha_0 = 0, \ \alpha_1 = 2n, \ \beta_0 = n(n + 1), \ r_1(z) = -2nz, \ r_2(z) = -n(n + 1),$$

leading to $g(z) = z^n + z^{n-1} - \alpha^n$ show that the conditions (3) and (4) are satisfied; thus the case $\alpha_0 = 0$, $\beta_0 \ne 0$ applies. □

Proof of 2. The case $n = 1$ can be resolved using

$$\alpha_0 = -2\beta, \ \alpha_1 = 2, \ \beta_0 = 2, \ r_1(z) = -2(z - \beta), \ r_2(z) = -2$$

leading to $g(z) = z - \alpha + 1$.
For $n \ge 2$ take:

$$\alpha_0 = 0, \ \alpha_1 = 2(\alpha^n - \beta^n), \ \beta_0 = (3n - 1)\alpha^n - (n - 1)\beta^n,$$
$$r_1(z) = 2nz^{n-1}, \ r_2(z) = n(3n - 1)z^{n-2}$$

leading to $g(z) = nz^{n-2}$. □

Proof of 3. Use

$$\alpha_0 = 0, \ \alpha_1 = 2, \ \beta_0 = nk - 1, \ g(z) = nkz^{nk-2}$$

and

$$r_1(z) = \left[2nz^{n-1} \frac{z^{kn} - \alpha^{kn}}{z^n - \alpha^n} - 2knz^{kn-1} \right] / (z^n - \alpha^n),$$

$$r_2(z) = \left[n(n-1)z^{n-2} \frac{z^{kn} - \alpha^{kn}}{z^n - \alpha^n} + 2nz^{n-1} \times \right.$$

$$\left. \times \frac{d}{dz} \left(\frac{z^{kn} - \alpha^{kn}}{z^n - \alpha^n} \right) - kn(kn-1)z^{kn-2} \right] / (z^n - \alpha^n).$$

The conditions are easily checked.   □

Proof of 4. Choose

$$\alpha_0 + \alpha_1 z = z - z_0, \ \beta_0 = 1, \ r_1(z) = -2nz^n \frac{z - z_0}{z_0}, \ r_2(z) = -2nz^{n-1} \frac{z - z_0}{z_0}.$$

The conditions follow from the requirements that p and q are co-prime and have simple zeros and from the condition (4) for

$$g(z) = nz \left\{ (3n^2 + 4n + 1)z^{2n} - (3n^2 + 2n - 1)z_0 z^{2n-1} \right. \tag{23}$$

$$\left. - (3n^2 + 4n + 1)\alpha^n z^n + (3n^2 - 2n - 1)z_0 \alpha^n z^{n-1} \right\} / ((n+1)z_0). \quad □$$

Proof of 5. The proof uses

$$\alpha_0 + \alpha_1 z + \alpha_2 z^2 = z^2 - \xi^2, \ \beta_0 + \beta_1 z = 2z,$$

$$r_1(z) = -n(3n+1)z^n(z^2 - \xi^2)^2/(2\xi^2), \ r_2(z) = -n(3n+1)z^{n-1}(z^2 - \xi^2)^2/\xi^2$$

and $g(z)$ given by

$$\frac{n[(3n^2 + 4n + 1)(z^n - \alpha^n)z^{n+2} + \xi^2 z^n \left\{ (3n^2 - 8n - 3)\alpha^n - 3(n-1)z^n) \right\}]}{2(n+1)\xi^2}.$$

The conditions a.o come in to ensure that q has simple zeros. The condition (6) is automatically satisfied because $g(z_j) = -4n^2\alpha^{2n}/(n+1)$.

The condition (8), in the integral $A = B = 1$, is automatically satisfied for n odd and leads to $\xi^n \neq (n+1)\alpha^n$ for n even.   □

Discussion

From the proofs it has become clear that method **A** necessitates a very special form for the node generating polynomials, while method **B** depends on the possibility of degree reduction in order to find a differential equation that can be solved without too many difficulties.

The fact that there does not appear to be much difference between the applicability of the methods **B** and **C** lies in the fact that both methods exploit the method of reducing the differential equation for the intermediary Q to a simple one.

The advantage of method **C** then lies in the fact that it enables one to find 'companion' node generating polynomials $q(z)$ to a *given* node generating polynomial $p(z)$ that lead to a regular $(0, 2)$ Pál-type interpolation problem on the nodes $\{p(z)q(z), p(z)\}$.

References

[1] L.G. Pál, A new modification of the Hermite-Fejér interpolation, Analysis Math. **1** (1975), 197–205.

[2] G.G. Lorentz, S.D. Riemenschneider and K. Jetter, *Birkhoff Interpolation*, Addison Wesley Pub. Co., Mass. USA, 1983.

[3] Y. Bao, *On simultaneous approximation to a differentiable function and its derivative by inverse Pál-type interpolation polynomials*, Approx. Theory Appl. (N.S.) **11** (1995), 15–23.

[4] T.F. Xie and S.P. Zhou, *On simultaneous approximation to a differentiable function and its derivative by Pál-type interpolation polynomials*, Acta Math. Hungar. **69** (1995), 135–147.

[5] Marcel G. de Bruin, *Interpolation on non-uniformly distributed nodes*, Nonlinear Analysis **17** (2001), 1931–1940.

[6] Marcel G. de Bruin, *Regularity of some incomplete Pál-type interpolation problems*, J. of Computational and Applied Mathematics **145** (2002), 407–415.

[7] Marcel G. de Bruin, Detlef H. Mache, $(0, 1)$ *Pál-type interpolation: a general method for regularity*, in "Advanced Problems in Constructive Approximation", M.D. Buhmann and D.H. Mache (eds.), Birkhäuser Verlag Basel, International Series of Numerical Mathematics **142** (2002), 21–26.

Marcel G. de Bruin
Delft University of Technology
Institute of Applied Mathematics
Mekelweg 4, 2628CD
Delft, Netherlands
e-mail: `M.G.deBruin@EWI.TUDelft.NL`

Detlef H. Mache
University of Applied Sciences TFH Bochum
FB3-Applied Mathematics (Constructive Approximation)
Herner Straße 45
D-44787 Bochum, Germany
e-mail: `Mache@TFH-Bochum.de`

Trends and Applications in Constructive Approximation
(Eds.) M.G. de Bruin, D.H. Mache & J. Szabados
International Series of Numerical Mathematics, Vol. 151, 71–75
© 2005 Birkhäuser Verlag Basel/Switzerland

Sufficient Convergence Conditions for Certain Accelerated Successive Approximations

Emil Cătinaş

Abstract. We have recently characterized the q-quadratic convergence of the perturbed successive approximations. For a particular choice of the parameters, these sequences resulted as accelerated iterations toward a fixed point.

We give here a Kantorovich-type result, which provides sufficient conditions ensuring the convergence of the accelerated iterates.

1. Introduction

Let $(X, \|\cdot\|)$ be a Banach space and $G : \Omega \subseteq X \to \Omega$ a nonlinear mapping having $x^* \in \mathrm{int}\,\Omega$ as fixed point:

$$x^* = G(x^*).$$

We are interested in the q-quadratic convergence toward x^* of the sequences of successive approximation type. Recall that an arbitrary sequence $(y_k)_{k \geq 0} \subset X$ converges $(q\text{-})$quadratically to its limit $\bar{y} \in X$ if [11], [12], [13]

$$\inf\left\{\alpha \in [1, +\infty) : Q_\alpha\{y_k\} = +\infty\right\} = 2,$$

where

$$
Q_\alpha\{y_k\} =
\begin{cases}
0, & \text{if } y_k = \bar{y}, \text{ for all but finitely many } k, \\
\displaystyle\limsup_{k \to \infty} \frac{\|y_{k+1} - \bar{y}\|}{\|y_k - \bar{y}\|^\alpha}, & \text{if } y_k \neq \bar{y}, \text{ for all but finitely many } k, \\
+\infty, & \text{otherwise.}
\end{cases}
$$

In the case when $0 < Q_2\{y_k\} < +\infty$, one obtains the well-known estimate of the form

$$\|y_{k+1} - \bar{y}\| \leq \left(Q_p\{y_k\} + \varepsilon\right)\|y_k - \bar{y}\|^2, \quad \text{for all } k \geq k_0$$

(in the sense that for all $\varepsilon > 0$ there exists $k_0 \geq 0$ such the above inequalities hold).

The successive approximations converging quadratically to x^* are characterized by the following result.

Theorem 1.1. [6] *Assume that G is differentiable on a neighborhood D of x^*, with the derivative G' Lipschitz continuous:*

$$\|G'(x) - G'(y)\| \leq L \|x - y\|, \quad \forall x, y \in D.$$

Suppose further that for a certain initial approximation $x_0 \in D$, the successive approximations

$$x_{k+1} = G(x_k), \quad k \geq 0,$$

converge to x^*, and $I - G'(x_k)$ are invertible starting from a certain step.

   Then the convergence is with order 2 if and only if G' has a zero eigenvalue and, starting from a certain step, the corrections $x_{k+1} - x_k$ are corresponding eigenvectors:

$$G'(x^*)(x_{k+1} - x_k) = 0, \quad \forall k \geq k_0.$$

This condition holds equivalently iff the errors $x_k - x^*$ are corresponding eigenvectors:

$$G'(x^*)(x_k - x^*) = 0, \quad \forall k \geq k_0,$$

or iff

$$x_k \in x^* + \operatorname{Ker} G'(x^*), \quad \forall k \geq k_0.$$

   This result implies that if G' has no eigenvalue 0, there exists no sequence of successive approximations converging to x^* with order 2. [1] In such a case, one may choose to consider for some $(\delta_k)_{k \geq 0} \subset X$ the perturbed successive approximations

$$x_{k+1} = G(x_k) + \delta_k, \quad k \geq 0. \tag{1}$$

Their quadratic convergence is characterized by the following result, which does not require the existence of the eigenvalue 0.

Theorem 1.2. [6] *Suppose that G satisfies the assumptions of Theorem 1.1, and that the sequence (1) of perturbed successive approximations converges to x^*. Then the convergence is with q-order 2 iff*

$$\|G'(x_k)(x_k - G(x_k)) + (I - G'(x_k))\delta_k\| = O(\|x_k - G(x_k)\|^2), \ as \ k \to \infty. \tag{2}$$

   In [5] we have shown that if we write

$$\delta_k = (I - G'(x_k))^{-1}\big(G'(x_k)(G(x_k) - x_k) + \gamma_k\big)$$

with $(\gamma_k)_{k \geq 0} \subset X$, then condition

$$\gamma_k = O(\|x_k - G(x_k)\|^2), \quad as \ k \to \infty,$$

is equivalent to (2).

   We have also noticed in [5] that, under the assumption $\|G'(x)\| \leq q < 1$ for all x in a certain neighborhood of x^*, and for a given $K > 0$, a natural choice (implied by the Banach lemma) for δ_k is:

$$\delta_k = \big(I + \cdots + G'(x_k)^{i_k}\big)G'(x_k)(G(x_k) - x_k),$$

---

[1]In this case, the successive approximations cannot converge faster than q-linearly [6].

with i_k such that

$$\frac{q^{i_k+2}}{1-q} \le K\|x_k - G(x_k)\|. \tag{3}$$

When applying Theorem 1.2 to characterize the quadratic convergence of the resulting sequence

$$x_{k+1} = G(x_k) + \left(I + \cdots + G'(x_k)^{i_k}\right)G'(x_k)(G(x_k) - x_k), \quad k \ge 0, \tag{4}$$

with i_k given by (3), we must assume that this sequence converges to the fixed point x^*. But is this assumption reasonable? The purpose of this note is to show that under certain natural conditions the sequence converges to x^*, so the answer is positive.

2. Main result

First of all, we remark that the fixed point problem is equivalent to solving

$$F(x) = 0, \quad \text{with } F(x) = x - G(x),$$

for which the Newton method generates the iterates

$$s_k^N = -F'(x_k)^{-1}F(x_k) \tag{5}$$

$$x_{k+1} = x_k + s_k^N, \quad k = 0, 1, \dots.$$

In this setting, iterations (4) may be rewritten as

$$x_{k+1} = x_k + (I + G'(x_k) + \cdots + G'(x_k)^{i_k+1})(G(x_k) - x_k) \tag{6}$$

$$:= x_k + s_k, \quad k = 0, 1, \dots,$$

$$\text{with } i_k \text{ s.t. } \frac{q^{i_k+2}}{1-q} \le K\|F(x_k)\|,$$

i.e., as quasi-Newton iterations (see, e.g., [11], [8], [7]).

We obtain the following sufficient Kantorovich-type conditions for the convergence to x^* of these iterates.

Theorem 2.1. *Assume that G is differentiable on the domain Ω, with G' bounded on Ω by*

$$\|G'(x)\| \le q < 1, \quad \forall x \in \Omega, \tag{7}$$

and Lipschitz continuous:

$$\|G'(x) - G'(y)\| \le L\,\|x - y\|, \quad \forall x, y \in \Omega.$$

Let $x_0 \in \Omega$ and $K > 0$ be chosen such that

$$\nu = \left(\frac{L}{2(1-q)^2} + K(1+q)\right)\|F(x_0)\| < 1, \tag{8}$$

and suppose that $\bar{B}_r(x_0) = \{x \in X : \|x - x_0\| \le r\} \subseteq \Omega$ for

$$r = \frac{1}{(1-\nu)(1-q)}\,\|F(x_0)\|.$$

Then the elements of the sequence defined by (6) remain in the ball $\bar{B}_r(x_0)$ and converge to a fixed point x^* of G, which is unique in this ball. According to Theorem 1.2, the convergence is quadratic.

Proof. Recall first [11, 3.2.12] that the Lipschitz hypothesis on G' implies that

$$\|G(y) - G(x) - G'(x)(y-x)\| \leq \tfrac{L}{2}\|y-x\|^2, \quad \forall x, y \in \Omega,$$

while (7) implies the existence of $(I - G'(x))^{-1} = I + G'(x) + \cdots + G'(x)^k + \cdots$ and the bound

$$\left\|(I - G'(x))^{-1}\right\| \leq \tfrac{1}{1-q}, \quad \forall x \in \Omega.$$

Our hypotheses imply the following inequalities:

$$\|s_0\| \leq \tfrac{1}{1-q}\|F(x_0)\|,$$

i.e., $x_1 \in \bar{B}_r(x_0)$ and also

$$
\begin{aligned}
\|F(x_1)\| = \left\|F(x_1) - F(x_0) - F'(x_0)s_0^N\right\| \quad \text{(by (5))} \\
\leq \left\|F(x_1) - F(x_0) - F'(x_0)s_0\right\| + \left\|F'(x_0)(s_0^N - s_0)\right\| \\
\leq \left\|G(x_1) - G(x_0) - G'(x_0)s_0\right\| + (1+q)\left\|s_0^N - s_0\right\| \\
\leq \tfrac{L}{2}\|s_0\|^2 + (1+q)\left\|G'(x_0)^{i_0+2}(I + G'(x_0) + \cdots)F(x_0)\right\| \\
\leq \tfrac{L}{2}(1 + q + \cdots + q^{i_0+1})^2\|F(x_0)\|^2 + \tfrac{q^{i_0+2}}{1-q}(1+q)\|F(x_0)\| \\
\leq \tfrac{L(1-q^{i_0+2})^2}{2(1-q)^2}\|F(x_0)\|^2 + K(1+q)\|F(x_0)\|^2 \leq \nu\|F(x_0)\|.
\end{aligned}
$$

In an analogous fashion, we obtain by induction that for all $k \geq 2$

$$
\begin{aligned}
\|F(x_k)\| &\leq \left(\tfrac{L}{2(1-q)^2} + K(1+q)\right)\|F(x_{k-1})\|^2 \\
&\leq \nu\|F(x_{k-1})\| \\
&\ \ \vdots \\
&\leq \nu^k\|F(x_0)\|,
\end{aligned}
$$

$$\|x_k - x_{k-1}\| = \|s_{k-1}\| \leq \tfrac{1}{1-q}\|F(x_{k-1})\| \leq \tfrac{\nu^{k-1}}{1-q}\|F(x_0)\|,$$

$$\|x_k - x_0\| \leq \|x_k - x_{k-1}\| + \cdots + \|x_1 - x_0\| \leq \tfrac{1}{(1-\nu)(1-q)}\|F(x_0)\| = r.$$

It follows that

$$
\begin{aligned}
\|x_{k+m} - x_k\| &\leq \|x_{k+m} - x_{k+m-1}\| + \cdots + \|x_{k+1} - x_k\| \\
&\leq \tfrac{\nu^{k+m-1} + \cdots + \nu^k}{(1-q)}\|F(x_0)\| \leq \tfrac{\nu^k}{(1-\nu)(1-q)}\|F(x_0)\|,
\end{aligned}
$$

which shows that $(x_k)_{k \geq 0}$ is a Cauchy sequence, and therefore converges to a certain $x^* \in \bar{B}_r(x_0)$. By the definition and continuity of F, x^* is a fixed point of G, which is unique in $\bar{B}_r(x_0)$ (and also in Ω) since G is a contraction. $\qquad\square$

We note that condition (8) contains certain natural demands: $\|F(x_0)\|$ is sufficiently small (which holds, e.g., when x_0 is sufficiently close to x^*), q is sufficiently small (in accordance with the results in [6]), the Lipschitz constant L is sufficiently small (the graph of G is close to a constant in case $X = \mathbb{R}$) and K is sufficiently small (the linear systems are solved with increasingly precision, the iterates approaching to those given by the Newton method – see the classical results of Dennis and Moré [8]).

Acknowledgement

The author would like to thank Professor Detlef Mache for his kind support regarding the participation to the IBoMat04 Conference.

References

[1] I. Argyros, *On the convergence of the modified contractions*, J. Comp. Appl. Math., **55** (1994), 183–189.

[2] C. Brezinski, *Dynamical systems and sequence transformations*, J. Phys. A, **34** (2001), pp. 10659–10669.

[3] C. Brezinski and J.-P. Chehab, *Nonlinear hybrid procedures and fixed point iterations*, Numer. Funct. Anal. Optimiz., **19** (1998), pp. 465–487.

[4] E. Cătinaş, *Inexact perturbed Newton methods and applications to a class of Krylov solvers*, J. Optim. Theory Appl., **108** (2001), 543–570.

[5] E. Cătinaş, *On accelerating the convergence of the successive approximations method*, Rev. Anal. Numér. Théor. Approx., **30** (2001) no. 1, pp. 3–8.

[6] E. Cătinaş, *On the superlinear convergence of the successive approximations method*, J. Optim. Theory Appl., **113** (2002) no. 3, pp. 473–485

[7] E. Cătinaş, *The inexact, inexact perturbed and quasi-Newton methods are equivalent models*, Math. Comp., **74** (2005) no. 249, 291–301.

[8] J.E. Dennis, Jr., and J.J. Moré, *A characterization of superlinear convergence and its application to quasi-Newton methods*, Math. Comp., **28** (1974), 549–560.

[9] Şt. Măruşter, *Quasi-nonexpansivity and two classical methods for solving nonlinear equations*, Proc. AMS, **62** (1977), 119–123.

[10] I. Moret, *On the behaviour of approximate Newton methods*, Computing, **37** (1986), pp. 185–193.

[11] J.M. Ortega, and W.C. Rheinboldt, *Iterative Solution of Nonlinear Equations in Several Variables*, Academic Press, New York, New York, 1970.

[12] F.A. Potra, *On Q-order and R-order of convergence*, J. Optim. Theory Appl., **63** (1989), 415–431.

[13] W.C. Rheinboldt, *Methods for Solving Systems of Nonlinear Equations*, SIAM, Philadelphia, 1998.

[14] D.A. Smith, W.F. Ford and A. Sidi, *Extrapolation methods for vector sequences*, SIAM Rev., **29** (1987), pp. 199–233.

Emil Cătinaş
"T. Popoviciu" Institute of
Numerical Analysis (Romanian Academy)
P.O. Box 68-1
Cluj-Napoca, Romania
e-mail: ecatinas@ictp.acad.ro, ecatinas@cs.ubbcluj.ro

M. de Bruin, P. Vertesi and J. Szabados in a seminar room

Trends and Applications in Constructive Approximation
(Eds.) M.G. de Bruin, D.H. Mache & J. Szabados
International Series of Numerical Mathematics, Vol. 151, 77–89

The Combined Shepard-Lidstone Bivariate Operator

Teodora Cătinaş

Abstract. We extend the Shepard operator by combining it with the Lidstone bivariate operator. We study this combined operator and give some error bounds.

1. Preliminaries

1.1. The Shepard bivariate operator

Recall first some results regarding the Shepard operator for the bivariate case [7], [17]. Let f be a real-valued function defined on $X \subset \mathbb{R}^2$, $(x_i, y_i) \in X$, $i = 0, \ldots, N$ some distinct points and $r_i(x, y)$, the distances between a given point $(x, y) \in X$ and the points (x_i, y_i), $i = 0, 1, \ldots, N$.

The Shepard interpolation operator is defined by

$$(Sf)(x, y) = \sum_{i=0}^{N} A_i(x, y) f(x_i, y_i),$$

where

$$A_i(x, y) = \frac{\prod_{\substack{j=0 \\ j \neq i}}^{N} r_j^{\mu}(x, y)}{\sum_{k=0}^{N} \prod_{\substack{j=0 \\ j \neq k}}^{N} r_j^{\mu}(x, y)}, \tag{1}$$

with $\mu \in \mathbb{R}_+$.

It follows that

$$\sum_{i=0}^{N} A_i(x, y) = 1. \tag{2}$$

^{0}This work has been supported by CNCSIS under Grant 8/139/2003.

Because of its small degree of exactness we are interested in extending the Shepard operator S by combining it with some other operators. Let $\Lambda := \{\lambda_i : i = 0, \ldots, N\}$ be a set of functionals and P the corresponding interpolation operator. We consider the subsets $\Lambda_i \subset \Lambda$, $i = 0, \ldots, N$ such that $\bigcup_{i=0}^{N} \Lambda_i = \Lambda$ and $\Lambda_i \bigcap \Lambda_j \neq \emptyset$, excepting the case $\Lambda_i = \{\lambda_i\}$, $i = 0, \ldots, N$, when $\Lambda_i \bigcap \Lambda_j = \emptyset$ for $i \neq j$. We associate the interpolation operator P_i to each subset Λ_i, for $i = 0, \ldots, N$.

The combined operator of S and P, denoted by S_P, is defined in [9] by

$$(S_P f)(x, y) = \sum_{i=0}^{N} A_i(x, y)(P_i f)(x, y).$$

Remark 1.1. [16] *If P_i, $i = 0, \ldots, N$, are linear and positive operators then S_P is a linear and positive operator.*

Remark 1.2. [16] *Let P_i, $i = 0, \ldots, N$, be some arbitrary linear operators. If $\mathrm{dex}(P_i) = r_i$, $i = 0, \ldots, N$, then*

$$\mathrm{dex}(S_P) = \min\{r_0, \ldots, r_N\}.$$

1.2. Two variable piecewise Lidstone interpolation

We recall first some results from [1] and [2]. Consider $a, b, c, d \in \mathbb{R}$, $a < b$, $c < d$ and let $\Delta : a = x_0 < x_1 < \ldots < x_{N+1} = b$ and $\Delta' : c = y_0 < y_1 < \ldots < y_{M+1} = d$ denote uniform partitions of the intervals $[a, b]$ and $[c, d]$ with stepsizes $h = (b-a)/(N+1)$ and $l = (d-c)/(M+1)$, respectively. Denote by $\rho = \Delta \times \Delta'$ the resulting rectangular partition of $[a, b] \times [c, d]$. For the univariate function f and the bivariate function g and each positive integer r we denote by $D^r f = d^r f/dx^r$, $D_x^r g = \partial^r g/\partial x^r$ and $D_y^r g = \partial^r g/\partial y^r$.

Definition 1.3. [2] *For each positive integer r and p, $1 \leq p \leq \infty$, let $PC^{r,p}[a, b]$ be the set of all real-valued functions f such that:*

(i) *f is $(r-1)$ times continuously differentiable on $[a, b]$;*

(ii) *there exist s_i, $0 \leq i \leq L+1$ with $a = s_0 < \ldots < s_{L+1} = b$, such that on each subinterval (s_i, s_{i+1}), $0 \leq i \leq L$, $D^{r-1}f$ is continuously differentiable;*

(iii) *the L_p-norm of $D^r f$ is finite, i.e.,*

$$\|D^r f\|_p = \left(\sum_{i=0}^{L} \int_{s_i}^{s_{i+1}} |D^r f(x)|^p \, dx \right)^{1/p} < \infty.$$

For the case $p = \infty$ it reduces to

$$\|D^r f\|_\infty = \max_{0 \leq i \leq L} \sup_{x \in (s_i, s_{i+1})} |D^r f(x)| < \infty.$$

Definition 1.4. [2] *For each positive integer r and p, $1 \leq p \leq \infty$, let $PC^{r,p}([a, b] \times [c, d])$ be the set of all real-valued functions f such that:*

(i) *f is $(r-1)$ times continuously differentiable on $[a, b] \times [c, d]$, i.e., $D_x^\mu D_y^\nu f$, $0 \leq \mu + \nu \leq r - 1$, exist and are continuous on $[a, b] \times [c, d]$;*

(ii) *there exist s_i, $0 \leq i \leq L+1$ and v_j, $0 \leq j \leq R+1$ with $a = s_0 < \ldots < s_{L+1} = b$ and $c = v_0 < v_1 < \ldots < v_{R+1} = d$, such that on each open subrectangle $(s_i, s_{i+1}) \times (v_j, v_{j+1})$, $0 \leq i \leq L$, $0 \leq j \leq R$ and for all $0 \leq \mu \leq r-1$, $0 \leq \nu \leq r-1$ with $\mu + \nu = r-1$, $D_x^\mu D_y^\nu f$ are continuously differentiable;*

(iii) *for all $0 \leq \mu \leq r$, $0 \leq \nu \leq r$ such that $\mu + \nu = r$, the L_p-norm of $D_x^\mu D_y^\nu f$ is finite, i.e.,*

$$\left\| D_x^\mu D_y^\nu f \right\|_p = \left(\sum_{i=0}^{L} \sum_{j=0}^{R} \int_{s_i}^{s_{i+1}} \int_{v_j}^{v_{j+1}} \left| D_x^\mu D_y^\nu f(x,y) \right|^p dx dy \right)^{1/p} < \infty.$$

For the particular case $p = \infty$ it reduces to

$$\left\| D_x^\mu D_y^\nu f \right\|_\infty = \max_{\substack{0 \leq i \leq L \\ 0 \leq j \leq R}} \sup_{(x,y) \in (s_i, s_{i+1}) \times (v_j, v_{j+1})} \left| D_x^\mu D_y^\nu f(x,y) \right| < \infty.$$

Definition 1.5. [2] *Let $PC^{r_1, r_2, p}([a, b] \times [c, d])$ be the set of all real-valued functions f such that:*

(i) *$D_x^\mu D_y^\nu f$, $0 \leq \mu \leq r_1-1$, $0 \leq \nu \leq r_2-1$ exist and are continuous on $[a,b] \times [c,d]$;*

(ii) *on each open subrectangle $(s_i, s_{i+1}) \times (v_j, v_{j+1})$, $0 \leq i \leq L$, $0 \leq j \leq R$ and for all $0 \leq \mu \leq r_1$, $0 \leq \nu \leq r_2$, $D_x^\mu D_y^\nu f$ exist and are continuous;*

(iii) *for all $0 \leq \mu \leq r_1$, $0 \leq \nu \leq r_2$ the L_p-norm of $D_x^\mu D_y^\nu f$ is finite.*

The Lidstone polynomial is the unique polynomial Λ_n of degree $2n+1$, $n \in \mathbb{N}$ on the interval $[0, 1]$, defined by (see, e.g., [1], [2])

$$\Lambda_0(x) = x,$$
$$\Lambda_n''(x) = \Lambda_{n-1}(x),$$
$$\Lambda_n(0) = \Lambda_n(1) = 0, \quad n \geq 1.$$

As in [1] and [2], for a fixed partition Δ denote the set $L_m(\Delta) = \{ h \in C[a, b] : h$ is a polynomial of degree at most $2m-1$ in each subinterval $[x_i, x_{i+1}]$, $0 \leq i \leq N \}$. Its dimension is $2m(N + 1) - N$.

Definition 1.6. [2] *For a given function $f \in C^{2m-2}[a, b]$ we say that $L_m^\Delta f$ is the Lidstone interpolant of f if $L_m^\Delta f \in L_m(\Delta)$ with*

$$D^{2k}(L_m^\Delta f)(x_i) = f^{(2k)}(x_i), \quad 0 \leq k \leq m-1, \ 0 \leq i \leq N+1.$$

According to [2], for $f \in C^{2m-2}[a, b]$ the Lidstone interpolant $L_m^\Delta f$ uniquely exists and on the subinterval $[x_i, x_{i+1}]$, $0 \leq i \leq N$ can be explicitly expressed as

$$(L_m^\Delta f)|_{[x_i, x_{i+1}]}(x) = \sum_{k=0}^{m-1} \left[\Lambda_k \left(\tfrac{x_{i+1}-x}{h} \right) f^{(2k)}(x_i) + \Lambda_k \left(\tfrac{x-x_i}{h} \right) f^{(2k)}(x_{i+1}) \right] h^{2k}. \quad (3)$$

It follows that

$$(L_m^\Delta f)(x) = \sum_{i=0}^{N+1} \sum_{j=0}^{m-1} r_{m,i,j}(x) f^{(2j)}(x_i),$$

where $r_{m,i,j}$, $0 \leq i \leq N+1$, $0 \leq j \leq m-1$ are the basic elements of $L_m(\Delta)$ satisfying

$$D^{2\nu} r_{m,i,j}(x_\mu) = \delta_{i\mu} \delta_{2\nu,j}, \quad 0 \leq \mu \leq N+1, 0 \leq \nu \leq m-1 \tag{4}$$

and

$$r_{m,i,j}(x) = \begin{cases} \Lambda_j\left(\frac{x_{i+1}-x}{h}\right)h^{2j}, & x_i \leq x \leq x_{i+1}, 0 \leq i \leq N \\ \Lambda_j\left(\frac{x-x_{i-1}}{h}\right)h^{2j}, & x_{i-1} \leq x \leq x_i, 1 \leq i \leq N+1 \\ 0, & \text{otherwise.} \end{cases}$$

Proposition 1.7. [1], [11] *The Lidstone operator L_m^Δ is exact for the polynomials of degree not greater than $2m-1$.*

We have the interpolation formula

$$f = L_m^\Delta f + R_m^\Delta f,$$

where $R_m^\Delta f$ denotes the remainder.

Taking into account Theorem 5.4.3. from [2] we have for $f \in PC^{2m-2,\infty}[a,b]$ the following estimation of the remainder:

$$\|R_m^\Delta f\|_\infty \leq \tag{5}$$

$$\leq d_{2m-2,0} \cdot h^{2m-2} \max_{0 \leq i \leq N} \sup_{x \in (x_i, x_{i+1})} \left| f^{(2m-2)}(x) - \frac{x_{i+1}-x}{h} f^{(2m-2)}(x_i) - \right.$$

$$\left. - \frac{x-x_i}{h} f^{(2m-2)}(x_{i+1}) \right|$$

$$\leq 2d_{2m-2,0} \cdot h^{2m-2} \|f^{(2m-2)}\|_\infty,$$

where $d_{2m,k}$, $0 \leq k \leq 2m-2$ are the numbers given by

$$d_{2m,k} = \begin{cases} \frac{(-1)^{m-i} E_{2m-2i}}{2^{2m-2i}(2m-2i)!}, & k = 2i, \ 0 \leq i \leq m \\ (-1)^{m-i+1} \frac{2(2^{2m-2i}-1)}{(2m-2i)!} B_{2m-2i}, & k = 2i+1, \ 0 \leq i \leq m-1 \\ 2, & k = 2m+1, \end{cases} \tag{6}$$

E_{2m} and B_{2m} being the $2m$th Euler and Bernoulli numbers (see, e.g., [2]). After some computations we get

$$d_{2m,0} = \begin{cases} 1, & m = 0 \\ \frac{4}{\pi^{2m+1}} \sum_{k=0}^{\infty} \frac{\sin(2k+1)\pi t}{(2k+1)^{2m+1}}, & m \geq 1. \end{cases} \tag{7}$$

For a fixed rectangular partition $\rho = \Delta \times \Delta'$ of $[a,b] \times [c,d]$ the set $L_m(\rho)$ is defined as follows (see, e.g., [1] and [2]):

$$L_m(\rho) = L_m(\Delta) \otimes L_m(\Delta')$$

$$= \text{Span} \left\{ r_{m,i,\mu}(x) r_{m,j,\nu}(y) \right\}_{i=0\ \mu=0\ j=0\ \nu=0}^{N+1\ m-1\ M+1\ m-1}$$

$$= \left\{ h \in C([a,b] \times [c,d]) : h \text{ is a two-dimensional polynomial} \right.$$

of degree at most $2m-1$ in each variable and subrectangle

$$\left. [x_i, x_{i+1}] \times [y_j, y_{j+1}];\ 0 \leq i \leq N,\ 0 \leq j \leq M \right\}$$

and its dimension is $(2m(N+1) - N)(2m(M+1) - M)$.

Definition 1.8. [2] *For a given function $f \in C^{2m-2,2m-2}([a,b] \times [c,d])$ we say that $L_m^\rho f$ is the two-dimensional Lidstone interpolant of f if $L_m^\rho f \in L_m(\rho)$ with*

$$D_x^{2\mu} D_y^{2\nu} (L_m^\rho f)(x_i, y_j) = f^{(2\mu, 2\nu)}(x_i, y_j),$$

$0 \leq i \leq N+1,\ 0 \leq j \leq M+1,\ 0 \leq \mu, \nu \leq m-1$.

According to [2], for $f \in C^{2m-2,2m-2}([a,b] \times [c,d])$, the Lidstone interpolant $L_m^\rho f$ uniquely exists and can be explicitly expressed as

$$(L_m^\rho f)(x,y) = \sum_{i=0}^{N+1} \sum_{\mu=0}^{m-1} \sum_{j=0}^{M+1} \sum_{\nu=0}^{m-1} r_{m,i,\mu}(x) r_{m,j,\nu}(y) f^{(2\mu,2\nu)}(x_i, y_j), \qquad (8)$$

where $r_{m,i,j},\ 0 \leq i \leq N+1,\ 0 \leq j \leq m-1$ are the basic elements of $L_m(\rho)$ satisfying (4).

Lemma 1.9. [2] *If $f \in C^{2m-2,2m-2}([a,b] \times [c,d])$ then*

$$(L_m^\rho f)(x,y) = (L_m^\Delta L_m^{\Delta'} f)(x,y) - (L_m^{\Delta'} L_m^\Delta f)(x,y).$$

Corollary 1.10. [2] *For a function $f \in C^{2m-2,2m-2}([a,b] \times [c,d])$, from Lemma 1.9 we have that*

$$f - L_m^\rho f = (f - L_m^\Delta f) + L_m^\Delta (f - L_m^{\Delta'} f) \qquad (9)$$

$$= (f - L_m^\Delta f) + \left[L_m^\Delta (f - L_m^{\Delta'} f) - (f - L_m^{\Delta'} f) \right] + (f - L_m^{\Delta'} f).$$

1.3. Estimation of the error for the Shepard-Lidstone univariate interpolation

We recall some results regarding error bounds for the Shepard-Lidstone univariate interpolation formula, obtained by us in [4].

With the previous assumptions we denote by $L_m^{\Delta,i} f$ the restriction of the Lidstone interpolation polynomial to the subinterval $[x_i, x_{i+1}]$, $0 \leq i \leq N$, given by (3), and in analogous way we obtain the expression of the restriction $L_m^{\Delta',i} f$ to the subinterval $[y_i, y_{i+1}] \subseteq [c,d]$, $0 \leq i \leq N$. We denote by S_L the univariate combined Shepard-Lidstone operator, given by

$$(S_L f)(x) = \sum_{i=0}^{N} A_i(x)(L_m^{\Delta,i} f)(x).$$

We have obtained the following result regarding the estimation of the remainder $R_L f$ of the univariate Shepard-Lidstone interpolation formula

$$f = S_L f + R_L f. \tag{10}$$

Theorem 1.11. [4] *If $f \in PC^{2m-2,\infty}[a,b]$ then*

$$\|R_L f\|_\infty \le \tag{11}$$

$$\le d_{2m-2,0} \cdot h^{2m-2} \max_{0 \le i \le N} \sup_{x \in (x_i, x_{i+1})} \left| f^{(2m-2)}(x) - \frac{x_{i+1}-x}{h} f^{(2m-2)}(x_i) - \right.$$

$$\left. - \frac{x-x_i}{h} f^{(2m-2)}(x_{i+1}) \right|$$

$$\le 2d_{2m-2,0} \cdot h^{2m-2} \|f^{(2m-2)}\|_\infty,$$

with $d_{2m,0}$ given by (7).

2. The combined Shepard-Lidstone bivariate interpolation

2.1. The combined Shepard-Lidstone bivariate operator

We consider $f \in C^{2m-2,2m-2}([a,b] \times [c,d])$ and the set of Lidstone functionals

$$\Lambda_{Li}^i = \{f(x_i, y_i), f(x_{i+1}, y_{i+1}), \dots, f^{(2m-2,2m-2)}(x_i, y_i), f^{(2m-2,2m-2)}(x_{i+1}, y_{i+1})\}$$

regarding each subrectangle $[x_i, x_{i+1}] \times [y_i, y_{i+1}], 0 \le i \le N$, with $\left| \Lambda_{Li}^i \right| = 4m$, $0 \le i \le N$. We denote by $L_m^{\rho,i} f$ the restriction of the polynomial given by (8) to the subrectangle $[x_i, x_{i+1}] \times [y_i, y_{i+1}], 0 \le i \le N$. This $2m-1$ degree polynomial in each variable solves the interpolation problem corresponding to the set Λ_{Li}^i, $0 \le i \le N$ and it uniquely exists.

We have

$$(L_m^{\rho,i} f)^{(2\nu,2\nu)}(x_k, y_k) = f^{(2\nu,2\nu)}(x_k, y_k),$$

$0 \le i \le N;\ 0 \le \nu \le m-1;\ k = i, i+1.$

We denote by S^{Li} the Shepard operator of Lidstone type, given by

$$(S^{Li} f)(x,y) = \sum_{i=0}^{N} A_i(x,y)(L_m^{\rho,i} f)(x,y), \tag{12}$$

where A_i, $i = 0, \dots, N$ are given by (1). We call S^{Li} the combined Shepard-Lidstone bivariate operator.

Theorem 2.1. *The operator S^{Li} is linear.*

Proof. For arbitrary $h_1, h_2 \in C^{2m-2,2m-2}([a,b] \times [c,d])$ and $\alpha, \beta \in \mathbb{R}$ one gets

$$S^{Li}(\alpha h_1 + \beta h_2)(x,y) = \sum_{i=0}^{N} A_i(x,y) L_m^{\rho,i}(\alpha h_1 + \beta h_2)(x,y)$$

$$= \alpha \sum_{i=0}^{N} A_i(x,y)(L_m^{\rho,i} h_1)(x,y) + \beta \sum_{i=0}^{N} A_i(x,y)(L_m^{\rho,i} h_2)(x,y)$$

$$= \alpha S^{Li}(h_1)(x,y) + \beta S^{Li}(h_2)(x,y). \qquad \square$$

Theorem 2.2. *The operator S^{Li} has the interpolation property:*

$$(S^{Li}f)^{(2\nu,2\nu)}(x_k,y_k) = f^{(2\nu,2\nu)}(x_k,y_k), \quad 0 \le \nu \le m-1, 0 \le k \le N+1, \quad (13)$$

for $\mu > 4m - 4$.

Proof. It is not difficult to show the following relations [8]

$$A_k^{(p,q)}(x_i,y_i) = 0, \ 0 \le i \le N; \ 0 \le p,q \le 2m-2, \ i \ne k,$$
$$A_k^{(p,q)}(x_k,y_k) = 0, \ p+q \ge 1,$$

for all $k = 0,\ldots,N$ and $\mu > \max\{p+q \mid 0 \le p,q \le m-1\}$.
From

$$(S^{Li}f)^{(2\nu,2\nu)}(x_k,y_k) = \sum_{i=0}^{N} \left(A_i(x,y)(L_m^{\rho,i}f)\right)^{(2\nu,2\nu)}(x_k,y_k),$$

we obtain

$$(S^{Li}f)^{(2\nu,2\nu)}(x_k,y_k) = \sum_{i=0}^{N} A_i(x_k,y_k)(L_m^{\rho,i}f)^{(2\nu,2\nu)}(x_k,y_k),$$

and taking into account the cardinality property of A_i's we get (13). $\qquad \square$

Theorem 2.3. *The degree of exactness of the combined operator S^{Li} is* $\text{dex}(S^{Li}) = 2m-1$.

Proof. By Proposition 1.7 we have that $\text{dex}(L_m^{\rho,i}) = 2m-1$. This implies $L_m^{\rho,i}e_{pq} = e_{pq}$, where $e_{pq}(x,y) = x^p y^q$, for $p,q \in \mathbb{N}$, with $p+q \le 2m-1$. Taking into account (2), we get

$$(S^{Li}e_{pq})(x,y) = \sum_{i=0}^{N} A_i(x,y)(L_m^{\rho,i}e_{pq})(x,y)$$

$$= \sum_{i=0}^{N} A_i(x,y)e_{pq}(x,y)$$

$$= e_{pq}(x,y)\sum_{i=0}^{N} A_i(x,y) = e_{pq}(x,y), \quad \text{for } p+q \le 2m-1.$$

Therefore the result is proved. $\qquad \square$

2.2. Estimation of the error for the Shepard-Lidstone bivariate interpolation

We obtain the bivariate Shepard-Lidstone interpolation formula,

$$f = S^{Li}f + R^{Li}f,$$

where $S^{Li}f$ is given by (12) and $R^{Li}f$ denotes the remainder term.

Theorem 2.4. *If* $f \in PC^{2m-2,2m-2,\infty}([a,b] \times [c,d])$ *then*

$$\left\| R^{Li}f \right\|_{\infty} \leq 4d_{2m-2,0} \cdot h^{2m-2} \left\| f^{(2m-2)} \right\|_{\infty} +$$
$$+ 2d_{2m-2,0} \cdot h^{2m-2} \max_{0 \leq i \leq N} \left\| (f - L_m^{\Delta',i}f)^{(2m-2)} \right\|_{\infty} \quad (14)$$

$$\leq 2d_{2m-2,0} \cdot h^{2m-2} \left[2 \left\| f^{(2m-2)} \right\|_{\infty} + \left\| (f - L_m^{\Delta'}f)^{(2m-2)} \right\|_{\infty} \right],$$

with $d_{2m-2,0}$ *given by (7).*

Proof. Taking into account (12) and (2) we get

$$(R^{Li}f)(x,y) = f(x,y) - (S^{Li}f)(x,y)$$

$$= f(x,y) - \sum_{i=0}^{N} A_i(x,y)(L_m^{\rho,i}f)(x,y)$$

$$= \sum_{i=0}^{N} A_i(x,y)f(x,y) - \sum_{i=0}^{N} A_i(x,y)(L_m^{\rho,i}f)(x,y)$$

$$= \sum_{i=0}^{N} A_i(x,y)\left[f(x,y) - (L_m^{\rho,i}f)(x,y) \right].$$

Next, applying formulas (9) and (2) we get

$$(R^{Li}f)(x,y) = \sum_{i=0}^{N} A_i(x,y) \left\{ (f - L_m^{\Delta,i}f)(x,y) + \left[L_m^{\Delta,i}(f - L_m^{\Delta',i}f)(x,y) - \right. \right.$$
$$\left. \left. - (f - L_m^{\Delta',i}f)(x,y) \right] + (f - L_m^{\Delta',i}f)(x,y) \right\}$$

$$= \sum_{i=0}^{N} A_i(x,y)(f - L_m^{\Delta,i}f)(x,y)$$

$$+ \sum_{i=0}^{N} A_i(x,y)\left[L_m^{\Delta,i}(f - L_m^{\Delta',i}f)(x,y) - (f - L_m^{\Delta',i}f)(x,y) \right]$$

$$+ \sum_{i=0}^{N} A_i(x,y)(f - L_m^{\Delta',i}f)(x,y)$$

$$= \left[f(x,y) \sum_{i=0}^{N} A_i(x,y) - \sum_{i=0}^{N} A_i(x,y)(L_m^{\Delta,i}f)(x,y) \right]$$

$$- \sum_{i=0}^{N} A_i(x,y) \left[(f - L_m^{\Delta',i}f)(x,y) - L_m^{\Delta,i}(f - L_m^{\Delta',i}f)(x,y) \right]$$

$$+ \left[f(x,y) \sum_{i=0}^{N} A_i(x,y) - \sum_{i=0}^{N} A_i(x,y)(L_m^{\Delta',i}f)(x,y) \right]$$

$$= \left[f(x,y) - \sum_{i=0}^{N} A_i(x,y)(L_m^{\Delta,i}f)(x,y) \right]$$

$$- \sum_{i=0}^{N} A_i(x,y) \left[(f - L_m^{\Delta',i}f)(x,y) - L_m^{\Delta,i}(f - L_m^{\Delta',i}f)(x,y) \right]$$

$$+ \left[f(x,y) - \sum_{i=0}^{N} A_i(x,y)(L_m^{\Delta',i}f)(x,y) \right],$$

whence it follows that

$$\left\| R^{Li}f \right\|_\infty \leq \left\| f - \sum_{i=0}^{N} A_i L_m^{\Delta,i}f \right\|_\infty$$

$$+ \max_{0 \leq i \leq N} \sup_{x \in (x_i, x_{i+1})} \left| (f - L_m^{\Delta',i}f) - L_m^{\Delta,i}(f - L_m^{\Delta',i}f) \right|$$

$$+ \left\| f - \sum_{i=0}^{N} A_i (L_m^{\Delta',i}f) \right\|_\infty .$$

We have $f(\cdot,y) \in PC^{2m-2,\infty}[a,b]$, $(f - L_m^{\Delta',i}f)(\cdot,y) \in PC^{2m-2,\infty}[a,b]$, for all $y \in [c,d]$ and $f(x,\cdot) \in PC^{2m-2,\infty}[c,d]$, for all $x \in [a,b]$. From (11) we get that

$$\left\| R^{Li}f \right\|_\infty \leq 4d_{2m-2,0} \cdot h^{2m-2} \left\| f^{(2m-2)} \right\|_\infty$$

$$+ \max_{0 \leq i \leq N} \sup_{x \in (x_i, x_{i+1})} \left| (f - L_m^{\Delta',i}f) - L_m^{\Delta,i}(f - L_m^{\Delta',i}f) \right| \qquad (15)$$

and from (5) we obtain

$$\max_{0 \leq i \leq N} \sup_{x \in (x_i, x_{i+1})} \left| (f - L_m^{\Delta',i}f) - L_m^{\Delta,i}(f - L_m^{\Delta',i}f) \right| \leq \qquad (16)$$

$$\leq 2d_{2m-2,0} \cdot h^{2m-2} \max_{0 \leq i \leq N} \left\| (f - L_m^{\Delta',i}f)^{(2m-2)} \right\|_\infty$$

$$\leq 2d_{2m-2,0} \cdot h^{2m-2} \left\| (f - L_m^{\Delta'}f)^{(2m-2)} \right\|_\infty .$$

Finally, replacing (16) in (15) we are led to (14). $\qquad\square$

Next, we give an estimation of the approximation error in terms of the mesh length and using the modulus of smoothness of order k.

Recall that the k−th modulus of smoothness of $f \in L_p[a,b]$, $0 < p < \infty$, or of $f \in C[a,b]$, if $p = \infty$, is defined by (see, e.g., [13]):

$$\omega_k(f;t)_p = \sup_{0 < h \leq t} \left\| \Delta_h^k f(x) \right\|_p,$$

where

$$\Delta_h^k f(x) = \sum_{i=0}^{k} (-1)^{k+i} \binom{k}{i} f(x + ih).$$

We will use some results for spline approximation given in [12].

Definition 2.5. [12, p.134] *Let $T := (t_i)_1^s$ or $T := (t_i)_{-\infty}^{+\infty}$ be a finite or infinite strictly increasing sequence of points of $\mathbb{R}$; in the second case, we assume that $|t_i| \to \infty$ for $i \to \pm\infty$. A function S on $\mathbb{R}$ is a spline of order r ($r = 1, 2, \dots$), equivalently of degree $m = r-1$, with the breakpoints T if on each interval (t_i, t_{i+1}), and on the intervals $(-\infty, t_1)$, $(t_s, +\infty)$ if T is finite, it is a polynomial of degree $\leq m$, and on one of them of degree exactly m. At the breakpoints t_i, S and its derivatives (which are also splines) are defined by continuity.*

Definition 2.6. [12, p. 135] *For given $A = [a,b]$ and $T = (t_i)_1^s$ (we assume that $a < t_i < b$, $i = 1, \dots, s$) we form the Schoenberg space, denoted by $\mathcal{S}_r(T, A)$, which is the space of all splines of order $\leq r$ on A whose breakpoints are contained in T, and of smoothness $\geq m_i$ at t_i $(0 \leq m_i \leq r)$, $i = 1, \dots, s$.*

A Schoenberg space $\mathcal{S}_r(T, A)$ contains the space $\mathbb{P}_{r-1}$, of polynomials of degree $\leq r - 1$ [12, p. 135].

Definition 2.7. [12, p. 144] *A projection operator Q from L_1 onto the Schoenberg space $\mathcal{S}_r := \mathcal{S}_r(T, A)$, and thereby from each L_p onto $\mathcal{S}_r$, for each $1 \leq p \leq \infty$, is called a quasi-interpolant of order r.*

Theorem 2.8. [12, Th. 7.3., p. 225] *Given a quasi-interpolant Q of order r, for each $f \in C[a,b]$, one has the following estimation:*

$$\|f - Q(f)\|_\infty \leq C_r \omega_r (f; \delta)_\infty,$$

where C_r is a constant and δ is defined by:

$$\delta := \max_{0 \leq j \leq N} (x_{i+1} - x_i).$$

By Definition 2.7, it follows that the operators L_m^Δ and $L_m^{\Delta'}$, given in Subsection 1.3 are quasi-interpolants of order $2m$. Therefore, we can apply Theorem 2.8 for $f \in C[a,b]$ and $g \in C[c,d]$ and we obtain

$$\left\| f - L_m^\Delta(f) \right\|_\infty \leq C_{2m} \omega_{2m} (f; \delta_1)_\infty, \tag{17}$$

$$\left\| g - L_m^{\Delta'}(g) \right\|_\infty \leq C'_{2m} \omega_{2m} (g; \delta_2)_\infty,$$

where

$$\delta_1 = \max_{0 \le j \le N} (x_{i+1} - x_i), \qquad (18)$$

$$\delta_2 = \max_{0 \le j \le N} (y_{j+1} - y_j),$$

and C_{2m}, C'_{2m} are some constants.

We obtain an estimation of $R_L f$ from (10), in terms of the modulus of smoothness of high order.

Theorem 2.9. *If $f \in PC^{2m-2,\infty}[a,b]$ then*

$$\|R_L f\|_\infty \le C_{2m} \omega_{2m} (f; \delta_1)_\infty, \qquad (19)$$

with

$$\delta_1 = \max_{0 \le j \le N} (x_{i+1} - x_i).$$

Proof. We have

$$(R_L f)(x) = f(x) - \sum_{i=0}^{N} A_i(x)(L_m^{\Delta,i} f)(x)$$

$$= \sum_{i=0}^{N} A_i(x) f(x) - \sum_{i=0}^{N} A_i(x)(L_m^{\Delta,i} f)(x)$$

$$= \sum_{i=0}^{N} A_i(x)[f(x) - (L_m^{\Delta,i} f)(x)],$$

and taking into account that $\sum_{i=0}^{N} |A_i(x)| = 1$ and (17), the conclusion follows. $\square$

We obtain an estimation of the remainder for the bivariate Shepard-Lidstone formula, in terms of the modulus of smoothness of high order.

Theorem 2.10. *If $f \in PC^{2m-2,2m-2,\infty}([a,b] \times [c,d])$ then*

$$\|R^{Li} f\|_\infty \le C_{2m} \max_{y \in [c,d]} \omega_{2m} (f(\cdot,y); \delta_1)_\infty$$

$$+ C_{2m} \max_{y \in [c,d]} \omega_{2m} \left((f - L_m^{\Delta'} f)(\cdot,y); \delta_1\right)_\infty$$

$$+ C'_{2m} \max_{x \in [a,b]} \omega_{2m} (f(x,\cdot); \delta_2)_\infty,$$

where δ_1 and δ_2 are given in (18) and C_{2m}, C'_{2m} are some constants.

Proof. This result follows using the same procedure as in proof of Theorem 2.4 and applying formula (17) and two times Theorem 2.9. $\square$

Example 2.11. *Let $f : [-2,2] \times [-2,2] \to \mathbb{R}$,*

$$f(x,y) = xe^{-(x^2+y^2)}$$

and consider the nodes $z_1 = (-1,-1)$, $z_2 = (-0.5,-0.5)$, $z_3 = (-0.3,-0.1)$, $z_4 = (0,0)$, $z_5 = (0.5,0.8)$, $z_6 = (1,1)$. Below we plot the graphics of f and $S^{Li} f$.

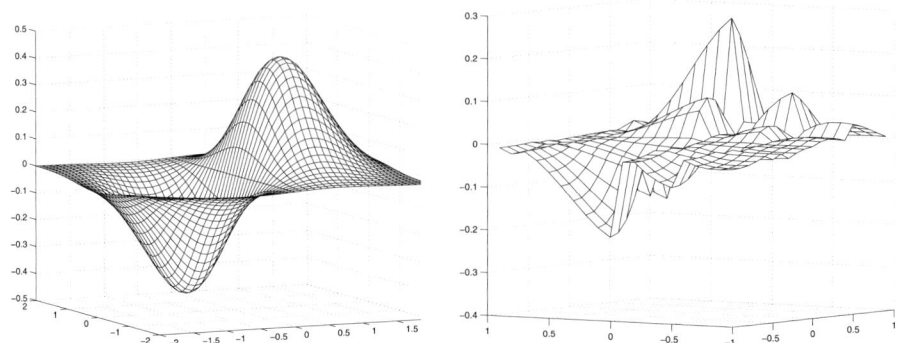

FIGURE 1. The graphic of $f(x,y) = xe^{-(x^2+y^2)}$ (left) and $S^{Li}f$ for $\mu = 1$ (right).

References

[1] R. Agarwal, P.J.Y. Wong, *Explicit error bounds for the derivatives of piecewise-Lidstone interpolation*, J. Comp. Appl. Math., **58** (1993), pp. 67–88.

[2] R. Agarwal, P.J.Y. Wong, *Error Inequalities in Polynomial Interpolation and their Applications*, Kluwer Academic Publishers, Dordrecht, 1993.

[3] T. Cătinaş, *The combined Shepard-Abel-Goncharov univariate operator*, Rev. Anal. Numér. Théor. Approx., **32** (2003) no. 1, pp. 11–20.

[4] T. Cătinaş, *The combined Shepard-Lidstone univariate operator*, Tiberiu Popoviciu Itinerant Seminar of Functional Equations, Approximation and Convexity, Cluj-Napoca, May 21–25, 2003, pp. 3–15.

[5] W. Cheney and W. Light, *A Course in Approximation Theory*, Brooks/Cole Publishing Company, Pacific Grove, 2000.

[6] Gh. Coman, *The remainder of certain Shepard type interpolation formulas*, Studia Univ. "Babeş-Bolyai", Mathematica, **32** (1987) no. 4, pp. 24–32.

[7] Gh. Coman, *Shepard-Taylor interpolation*, Itinerant Seminar on Functional Equations, Approximation and Convexity, Cluj-Napoca, (1988), pp. 5–14.

[8] Gh. Coman, *Shepard operators of Birkhoff type*, Calcolo, **35** (1998), pp. 197–203.

[9] Gh. Coman and R. Trîmbiţaş, *Combined Shepard univariate operators*, East Jurnal on Approximations, **7** (2001) no. 4, pp. 471–483.

[10] Gh. Coman and R. Trîmbiţaş, *Univariate Shepard-Birkhoff interpolation*, Rev. Anal. Numér. Théor. Approx., **30** (2001) no. 1, pp. 15–24.

[11] F.A. Costabile and F. Dell'Accio, *Lidstone approximation on the triangle*, (2002), technical report (http://lan.unical.it/Ricerca/Rapporti_interni.htm), Recondiconti di Matematica e delle sue Aplicazioni, Univ. "La Sapienza", Roma.

[12] R.A. DeVore and G.G. Lorentz, *Constructive Approxiamtion*, Springer-Verlag, 1993.

[13] Z. Ditzian and V. Totik, *Moduli of Smoothness*, Springer-Verlag, Berlin-Heidelberg-New York, Series in Computational Mathematics, vol. 9, 1987.

[14] R. Farwig, *Rate of convergence of Shepard's global interpolation formula*, Math. Comp., **46** (1986) no. 174, pp. 577–590.

[15] B. Sendov and A. Andreev, *Approximation and Interpolation Theory*, in Handbook of Numerical Analysis, vol. III, ed. P.G. Ciarlet and J.L. Lions, 1994.

[16] D.D. Stancu, Gh. Coman, O. Agratini, R. Trîmbiţaş, *Numerical Analysis and Approximation Theory*, vol. I, Presa Universitară Clujeană, 2001 (in Romanian).

[17] D.D. Stancu, Gh. Coman, P. Blaga, *Numerical Analysis and Approximation Theory*, vol. II, Presa Universitară Clujeană, 2002 (in Romanian).

[18] J. Szabados and P. Vértesi, *Interpolation of Functions*, World Scientific, Singapore, 1990.

[19] P. Vértesi, *Lower estimations for some interpolating processes*, Stud. Sci. Math. Hungar., **5** (1970), pp. 401–410.

[20] P. Vértesi, *Saturation of the Shepard operator*, Acta Math. Hungar., **72** (1996) no. 4, pp. 307–317.

Teodora Cătinaş
Babes Bolyai University
Faculty of Mathematics and Computer Science
str. M. Kogalniceanu 1
RO-3400 Cluj-Napoca, Romania
e-mail: `tcatinas@math.ubbcluj.ro`

"Lunch-Time" in Haus Bommerholz, Witten

Trends and Applications in Constructive Approximation
(Eds.) M.G. de Bruin, D.H. Mache & J. Szabados
International Series of Numerical Mathematics, Vol. 151, 91–102
© 2005 Birkhäuser Verlag Basel/Switzerland

Local RBF Approximation for Scattered Data Fitting with Bivariate Splines

Oleg Davydov, Alessandra Sestini and Rossana Morandi

Abstract. In this paper we continue our earlier research [4] aimed at developing efficient methods of local approximation suitable for the first stage of a spline based two-stage scattered data fitting algorithm. As an improvement to the pure polynomial local approximation method used in [5], a hybrid polynomial/radial basis scheme was considered in [4], where the local knot locations for the RBF terms were selected using a greedy knot insertion algorithm. In this paper standard radial local approximations based on interpolation or least squares are considered and a faster procedure is used for knot selection, significantly reducing the computational cost of the method. Error analysis of the method and numerical results illustrating its performance are given.

1. Introduction

Let $\mathbf{X} \subset \Omega$, be a set of scattered distinct sites and $\{(\mathbf{x}, f_{\mathbf{x}}) : \mathbf{x} \in \mathbf{X}, f_{\mathbf{x}} \in \mathbb{R}\}$ the set of data points to be approximated, $\Omega \subset \mathbb{R}^d$. The idea of the two-stage method [16] is to compute in the first stage a large number of *local approximations* to the data and use them in the second stage as a source of information (e.g., function values and gradients at vertices of a triangulation) for building a *global spline approximation* of the full data set using a localized *quasi-interpolation* type operator. This helps to avoid solving large linear systems and large scale optimization problems arising if the interpolating, smoothing or minimal energy spline is directly computed from the data. For a long time it has been believed that two-stage methods cannot produce approximations of the same quality as the above-mentioned global methods.

Recently, a promising two-stage bivariate spline algorithm has been developed and tested in [5, 9]. Convincing numerical evidence has been provided that the new method is efficient, robust and avoids the drawbacks usually associated with the two-stage methods. One of the goals of [4] and this paper is to improve the performance of this method at the first stage by achieving the approximation quality of the radial basis function (RBF) methods [2], in the same time also avoiding their well-known computational difficulties, by applying them only to small subsets of the data.

In the original approach (see [5]) the local polynomial approximations are computed as discrete least squares, with the polynomial degree automatically adjusted to local data by taking into account the estimates of the approximation power of local least squares [3]. In [4] and here we consider local approximation schemes defining non-polynomial approximations which are later converted into polynomials and then extended to a spline by the same method as in [5]. In [4] we have provided numerical evidence that better accuracy of the approximation may be achieved if local polynomials are augmented by linear combinations of radial basis functions, so defining *hybrid* approximations which are still computed by discrete least squares. The knot set used for each local hybrid approximation is chosen using an adaptive greedy algorithm based on successive knot insertion and estimates from [3].

In this paper we consider the standard radial approximations in the local stage that are computed by interpolation or by the least-squares method, with the local knots selected using a thinning algorithm similar to that suggested in [7] in the context of multiresolution. (Note that our motivation for thinning is entirely based on the computational considerations since the condition numbers of the matrices arising in the RBF method depend on the so-called *separation distance* of the knots.)

The paper is organized as follows. In Section 2 we introduce the local approximation scheme. In Section 3 we provide an error analysis of this version of the two-stage method based on available estimates for the RBF interpolants. Sections 4 and 5 are devoted to extensive numerical tests with two goals: to verify the approximation order of the method, and to compare the performance of this new method with the method of [4] for some real world data sets.

2. Local RBF approximation

At the first stage of a two-stage method, the local approximations are needed for each cell T of a partition of Ω associated with the spline method used. (Such a cell is usually a d-dimensional simplex or cube.) The task of the first stage is to find a good approximation of the underlying function on T. To this end, the data from some domain ω, where $T \subset \omega \subset \Omega$, are used. As in [4, 5, 9], we select the domain ω initially as a ball with center at the barycenter of T and of radius equal to the diameter of T. If the number of data points located in this ball is smaller than a user specified number $M_{\min}$, then the radius of the ball ω is enlarged until this number is achieved. Another user specified parameter, $M_{\max}$, controls the maximal number of points to be used, and a uniform type thinning is employed, if needed. Thus, this *data selection* procedure delivers a set of data sites $\mathbf{X}_\omega = \{\mathbf{x}_1, \ldots, \mathbf{x}_{N_\omega}\} \subset \mathbf{X} \cap \omega$, where

$$N_\omega \leq M_{\max}. \tag{1}$$

The local RBF approximation has the following form

$$\ell_\omega(\cdot) = \sum_{j=1}^{m} a_j \, p_j(\cdot) + \sum_{j=1}^{n_\omega} b_j \, \phi_\omega(\| \cdot - \mathbf{y}_j \|_2), \qquad (2)$$

where the set of *knots* $\mathbf{Y}_\omega = \{\mathbf{y}_j : j = 1, \ldots, n_\omega\}$ is a subset of $\mathbf{X}_\omega$, $\{p_1, \ldots, p_m\}$, $m = \binom{d+q}{d}$, is a suitable basis for the space Π_q^d of d-variate polynomials of total degree $q \geq 0$, and $\phi_\omega : \mathbb{R}_{\geq 0} \to \mathbb{R}$ is a *radial basis function*, i.e., a positive definite function or a conditionally positive definite function of order $s \leq q + 1$ on $\mathbb{R}^d$ [2], adjusted to the size of ω by scaling. Thus, we take

$$\phi_\omega(r) = \phi\left(\frac{r}{\delta d_\omega}\right), \qquad r \geq 0, \qquad (3)$$

where ϕ is a fixed radial basis function, d_ω is the diameter of ω, and δ is a user specified parameter.

In this paper we consider only positive definite radial basis functions or conditionally positive definite radial basis functions of order 1. Therefore, it is sufficient to take

$$q = 0.$$

The function ℓ_ω of the form (2) is selected by using *interpolation* on the coarse set $\mathbf{Y}_\omega$, i.e. requiring

$$\ell_\omega(\mathbf{y}_j) = f_{\mathbf{y}_j}, \qquad j = 1, \ldots, n_\omega, \qquad (4)$$

with additional *orthogonality condition*

$$\sum_{j=1}^{n_\omega} b_j = 0. \qquad (5)$$

The existence and uniqueness of such a function is guaranteed for any $\mathbf{Y}_\omega$ (see, e.g., [2]). In particular, the matrix of the corresponding linear system,

$$\begin{bmatrix} e^T & A_{\mathbf{Y}_\omega} \\ 0 & e \end{bmatrix},$$

where $e := (1, \ldots, 1)$,

$$A_{\mathbf{Y}_\omega} := \begin{bmatrix} \phi_\omega(\|\mathbf{y}_1 - \mathbf{y}_1\|_2) & \cdots & \phi_\omega(\|\mathbf{y}_1 - \mathbf{y}_{n_\omega}\|_2) \\ \vdots & & \vdots \\ \phi_\omega(\|\mathbf{y}_{n_\omega} - \mathbf{y}_1\|_2) & \cdots & \phi_\omega(\|\mathbf{y}_{n_\omega} - \mathbf{y}_{n_\omega}\|_2) \end{bmatrix},$$

is nonsingular as soon as all knots $\mathbf{y}_1, \ldots, \mathbf{y}_{n_\omega}$ are distinct.

Since the linear system arising in this interpolation problem is of the size $n_\omega + 1 \leq M_{\max} + 1$, its solution can be easily computed if $M_{\max}$ is not large, and if the matrix $A_{\mathbf{Y}_\omega}$ is well conditioned.

To complete the description of the method we now explain how we choose $\mathbf{Y}_\omega$. It is known that the condition number of $A_{\mathbf{Y}_\omega}$ can be bounded in terms of the reciprocal of the *separation distance*

$$s(\mathbf{Y}_\omega) = \frac{1}{2} \min_{1 \le i < j \le n_\omega} \|\mathbf{y}_i - \mathbf{y}_j\|_2.$$

Therefore, we choose $\mathbf{Y}_\omega$ such that

$$d_\omega / s(\mathbf{Y}_\omega) \le S, \tag{6}$$

where S is again a user specified number. To guarantee (6), the thinning algorithm from [7] is adapted.

As an alternative to interpolation, the *discrete least squares approach* [10, 12] can also be considered, i.e., ℓ_ω of the form (2) can be selected via the minimization of the *least-squares error* (the ℓ_2-norm of the residual on $\mathbf{X}_\omega$),

$$\left(\sum_{i=1}^{N_\omega} (f_i - \ell_\omega(\mathbf{x}_i))^2 \right)^{1/2}, \tag{7}$$

using the orthogonality condition (5) as a linear equality constraint. The existence and uniqueness of the least squares approximation follows from the theory of constrained least squares, see [1].

Regardless whether we use interpolation or least squares, and besides the choice of the radial basis function ϕ, the scheme depends on the following *parameters* that are supposed to be specified by the user globally, i.e., the same values are used for all local approximations:

$$M_{\min}, M_{\max}, \delta, S. \tag{8}$$

In real world applications these parameters have to be adjusted to a particular type of data by some *calibration* procedure. The local error estimates discussed in the next section can also be useful for this.

3. Error bounds

To facilitate a correct comparison to the approximation results for global methods, we mention that the approximation order of a two-stage method is the minimum of the order of the spline operator and that of the local scheme [16]. More precisely, let us assume for simplicity that the subdomains where local approximations are needed are the *cells* T of a *uniform partition* $\triangle$ of Ω associated with the spline space, which is the case for the splines used in [5]. Then the approximation error of the two-stage scheme in the uniform norm for a sufficiently smooth function can be estimated by

$$C_1 h^{p+1} + C_2 \max\{e_T : T \in \triangle\}, \tag{9}$$

where h is the diameter of the cells, $p+1$ is the approximation order of the spline quasi-interpolation operator, e_T is the error of local approximation, and C_1, C_2 are some positive constants.

To assess the approximation error of the first stage of the two-stage method we invoke some results from the approximation theory of radial basis functions on bounded domains.

Let $\ell_\omega(f)$ be the sum (2) determined by the conditions (4) and (5) with $f_{\mathbf{y}_j} = f(\mathbf{y}_j)$, $j = 1, \ldots, n_\omega$, for a function $f : \mathbb{R}^d \to \mathbb{R}$. We assume that f is smooth enough to belong to the *native space* $\mathcal{F}_{\phi_\omega}$ associated with the radial basis function ϕ_ω,

$$\mathcal{F}_{\phi_\omega} = \{f \in L_2(\mathbb{R}^d) : |f|_{\phi_\omega} < \infty\},$$

where

$$|f|_{\phi_\omega} := \left(\int_{\mathbb{R}^d} \frac{|\hat{f}(\mathbf{x})|^2}{\hat{\Phi}_\omega(\mathbf{x})} \, d\mathbf{x} \right)^{1/2}, \qquad \Phi_\omega(\cdot) := \phi_\omega(\| \cdot \|_2),$$

and $\hat{f}$ denotes the generalized Fourier transform.

Well known error bounds for the interpolation with radial basis functions (see, e.g., [2, 11, 13, 17]) lead in the case $q = 0$ to the estimate

$$|f(\mathbf{x}) - \ell_\omega(f, \mathbf{x})| \le 2 \sqrt{E_0(\Phi_\omega)_{C(B_{h(\mathbf{x}, \mathbf{Y}_\omega)})}} \, |f|_{\phi_\omega}, \quad \mathbf{x} \in \mathbb{R}^d, \tag{10}$$

where $h(\mathbf{x}, \mathbf{Y}_\omega)$ is the distance between $\mathbf{x}$ and $\mathbf{Y}_\omega$,

$$h(\mathbf{x}, \mathbf{Y}_\omega) := \inf_{\mathbf{y} \in \mathbf{Y}_\omega} \|\mathbf{x} - \mathbf{y}\|_2,$$

$E_0(\Phi_\omega)_{C(B_{h(\mathbf{x}, \mathbf{Y}_\omega)})}$ is the error of the best constant approximation of Φ_ω,

$$E_0(\Phi_\omega)_{C(B_{h(\mathbf{x}, \mathbf{Y}_\omega)})} = \inf_{p \in \Pi_0^d} \|f - p\|_{C(B_{h(\mathbf{x}, \mathbf{Y}_\omega)})},$$

and B_r denotes the ball in $\mathbb{R}^d$ with center $\mathbf{0}$ and radius r.

Assuming that ϕ is monotone (which is true for all available radial basis functions at least in a neighborhood of zero) and considering the *fill distance* of $\mathbf{Y}_\omega$ with respect to T,

$$h(T, \mathbf{Y}_\omega) = \sup_{\mathbf{x} \in T} h(\mathbf{x}, \mathbf{Y}_\omega),$$

we have for all $x \in T$,

$$E_0(\Phi_\omega)_{C(B_{h(\mathbf{x}, \mathbf{Y}_\omega)})} \le E_0(\Phi_\omega)_{C(B_{h(T, \mathbf{Y}_\omega)})} = \frac{1}{2} |\phi_\omega(h(T, \mathbf{Y}_\omega)) - \phi_\omega(0)|,$$

which leads to the estimate

$$\|f - \ell_\omega(f)\|_{C(T)} \le \sqrt{2|\phi_\omega(h(T, \mathbf{Y}_\omega)) - \phi_\omega(0)|} \cdot |f|_{\phi_\omega}. \tag{11}$$

Obviously, $h(T, \mathbf{Y}_\omega) \le d_\omega$, and taking into account (3) we have

$$|\phi_\omega(h(T, \mathbf{Y}_\omega)) - \phi_\omega(0)| = \left| \phi\left(\frac{h(T, \mathbf{Y}_\omega)}{\delta d_\omega} \right) - \phi(0) \right| \le |\phi(1/\delta) - \phi(0)|, \tag{12}$$

which shows that increasing the value of the parameter δ may have a positive effect on the error. It should, however, be taken into account that the seminorm $|f|_{\phi_\omega}$ also depends on δ, in view of (3).

Among the most commonly used radial basis functions are the *thin plate splines*

$$\phi^{TP,\beta}(r) = \begin{cases} (-1)^{\lceil \beta/2 \rceil} r^{\beta}, & \beta \in \mathbb{R}_{>0} \setminus 2\mathbb{N}, \\ (-1)^{\beta/2+1} r^{\beta} \log r, & \beta \in 2\mathbb{N}, \end{cases} \tag{13}$$

that are conditionally positive definite of order $\lceil \beta/2 \rceil$ if $\beta \in \mathbb{R}_{>0} \setminus 2\mathbb{N}$, and $\beta/2+1$ if $\beta \in 2\mathbb{N}$. (Here $\lceil x \rceil$ denotes the smallest integer greater or equal to $x \in \mathbb{R}$.) Therefore they can be used in our scheme (where the polynomial degree is $q = 0$) if $0 < \beta < 2$.

The approximation order of the thin plate splines is understood better than that of the other available RBFs because their Fourier transform is *homogeneous*, $\hat{\Phi}^{TP,\beta}(\mathbf{x}) = K\|\mathbf{x}\|_2^{-\beta-d}$, with some constant K independent of $\mathbf{x}$. Therefore,

$$|f|^2_{\phi^{TP,\beta}_\omega} = (\delta d_\omega)^\beta |f|^2_{\phi^{TP,\beta}},$$

and we obtain from (11) and (12),

$$\|f - \ell_\omega(f)\|_{C(T)} \le \sqrt{2\, h(T, \mathbf{Y}_\omega)^\beta} \, |f|_\phi, \qquad 0 < \beta < 2. \tag{14}$$

By our algorithm (see Section 2), we always have $d_\omega \ge 2h$, where h is the diameter of the cell T, which is the same for all cells in our setting. On the other hand, we may assume that there is enough data so that $d_\omega \le ch$, for a constant c. Taking into account (6) and the obvious inequality $s(\mathbf{Y}_\omega) \le h(T, \mathbf{Y}_\omega)$, we have

$$2h/S \le d_\omega/S \le h(T, \mathbf{Y}_\omega) \le d_\omega \le ch.$$

Therefore, (9) and (14) suggest that the approximation order of the two-stage method with thin plate splines in the local stage should be $\mathcal{O}(h^{\min\{\beta/2,p+1\}})$ or, assuming that p is high enough, $\mathcal{O}(h^{\beta/2})$. Note that since the cell T where we use the local approximations covers only the central part of ω, the deterioration of the error near the boundary of ω only affects the quality of the local approximations at the boundary of the entire domain Ω.

Finally, we mention that the above estimates can be improved if f satisfies some more stringent requirements than $f \in \mathcal{F}_{\phi_\omega}$, see [2, 14, 15]. The improvement amounts basically (up to a constant factor) to removing the square root sign in (10), (11) and (14), and replacing the seminorm $|f|_{\phi_\omega}$ with a stronger seminorm $|f|$, whose boundedness requires "higher smoothness" of f.

In particular, for the thin plate splines the approximation order becomes $\mathcal{O}(h^\beta)$. Moreover, in this case the order $\mathcal{O}(h^{\beta+d/2})$ for scattered data and $\mathcal{O}(h^{\beta+d})$ for grid data has been proved (see [2]).

4. Numerical results: Approximation order

In our numerical experiments we restrict ourselves to the two-dimensional case $d = 2$. This section is devoted to numerical tests with randomly generated data for the well-known Franke test function [8]. The goals of the tests are to measure the approximation order of the two-stage method, compare it with the theoretical error bounds, and get hints for the selection of good values of the parameters (8) for the local approximation.

More precisely, 40 different random data sets $\mathbf{X}_i$, $i = 1, \ldots, 40$, of cardinality $\#\mathbf{X}_i = N$ were generated in the reference square $[0, 1]^2$, for $N = 10^2, 10^3, 10^4, 10^5$. For the second stage of the two-stage method we have chosen the method SQ_2^{av} of [5] which produces C^2 piecewise sextic splines on the four directional mesh. Based on our experiments with the Franke test function in [4], we take the grid size for the spline space to be $n \times n$, where n is the closest integer to $\sqrt{N}/2$. The experiments have been performed using the implementation of the spline operators in [6].

To measure the approximation error, we compute the maximum error ϵ_i of the spline relative to the exact function values on a dense $(10n + 1) \times (10n + 1)$ grid in a suitably reduced window $([0.2, 0.8]^2)$ for every data set $\mathbf{X}_i$ and take the geometric average max $= \exp(\frac{1}{40} \sum_{i=1}^{40} \ln \epsilon_i)$ of these errors. We think that the geometric average is the most appropriate way of averaging for the approximation order tests. The motivation for using the reduced window is our desire to avoid boundary effects.

In the local stage we use the interpolation method described in Section 2 above and choose 1) the *thin plate spline* $\phi(r) = -r^\beta$, $\beta = 3/2$ or $\beta = 7/4$, and 2) the *multiquadrics* $\phi(r) = -\sqrt{1 + r^2}$ for the experiments. We have chosen a high value $M_{max} = 400$ to eliminate the influence of this parameter and tried to find nearly optimal values for M_{min}, S and δ. The results are presented in Tables 1 and 2.

N	spline grid	max ($\beta = 3/2$)	max ($\beta = 7/4$)
10^2	5×5	$8.55 \cdot 10^{-2}$	$6.92 \cdot 10^{-2}$
10^3	16×16	$5.22 \cdot 10^{-3}$	$3.37 \cdot 10^{-3}$
10^4	50×50	$2.60 \cdot 10^{-4}$	$1.17 \cdot 10^{-4}$
10^5	158×158	$2.37 \cdot 10^{-5}$	$7.09 \cdot 10^{-6}$

TABLE 1. Maximum error using the local RBF interpolation scheme based on $\phi(r) = -r^\beta$, $\beta = 3/2$ and $7/4$. Parameter values: $M_{min} = 100$, $S = 100$, $\delta = 1$.

For the thin plate spline (Table 1), the experiments confirm that the parameter δ does not influence the error significantly. Therefore, we have chosen a nominal value $\delta = 1$. Note that the average number of RBF knots in the local approximations approaches 140 for the larger data sets, which makes these tests particularly slow. Although an increase of M_{min} was always profitable for $\phi(r) = -r^\beta$, fewer knots were sufficient to achieve nearly optimal errors for $N < 10^5$. In this sense nearly optimal values of the parameter M_{min} are: $M_{min} = 20$ for $N = 10^2$ (28 knots), $M_{min} = 30$ for $N = 10^3$ (45 knots), and $M_{min} = 60$ for $N = 10^4$ (65 knots). (We have taken $S = M_{min}$ in these tests.)

Table 1 suggests the approximation order about $h^{\beta+1}$, which conforms nicely to the available theoretical results, see the comments at the end of Section 3. Note that the approximation error of the spline operator SQ_2^{av} is $\mathcal{O}(h^7)$ [5] and hence it is negligible for this test.

N	$\delta = 0.4$ $S = 40$	$\delta = 0.8$ $S = 20$	$\delta = 1.2$ $S = 40/3$	$\delta = 1.6$ $S = 10$
10^2	$2.27 \cdot 10^{-2}$	$2.81 \cdot 10^{-2}$	$3.66 \cdot 10^{-2}$	$4.48 \cdot 10^{-2}$
10^3	$1.26 \cdot 10^{-5}$	$4.44 \cdot 10^{-6}$	$6.13 \cdot 10^{-5}$	$5.42 \cdot 10^{-4}$
10^4	$4.20 \cdot 10^{-6}$	$1.98 \cdot 10^{-7}$	$1.00 \cdot 10^{-7}$	$2.36 \cdot 10^{-7}$
10^5	$2.03 \cdot 10^{-6}$	$9.28 \cdot 10^{-8}$	$3.54 \cdot 10^{-8}$	$5.60 \cdot 10^{-8}$
#*knots* ($N = 10^5$)	122.2	87.2	60.4	42.8

TABLE 2. Maximum error using the local RBF interpolation scheme based on $\phi(r) = -\sqrt{1 + r^2}$ for different values of δ. The spline grid is the same as in Table 1. Other parameters: $M_{\min} = 20$ if $N = 10^2$ and $M_{\min} = 100$ otherwise.

It is clear from Table 2 that in the case of multiquadrics the correct choice of the parameter δ (which clearly is related to the reciprocal of the classical multiquadric coefficient c in $\sqrt{c^2 + r^2}$) is important. However, we had to choose the separation parameter S such that $\delta S \leq 16$ since otherwise the computation with multiquadrics turned out numerically instable. (Note that for the real world data, like those tested below in Section 5, δS must be even smaller.) The values of $M_{\min}$ lower than 100 were disadvantageous in our experiments for all N except $N = 10^2$. In the case $N = 10^2$, however, $M_{\min} = 100$ delivers relatively high errors: $1.94 \cdot 10^{-2}$ for $\delta = 0.2, S = 80$, $3.75 \cdot 10^{-2}$ for $\delta = 0.4, S = 40$, $6.59 \cdot 10^{-2}$ for $\delta = 0.8, S = 20$, $1.08 \cdot 10^{-1}$ for $\delta = 1.2, S = 40/3$, $1.65 \cdot 10^{-1}$ for $\delta = 1.6, S = 10$. (Note that $N = M_{\min} = 100$ means, in fact, that all local approximations are the same, and, hence, our spline does not differ much from the corresponding *global* multiquadric approximation.) Therefore, we use $M_{\min} = 20$ if $N = 10^2$, and $M_{\min} = 100$ for other N. The value $M_{\min} > 100$ may be advantageous for smaller δ. For example, for $N = 10^3$ and $M_{\min} = 200$ we have $3.36 \cdot 10^{-6}$ if $\delta = 0.4, S = 40$, and $1.26 \cdot 10^{-5}$ if $\delta = 0.8, S = 20$.

The results in Table 2 confirm that greater values of δ tend to provide better errors. Indeed, for higher N we have to increase δ in order to obtain the best errors, even though numerical stability considerations force us to take smaller S, which in turn leads to the reduction of the number of RBF knots (see the last row of the table). For any fixed δ, however, Table 2 shows a substantial deterioration of the approximation order as N increases. The estimates of Section 2 do not provide a full theoretical explanation for this behavior. In particular, (11) includes the term $|f|_{\phi_\omega}$, whose behavior for $d_\omega \to 0$ is not clear to us in the case of multiquadrics.

5. Numerical results: Real world data

The second group of our experiments is aimed at verifying the performance of the proposed scheme compared with the hybrid approach introduced in [4].

To this end we consider the same real-world data sets as in [4, 5], namely, the Glacier data (GL, 8345 points), the Black Forest data (BF, 15885 points) and the Rotterdam Port data (RP, 621624 points after cleaning, see [5]). Referring to [4, 5] for the description of these data sets, we only mention that GL is available from [8] and that RP has been provided by Quality Positioning Services (Zeist, The Netherlands), and it has been recorded using the QINSy software.

Note that in the first stage we use the *least-squares* method as described at the end of Section 2 since it consistently produced better results than interpolation for the real world data in our tests. To solve the constrained least squares problems we employ the routine DGESDD from LAPACK. (Note that the interpolation method in our implementation is also treated as special case of least squares.)

The results are reported in Table 3, where maximum (max), mean (*mean*) and root mean square (*rms*) errors at the data points, the average number of RBF knots (#*knots*) used for the local approximations, and the computational time (*time*) are shown. Results obtained with the method suggested in this paper (R) are compared with the hybrid approach (H) of [4]. As in [4], we use *multiquadric* RBF in these tests. The degree q of the polynomial term is 0 in all tests except RP/H, where $q = 1$. In the second stage we use the spline methods RQ_2^{av} (piecewise sextic) for GL and BF and Q_1^{av} (piecewise cubic) for RP, as in the respective tests in [4, 5]. The computer used for these experiments is a Pentium 4m / 1.9 GHz / 768 MB RAM.

	GL/H	GL/R	BF/H	BF/R	RP/H	RP/R
max	15.6 m	17.8 m	32.0 m	30.0 m	92.5 cm	90.2 cm
mean	1.57 m	1.49 m	1.39 m	1.25 m	5.46 cm	5.23 cm
rms	2.26 m	2.19 m	2.17 m	2.00 m	7.46 cm	7.22 cm
#*knots*	14.9	20.8	12.2	8.2	5.4	11.6
time	33.0 sec	3.7 sec	134 sec	7.11 sec	316 sec	97.3 sec

TABLE 3. Results for the data sets GL, BF and RP. Parameters (see [4] for the meaning of κ_P and κ_H): 1) GL. Spline grid 20×24, $M_{min} = 60$, $M_{max} = 160$, $\delta = 0.4$ for both H and R methods, $\kappa_H = 10^5$ for H, and $S = 8$ for R. 2) BF. Spline grid 80×80, $M_{max} = 100$ for both H und R methods, $M_{min} = 12$, $\delta = 0.3$, $\kappa_H = 10^4$ for H, and $M_{min} = 3$, $\delta = 0.4$, $S = 7$ for R. 3) RP. Spline grid 100×281, $M_{min} = 3$, $M_{max} = 100$, $\delta = 0.4$ for both H and R, $\kappa_P = 100$, $\kappa_H = 2 \cdot 10^4$ for H, and $S = 5$ for R.

In addition to Table 3, we provide Figures 1–3 that present zooms into the same subareas of the surfaces as ones used in [4]. They are produced with MATLAB using dense grid evaluations of the spline surfaces (see [4]). The figures confirm the high visual quality of our approximations, as they show no artifical oscillation or other unnatural behaviour. Table 3 also shows that the errors for both H and R methods are comparable, whereas the computational time for the method of this paper is substantially lower.

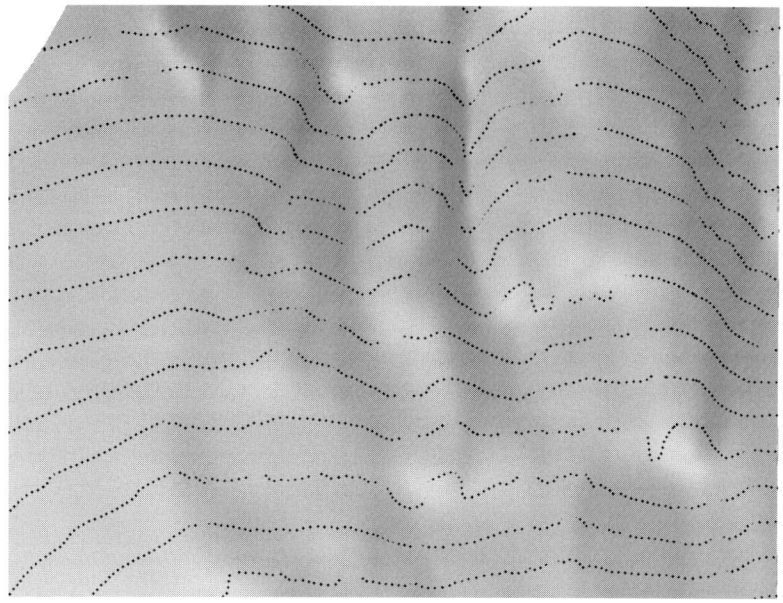

FIGURE 1. Glacier test.

References

[1] A. Björck, Numerical methods for least squares problems, SIAM, Philadelphia, 1996.

[2] M D. Buhmann, Radial basis functions, Cambridge University Press, 2003.

[3] O. Davydov, On the approximation power of local least squares polynomials, in algorithms for approximation IV, J. Levesley, I.J. Anderson and J.C. Mason (eds), 2002, 346–353.

[4] O. Davydov, R Morandi and A. Sestini, Local hybrid approximation for scattered data fitting with bivariate splines, manuscript, 2003.
http://www.uni-giessen.de/~gcn5/davydov/

[5] O. Davydov and F. Zeilfelder, Scattered data fitting by direct extension of local polynomials to bivariate splines, Adv. Comp. Math. 21 (2004), 223–271.

[6] O. Davydov and F. Zeilfelder, Toolbox for two-stage scattered data fitting, in preparation.

[7] M.S. Floater and A. Iske, Thinning algorithms for scattered data interpolation, BIT 38 (1998), 705–720.

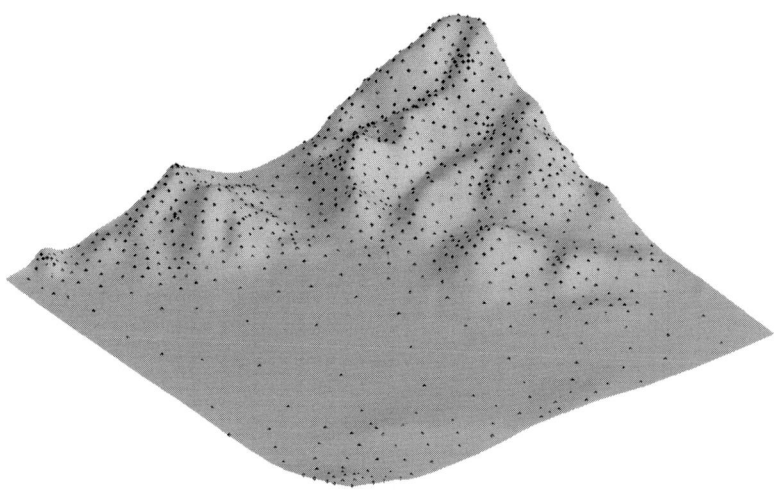

FIGURE 2. Black Forest test.

FIGURE 3. Rotterdam Port test.

[8] R. Franke, Homepage, http://www.math.nps.navy.mil/~rfranke/, Naval Postgraduate School.

[9] J. Haber, F. Zeilfelder, O. Davydov, H.-P. Seidel, Smooth approximation and rendering of large scattered data sets, in: Proceedings of IEEE Visualisation 2001 eds. T. Ertl, K. Joy and A. Varshney, 2001, pp. 341–347, 571.

[10] A. Iske, Reconstruction of smooth signals from irregular samples by using radial basis function approximation, in: Proceedings of the 1999 International Workshop on Sampling Theory and Applications, The Norwegian University of Science and Technology, Trondheim, 1999, pp. 82–87.

[11] K. Jetter, J. Stöckler and J.D. Ward, Error estimates for scattered data interpolation on spheres, Math. Comp. 68 (1999), 733–747.

[12] J.R. McMahon and R. Franke, Knot selection for least squares thin plate splines, SIAM J. Sci. Stat. Comput. 13 (1992), 484–498.

[13] R. Schaback, Reconstruction of multivariate functions from scattered data, manuscript, 1997. http://www.num.math.uni-goettingen.de/schaback/

[14] R. Schaback, Improved error bounds for scattered data interpolation by radial basis functions, Math. Comp. 68 (1999), 201–216.

[15] R. Schaback and H. Wendland, Inverse and saturation theorems for radial basis function interpolation, Math. Comp. 71 (2002), 669–681.

[16] L.L. Schumaker, Fitting surfaces to scattered data, in: Approximation Theory II eds. G.G. Lorentz, C.K. Chui, and L.L. Schumaker, Academic Press, New York, 1976, pp. 203–268.

[17] H. Wendland, Gaussian interpolation revisited, in: Trends in Approximation Theory eds. K. Kopotun, T. Lyche and M. Neamtu, Vanderbilt University Press, Nashville, TN, 2001, pp. 427–436.

Oleg Davydov
Mathematisches Institut
Justus-Liebig-Universität Giessen
D-35392 Giessen
Germany
e-mail: oleg.davydov@math.uni-giessen.de

Alessandra Sestini and Rossana Morandi
Dipartimento di Energetica
Università di Firenze
Via Lombroso 6/17
I-50134 Firenze, Italy
e-mail: morandi@de.unifi.it, sestini@de.unifi.it

Trends and Applications in Constructive Approximation
(Eds.) M.G. de Bruin, D.H. Mache & J. Szabados
International Series of Numerical Mathematics, Vol. 151, 103–123
© 2005 Birkhäuser Verlag Basel/Switzerland

Evolutionary Optimization of Neural Systems: The Use of Strategy Adaptation

Christian Igel, Stefan Wiegand and Frauke Friedrichs

Abstract. We consider the synthesis of neural networks by evolutionary algorithms, which are randomized direct optimization methods inspired by neo-Darwinian evolution theory. Evolutionary algorithms in general as well as special variants for real-valued optimization and for search in the space of graphs are introduced. We put an emphasis on strategy adaptation, a feature of evolutionary methods that allows for the control of the search strategy during the optimization process.

Three recent applications of evolutionary optimization of neural systems are presented: topology optimization of multi-layer neural networks for face detection, weight optimization of recurrent networks for solving reinforcement learning tasks, and hyperparameter tuning of support vector machines.

1. Introduction

The information processing capabilities of vertebrate brains outperform technical systems in many respects. Abstract models of neural networks (NNs) exist, which can – in principle – simulate all Turing machines and exhibit universal approximation properties (e.g., [40, 42, 44]). However, the general question of how to efficiently design an appropriate neural system for a given problem remains open and complexity theory reveals the need for using heuristics (e.g., [41]). The answer is likely to be found by investigating the three major organization principles of biological NNs: evolution, self-organization, and learning.

In the following, we consider the synthesis of NNs by evolutionary algorithms (EAs), which are randomized direct optimization methods inspired by neo-Darwinian evolution theory. We focus on strategy adaptation, a feature of evolutionary methods that enables control of the search strategy during the optimization process. Two related ways of adapting search strategies in evolutionary computation are presented. First, we describe the CMA evolution strategy [19], an efficient method for adjusting the covariance matrix of the search distribution in real-valued optimization. Second, we introduce an algorithm that adapts the probabilities of variation operators [24]. We sketch applications of these methods to the design of different types of neural systems: topology optimization of multi-layer NNs for face detection [49], weight optimization of recurrent networks for solving reinforcement learning tasks [21], and model selection for support vector machines [14].

An introduction to NNs is beyond the scope of this article, the reader is referred to the standard literature, for example the collection [1]. The articles [12, 4] provide starting points for reading about the theory of evolutionary computation. General surveys of evolutionary optimization of neural networks can be found in [29, 30, 39, 52].

2. Evolutionary computation

Evolutionary algorithms (EAs) can be considered as a special class of global random search algorithms. Let the search problem under consideration be described by a quality function $f : \mathcal{G} \to \mathcal{F}$ to be optimized, where $\mathcal{G}$ denotes the search space (i.e., the space of candidate solutions) and $\mathcal{F}$ the (at least partially) ordered space of cost values. The general global random search scheme can be described – with slight modifications – as follows [54, 25]:

$\boxed{1}$ Choose a joint probability distribution $P_{\mathcal{G}^\lambda}^{(t)}$ on $\mathcal{G}^\lambda$. Set $t \leftarrow 1$.

$\boxed{2}$ Obtain λ points $\boldsymbol{g}_1^{(t)}, \dots, \boldsymbol{g}_\lambda^{(t)}$ by sampling from the distribution $P_{\mathcal{G}^\lambda}^{(t)}$. Evaluate these points using f.

$\boxed{3}$ According to a fixed (algorithm dependent) rule construct a new probability distribution $P_{\mathcal{G}^\lambda}^{(t+1)}$ on $\mathcal{G}^\lambda$.

$\boxed{4}$ Check for some appropriate stopping condition; if the algorithm has not terminated, substitute $t \leftarrow t + 1$ and return to step [2].

Random search algorithms can differ fundamentally in the way they describe (parameterize) and alter the joint distribution $P_{\mathcal{G}^\lambda}^{(t)}$, which is typically represented by a semi-parametric model.

The scheme of a canonical EA is shown in figure 1. In evolutionary computation, the iterations of the algorithm are called *generations*. The search distribution of an EA is given by the *parent population*, the *variation operators*, and the *strategy parameters*.

The parent population is a multiset of μ points $\tilde{\boldsymbol{g}}_1^{(t)}, \dots, \tilde{\boldsymbol{g}}_\mu^{(t)} \in \mathcal{G}$. Each point corresponds to the *genotype* of an *individual*. In each generation, λ *offspring* $\boldsymbol{g}_1^{(t)}, \dots, \boldsymbol{g}_\lambda^{(t)} \in \mathcal{G}$ are created by the following procedure: Individuals for reproduction are chosen from $\tilde{\boldsymbol{g}}_1^{(t)}, \dots, \tilde{\boldsymbol{g}}_\mu^{(t)}$. This is called *mating selection* and can be deterministic or stocastic (where the sampling can be with or without replacement). The offspring's genotypes result from applying variation operators to these selected parents. Variation operators are deterministic or partially stochastic mappings from $\mathcal{G}^k$ to $\mathcal{G}^l$, $1 \leq k \leq \mu, 1 \leq l \leq \lambda$. An operator with $k = l = 1$ is called *mutation*, whereas *recombination* operators involve more than one parent and can lead to more than one offspring. Multiple operators can be applied consecutively to generate offspring. For example, an offspring $\boldsymbol{g}_i^{(t)}$ can be the product of applying recombination $o_{\mathrm{rec}} : \mathcal{G}^2 \to \mathcal{G}$ to two randomly selected parents $\tilde{\boldsymbol{g}}_{i_1}^{(t)}$ and $\tilde{\boldsymbol{g}}_{i_2}^{(t)}$ followed by mutation $o_{\mathrm{mut}} : \mathcal{G} \to \mathcal{G}$, that is, $\boldsymbol{g}_i^{(t)} = o_{\mathrm{mut}}\left(o_{\mathrm{rec}}\left(\tilde{\boldsymbol{g}}_{i_1}^{(t)}, \tilde{\boldsymbol{g}}_{i_2}^{(t)}\right)\right)$. Evolutionary algorithms allow for incorporation of *a priori* knowledge about the problem by using tailored variation operators combined with an appropriate encoding of the candidate solutions.

Let the probability that parents $\tilde{\boldsymbol{g}}_1^{(t)}, \ldots, \tilde{\boldsymbol{g}}_\mu^{(t)}$ lead to offspring $\boldsymbol{g}_1^{(t)}, \ldots, \boldsymbol{g}_\lambda^{(t)}$ be described by the conditional probability distribution

$$P_{\mathcal{G}^\lambda}\left(\boldsymbol{g}_1, \ldots, \boldsymbol{g}_\lambda \mid \tilde{\boldsymbol{g}}_1^{(t)}, \ldots, \tilde{\boldsymbol{g}}_\mu^{(t)}; \boldsymbol{\theta}^{(t)}\right) = P_{\mathcal{G}^\lambda}^{(t)}(\boldsymbol{g}_1, \ldots, \boldsymbol{g}_\lambda) . \tag{1}$$

This distribution is additionally parameterized by some *external strategy parameters* $\boldsymbol{\theta}^{(t)} \in \Theta$, which may vary over time. In some EAs, the offspring are created independently of each other based on the same distribution. In this case, the joint distribution $P_{\mathcal{G}^\lambda}^{(t)}$ can be factorized as

$$P_{\mathcal{G}^\lambda}^{(t)}(\boldsymbol{g}_1, \ldots, \boldsymbol{g}_\lambda) = P_{\mathcal{G}}^{(t)}(\boldsymbol{g}_1) \cdot \, \cdots \, \cdot P_{\mathcal{G}}^{(t)}(\boldsymbol{g}_\lambda) . \tag{2}$$

Evaluation of an individual corresponds to determining its fitness by assigning the corresponding cost value given by the quality function f. Evolutionary algorithms can handle optimization problems that are non-differentiable, non-continuous, multi-modal, and noisy. They are easy to parallelize by distributing the fitness evaluations of the offspring.

Updating the search distribution consists of two steps, *environmental selection* and sometimes *strategy adaptation* of external strategy parameters: A selection method chooses μ new parents $\tilde{\boldsymbol{g}}_1^{(t+1)}, \ldots, \tilde{\boldsymbol{g}}_\mu^{(t+1)}$ from $\tilde{\boldsymbol{g}}_1^{(t)}, \ldots, \tilde{\boldsymbol{g}}_\mu^{(t)}$ and $\boldsymbol{g}_1^{(t)}, \ldots, \boldsymbol{g}_\lambda^{(t)}$. This second selection process is called environmental selection and may be deterministic or stochastic. Either the mating or the environmental selection must be based on the objective function values of the individuals and must prefer those with better fitness – this is the driving force of the evolutionary adaptation process. An example of fitness-dependent environmental selection is choosing the μ best individuals out of $\lambda > \mu$ offspring. In addition, the EA may update external strategy parameters as discussed in the following section.

2.1. Strategy adaptation

Strategy adaptation, that is the automatic adjustment of the search strategy during the optimization process, is a key concept to improve the performance in evolutionary computation [11, 24, 43]. It is necessary because the best search strategy for a problem is usually not known in advance and typically changes (e.g., from coarse to fine) during optimization. Examples of strategy parameters that can be

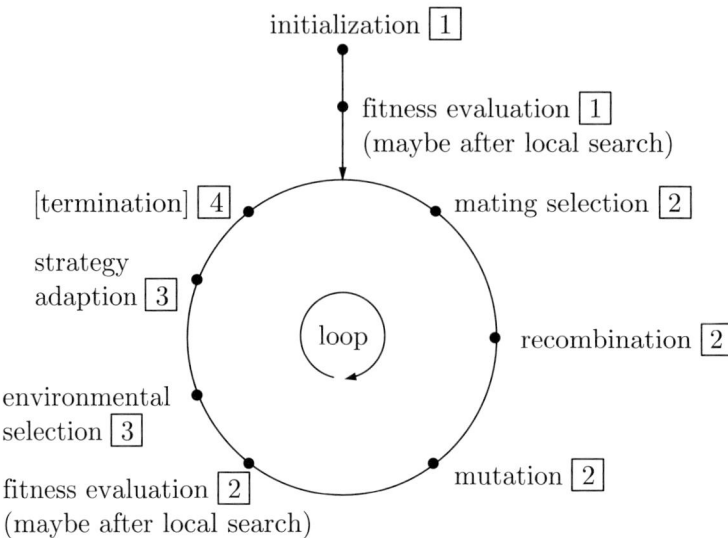

FIGURE 1. Basic EA loop. The numbers indicate the corresponding steps in the random search scheme. When optimizing adaptive systems, the local search usually corresponds to some learning process.

controlled externally include population sizes, the probabilities that certain variation operators are applied, and parameters that determine the mutation strength.

In the following section, we describe an efficient algorithm for adjusting the covariance matrix of Gaussian mutations in EAs. Thereafter, we present a way for adaptation of the application probabilities of variation operators. Both methods are deterministic and monitor the effects of the variation operators over the generations. They are based on the same rule of thumb that recent beneficial mutations are also likely to be beneficial in the following generations.

2.1.1. The CMA evolution strategy. Evolution strategies (ES, [3, 32, 38]) are one of the main branches of EAs. In the following, we describe the covariance matrix adaptation ES (CMA-ES) proposed in [18, 19], which performs efficient real-valued optimization. Each individual represents an n-dimensional real-valued object variable vector. These variables are altered by two variation operators, intermediate recombination and additive Gaussian mutation. The former corresponds to computing the center of mass of the μ individuals in the parent population. Mutation is realized by adding a normally distributed random vector with zero mean. In the CMA-ES, the complete covariance matrix of the Gaussian mutation distribution is adapted during evolution to improve the search strategy. More formally, the object parameters $g_k^{(t)}$ of offspring $k = 1, \ldots, \lambda$ created in generation t are given by

$$g_k^{(t)} = \langle \tilde{g} \rangle^{(t)} + \sigma^{(t)} B^{(t)} D^{(t)} z_k^{(t)} , \qquad (3)$$

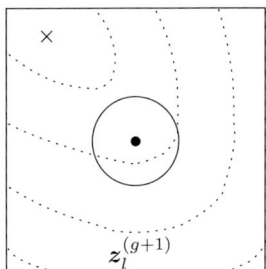
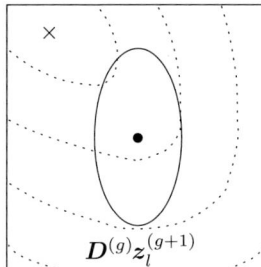
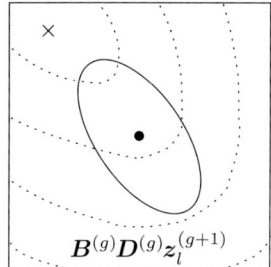

FIGURE 2. The dashed lines schematically visualize an error / fitness surface (landscape) for $\mathcal{G} \subset \mathbb{R}^2$, where each line represents points of equal fitness and the $\times$ symbol marks the optimum. The dot corresponds to the center of mass of the parent population and the solid lines indicate the mutation (hyper-) ellipsoids (i.e., surfaces of equal probability density to place an offspring) of the random vectors after the different transformations.

Evolution strategies that adapt only one global step size can only produce mutation ellipsoids as shown in the left plot. Algorithms that adapt n different step sizes, one for each object variable, can produce mutation ellipsoids scaled along the coordinate axes like the one shown in the center plot. Only if the complete covariance matrix is adapted, arbitrary normal distributions can be realized as shown in the right picture.

where $\langle \tilde{g} \rangle^{(t)} = \frac{1}{\mu} \sum_{i=1}^{\mu} \tilde{g}_i^{(t)}$ is the center of mass of the parent population in generation t and the $z_k^{(t)} \sim \mathcal{N}(0, I)$ are independent realizations of an n-dimensional normally distributed random vector with zero mean and covariance matrix equal to the identity matrix I. The covariance matrix $C^{(t)}$ of the random vectors

$$\sigma^{(t)} B^{(t)} D^{(t)} z_k^{(t)} \sim \mathcal{N}(0, C^{(t)}) \tag{4}$$

is a symmetric positive $n \times n$ matrix with

$$C'^{(t)} = C^{(t)} / \sigma^{{(t)}^2} = B^{(t)} D^{(t)} \left(B^{(t)} D^{(t)} \right)^T . \tag{5}$$

The columns of the orthogonal $n \times n$ matrix $B^{(t)}$ are the normalized eigenvectors of $C'^{(t)}$ and $D^{(t)}$ is a $n \times n$ diagonal matrix with the square roots of the corresponding eigenvalues. Figure 2 schematically shows the transformations of $z_k^{(t)}$ by $B^{(t)}$ and $D^{(t)}$.

The strategy parameters, both the matrix $C'^{(t)}$ and the so-called global step-size $\sigma^{(t)}$, are updated online using the covariance matrix adaptation (CMA) method. The key idea of the CMA is to alter the mutation distribution in a deterministic way such that the probability to reproduce steps in the search space that have led to the current population is increased. This enables the algorithm to detect correlations between object variables and to become invariant under orthogonal transformations of the search space (apart from the initialization). In order to use the information from previous generations efficiently, the search path of the population over a number of past generations is taken into account.

In the CMA-ES, rank-based (μ, λ)-selection is used for environmental selection. That is, the μ best of the λ offspring form the next parent population. After selection, the strategy parameters are updated:

$$\boldsymbol{s}^{(t+1)} = (1-c) \cdot \boldsymbol{s}^{(t)} + \underbrace{c_{\mathrm{u}} \cdot \sqrt{\mu} \, \boldsymbol{B}^{(t)} \boldsymbol{D}^{(t)} \langle \boldsymbol{z} \rangle_\mu^{(t)}}_{\frac{\sqrt{\mu}}{\sigma^{(t)}} \left(\langle \tilde{\boldsymbol{g}} \rangle^{(t+1)} - \langle \tilde{\boldsymbol{g}} \rangle^{(t)} \right)} \tag{6}$$

$$\boldsymbol{C}'^{(t+1)} = (1 - c_{\mathrm{cov}}) \cdot \boldsymbol{C}'^{(t)} + c_{\mathrm{cov}} \cdot \boldsymbol{s}^{(t+1)} \left(\boldsymbol{s}^{(t+1)} \right)^{\mathrm{T}} . \tag{7}$$

Herein, $\boldsymbol{s}^{(t+1)} \in \mathbb{R}^n$ is the evolution path – a weighted sum of the centers of the population over the generations starting from $\boldsymbol{s}^{(2)} = \sqrt{\mu} \, \boldsymbol{B}^{(1)} \boldsymbol{D}^{(1)} \langle \boldsymbol{z} \rangle_\mu^{(1)}$ (the factor $\sqrt{\mu}$ compensates for the loss of variance due to computing the center of mass). The parameter $c \in {]}0,1]$ controls the time horizon of the adaptation of $\boldsymbol{s}$; we set $c = 1/\sqrt{n}$. The constant $c_{\mathrm{u}} = \sqrt{c(2-c)}$ normalizes the variance of $\boldsymbol{s}$ (viewed as a random variable) as $1^2 = (1-c)^2 + c_{\mathrm{u}}^2$. The expression $\langle \boldsymbol{z} \rangle_\mu^{(t)} = \frac{1}{\mu} \sum_{i=1}^{\mu} \boldsymbol{z}_{i:\lambda}^{(t)}$ is the average of the realizations of the random vector that led to the new parent population, where $i{:}\lambda$ denotes the index of the offspring having the ith best fitness value of all offspring in the current generation, that is $\{\{\boldsymbol{g}_{1:\lambda}^{(t)}, \ldots, \boldsymbol{g}_{\mu:\lambda}^{(t)}\}\} = \{\{\tilde{\boldsymbol{g}}_1^{(t+1)}, \ldots, \tilde{\boldsymbol{g}}_\mu^{(t+1)}\}\}$. The parameter $c_{\mathrm{cov}} \in [0, 1[$ controls the update of $\boldsymbol{C}'^{(t)}$ and we set it to $c_{\mathrm{cov}} = 2/(n^2 + n)$.

The update rule (7) shifts $\boldsymbol{C}'^{(t)}$ towards the $n \times n$ matrix $\boldsymbol{s}^{(t+1)} \left(\boldsymbol{s}^{(t+1)} \right)^T$ making mutation steps in the direction of $\boldsymbol{s}^{(t+1)}$ more likely. The vector $\boldsymbol{s}$ does not only represent the last (adaptive) step of the parent population, but a time average over all previous adaptive steps. The influence of previous steps decays exponentially, where the decay rate is controlled by c.

The adaptation of the global step-size parameter σ is done separately on a shorter timescale (a single parameter can be estimated based on less samples compared to the complete covariance matrix). We keep track of a second evolution path $\boldsymbol{s}_\sigma$ without the scaling by $\boldsymbol{D}$:

$$\boldsymbol{s}_\sigma^{(t+1)} = (1 - c_\sigma) \cdot \boldsymbol{s}_\sigma^{(t)} + \underbrace{c_{\mathrm{u}_\sigma} \cdot \sqrt{\mu} \, \boldsymbol{B}^{(t)} \langle \boldsymbol{z} \rangle_\mu^{(t+1)}}_{\boldsymbol{B}^{(t)} \left(\boldsymbol{D}^{(t)} \right)^{-1} \left(\boldsymbol{B}^{(t)} \right)^{-1} \frac{\sqrt{\mu}}{\sigma^{(t)}} \left(\langle \boldsymbol{g} \rangle_\mu^{(t+1)} - \langle \boldsymbol{g} \rangle_\mu^{(t)} \right)} \tag{8}$$

$$\sigma^{(t+1)} = \sigma^{(t)} \cdot \exp\left(\frac{\|\boldsymbol{s}_\sigma^{(t+1)}\| - \hat{\chi}_n}{d \cdot \hat{\chi}_n} \right) , \tag{9}$$

where $\hat{\chi}_n$ is the expected length of a n-dimensional, normally distributed random vector with covariance matrix $\boldsymbol{I}$. The damping parameter $d \geq 1$ decouples the adaptation rate from the strength of the variation. We set $d = \sqrt{n}$ and start from $\boldsymbol{s}_\sigma^{(2)} = \sqrt{\mu}\,\boldsymbol{B}^{(1)}\langle\boldsymbol{z}\rangle_\mu^{(1)}$. The parameter $c_\sigma \in]0,1]$ controls the update of $\boldsymbol{s}_\sigma$. Here, we use $c_\sigma = c$. Setting $c_{u_\sigma} = \sqrt{c_\sigma(2 - c_\sigma)}$ normalizes the variance of $\boldsymbol{s}_\sigma$.

The evolution path $\boldsymbol{s}_\sigma$ is the sum of normally distributed random variables. Because of the normalization, its expected length would be $\hat{\chi}_n$ if there were no selection. Hence, the rule (9) basically increases the global step-size if the steps leading to selected individuals have been larger than expected and decreases the step size in the opposite case.

The CMA-ES needs only small population sizes. These are chosen according to the heuristic $\lambda = 4 + \lfloor 3\ln n \rfloor$ and $\mu = \lfloor \lambda/4 \rfloor$. Note that almost all the parameters of the algorithm can be set to the default values given in [18, 19]. The initializations of $\boldsymbol{C}'$ and σ allow for incorporation of prior knowledge about the scaling of the search space. In the following, we set $\boldsymbol{C}'^{(1)} = \boldsymbol{I}$ and choose $\sigma^{(1)}$ dependent on the problem.

2.1.2. Adaptation of operator probabilities. The search strategy is mainly determined by the variation operators and the probabilities of their application. In the CMA-ES, there is only one mutation and one recombination operator, which are both always applied, and strategy adaptation corresponds to adjusting the parameters of the mutation operator. Now we consider the case when there are several mutation operators and the strategy parameters to adapt are the probabilities of application of these operators.

The algorithm we propose combines concepts from [10] and from the CMA-ES. Let Ω denote a set of mutation operators. Each time an offspring $\boldsymbol{g}_i^{(t)}$ is created from a parent, first the number $v_i^{(t)}$ of variations is determined, then $v_i^{(t)}$ operators are randomly chosen from Ω and applied successively. Let $p_o^{(t)}$ be the probability that $o \in \Omega$ is chosen at generation t. Further, let $O_o^{(t)}$ contain all offspring produced in generation t by an application of the operator o. The case when an offspring is produced by applying more than one operator is treated as if the offspring has been generated several times, once by each of the operators involved. The operator probabilities are updated every τ generations. This period is called an adaptation cycle. The average performance achieved by an operator o over an adaptation cycle is measured by

$$q_o^{(t,\tau)} = \sum_{i=0}^{\tau-1} \sum_{\boldsymbol{g}\in O_o^{(t-i)}} \max(0, f(\boldsymbol{g}) - f(\text{parent}(\boldsymbol{g}))) \Big/ \sum_{i=0}^{\tau-1} \left|O_o^{(t-i)}\right| , \qquad (10)$$

where $\text{parent}(\boldsymbol{g})$ denotes the parent of offspring $\boldsymbol{g}$ (and we assume a fitness maximization task). The operator probabilities $p_o^{(t+1)}$ are adjusted every $\tau = 4$ generations according to

$$
s_o^{(t+1)} = \begin{cases} c_\Omega \cdot q_o^{(t,\tau)}/q_{\text{all}}^{(t,\tau)} + (1 - c_\Omega) \cdot s_o^{(t)} & \text{if } q_{\text{all}}^{(t,\tau)} > 0 \\ c_\Omega/|\Omega| + (1 - c_\Omega) \cdot s_o^{(t)} & \text{otherwise} \end{cases} \tag{11}
$$

and

$$
p_o^{(t+1)} = p_{\min} + (1 - |\Omega| \cdot p_{\min}) s_o^{(t+1)} \Big/ \sum_{o' \in \Omega} s_{o'}^{(t+1)} . \tag{12}
$$

The factor $q_{\text{all}}^{(t,\tau)} = \sum_{o' \in \Omega} q_{o'}^{(t,\tau)}$ is used for normalization and $s_o^{(t+1)}$ stores the weighted average of the quality of the operator o, where the influence of previous adaptation cycles decreases exponentially. The rate of this decay is controlled by $c_\Omega \in (0,1]$, which is set to $c_\Omega = 0.3$ in our experiments. The operator probability $p_o^{(t+1)}$ is computed from the weighted average $s_o^{(t+1)}$, such that all operator probabilities sum to one and are bounded from below by $p_{\min} < 1/|\Omega|$. Initially, $s_o^{(0)} = p_o^{(0)}$ for all $o \in \Omega$. Note that s_o has a similar function as the evolution paths in the CMA-ES.

A more detailed description and an empirical evaluation of the operator adaptation algorithm is given in [24].

3. Evolutionary optimization of neural networks

Learning of an adaptive (e.g., neural) system can be defined as goal-directed, data-driven changing of its behavior. The major components of an adaptive system can be described by a triple $(\mathcal{S}, \mathcal{A}, \mathcal{D})$, where $\mathcal{S}$ stands for the structure or architecture of the adaptive system, $\mathcal{A}$ is a learning algorithm that operates on $\mathcal{S}$ and adapts flexible parameters of the system, and $\mathcal{D}$ denotes the sample data.

Examples of learning algorithms for technical NNs include gradient-based heuristics, see Section 3.1, or quadratic program solvers, see Section 3.3. Such "classical" optimization methods are usually considerably faster than pure evolutionary optimization of NN parameters [22, 26, 45], although they might be more prone to getting stuck in local minima. However, there are cases where "classical" optimization methods are not applicable, for example when the neural model or the objective function is non-differentiable as in Section 3.2. Still, the main application of evolutionary optimization in the field of neurocomputing is adapting the structures of neural systems, that is, optimizing those parts that are not altered by the learning algorithm. Both in biological and technical neural systems the structure is crucial for the learning behavior – the evolved structures of brains are an important reason for their incredible learning performance: "development of intelligence requires a balance between innate structure and the ability to learn" [2]. Hence, it appears to be obvious to apply evolutionary methods for adapting the structure of neural systems for technical applications, a task for which generally no efficient "classical" methods exist. A prototypical example of evolutionary optimization of a neural architecture on which a learning algorithm operates is given

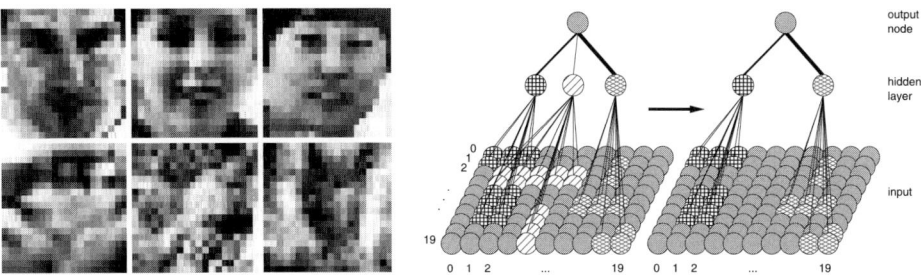

FIGURE 3. Left, the input data to the face detection network are pre-processed 20×20 pixel grayscale images showing either frontal, upright faces or nonface examples. Right, scheme of the *delete-node* operator. The linewidths indicate the magnitude of the corresponding weight values.

in the following Section 3.1, where the topology and the weights of multi-layer perceptron network are optimized. Adopting the extended definition of structure as that part of the adaptive system that cannot be optimized by the learning algorithm itself, Section 3.3 presents the optimization of the "structure" of support vector machines.

3.1. Evolving neural networks for face detection

Feed forward NNs have proven to be powerful tools in pattern recognition [53]. For example, they can be used to decide whether images of a fixed size contain a complete frontal upright face or not, see Fig. 3 (left). In the following, we discuss the evolutionary optimization of a feed forward multi-layer perceptron for such a face detection task. The optimization process has two objectives: improving the classification accuracy and the speed of processing.

3.1.1. Feed forward multi-layer perceptrons. The structure of a multi-layer perceptron (MLP, a good introduction is given in [5]) is given by a connected directed graph $G = (\mathcal{V}, \mathcal{E})$ with vertices $\mathcal{V}$ and edges $\mathcal{E} \subseteq \mathcal{V} \times \mathcal{V}$. We identify the vertices with neurons and edges with connecting synapses. If G is acyclic it describes a feed forward MLP. The n_{out} nodes without successors are called the output neurons. The $n_{\text{in}} + 1$ nodes without predecessors are the input neurons and an additional bias unit. The n_{hidden} nodes with at least one successor and at least one predecessor are called hidden neurons. Each feed forward MLP represents a static function that maps an input $\boldsymbol{x} \in \mathbb{R}^{n_{\text{in}}}$ to an output value $\boldsymbol{y} \in \mathbb{R}^{n_{\text{out}}}$. Traversing the graph from the input neurons, we compute the activation z_i of each neuron i. The activation of the input nodes is equal to the corresponding component of the input pattern $\boldsymbol{x}$, the activation of the additional node without predecessors (the bias node) is constantly equal to 1. For the other nodes the activation is given by

$$z_i = g \left(\sum_{j \in \mathrm{pred}(i)} w_{ij} z_j \right) \ , \tag{13}$$

where $\mathrm{pred}(i) \subset \mathcal{V}$ is the set of predecessors, w_{ij} is the weightening factor of the connection from neuron j to neuron i, and g is a non-linear transfer function, here the sigmoidal $g(u) = u/(|u| + 1)$. The activations of the output nodes correspond to the components of the output $\boldsymbol{y}$ of the network.

Learning of a MLP means adapting the weights w_{ij} based on some sample input-output patterns $\{(\boldsymbol{x}_1, \boldsymbol{y}_1), \ldots, (\boldsymbol{x}_\ell, \boldsymbol{t}_\ell)\}$. This is usually done by gradient-based minimization of the (squared) differences between the targets $\boldsymbol{t}_i$ and the corresponding outputs $\boldsymbol{y}_i$ of the NN given the input $\boldsymbol{x}_i$. The goal is not to learn the training patterns by heart, but to find a statistical model for the underlying relationship between input and output data. Such a model will generalize well, that is, will make good predictions for cases other than the training patterns. A critical issue is therefore to avoid overfitting during the learning process: The NN should just fit the signal and not the noise. This is usually achieved by restricting the effective complexity of the network, for example by regularization of the learning process.

3.1.2. Evolving neural face detectors. Real-time face recognition requires both fast and accurate face detection methods. Recognition speed may be crucial, for example when processing a huge amount of data from video streams.

As stated in a recent survey "The advantage of using neural networks for face detection is the feasibility of training a system to capture the complex class conditional density of face patterns. However, one drawback is that the network architecture has to be extensively tuned (number of layers, number of nodes, learning rates, etc.) to get exceptional performance" [51]. In the following, we show how evolutionary computation can help to overcome this drawback. We optimize the weights and the structure of an already existing neural face classifier, which is part of a complex face detection system similar to the one described in [35]. We consider an implementation in which the speed of classification scales approximately linearly with the number of hidden neurons. Therefore our goal is to reduce the number of neurons of a NN without loss of classification accuracy, whereas we tolerate an increase in the number of connections.

We apply an EA combined with gradient-based learning as schematically shown in Fig. 1. Each individual encodes a NN. In every generation, each parent creates one offspring, which inherits its parent's genotype. The offspring's genotype is then altered by elemental variation operators. These are chosen randomly from a set Ω of 8 different operators and are applied sequentially. The process of choosing and applying an operator is repeated $1 + \kappa$ times, where κ is an individual realization of a Poisson distributed random number with mean 1. There are 5 basic operators: *add-connection, delete-connection, add-node, delete-node*, and *jog-weights*. Their effects on the network graph represented by the genotype becomes obvious from their names, probably except for the operator *jog-weights*. The latter adds Gaussian noise to the weights in order to push the weight configuration out of local minima. The elemental deletion operators are based on the *magnitude based pruning* heuristic (see [33]), which assigns a higher probability to the deletion of small weights. As an example, the *delete-node* operator is schematically depicted in Fig. 3 (right).

In addition to the 5 basic operators, there are 3 task-specific mutations inspired by the concept of "receptive fields" (RFs). These RF-operators *add-RF-connection, delete-RF-connection*, and *add-RF-node* behave as their basic counterparts, but act on groups of connections. These groups are defined by rectangular regions of the input image, see the input layer in Fig. 3 (right). The RF-operators consider the topology of the image plane by taking into account that "isolated" processing of pixels is rarely useful for object detection. Not all operators might be necessary at all stages of the evolutionary process and questions such as when fine-tuning becomes more important than operating on receptive fields cannot be answered in advance. Hence, the application probabilities of the 8 variation operators are adapted using the method described in Section 2.1.2.

An inner loop of learning is embedded, just before fitness evaluation in order to fine-tune all weights. An iterative learning algorithm is used, namely the improve Rprop algorithm [23, 34]. Learning is done for a fixed number of iterations[1] and is stopped earlier when the generalization performance of the network, which is measured using a validation data set, decreases. The weight configuration with the smallest error on training and validation sample data found during network learning is regarded as the outcome of the learning process and is stored in the genome of the corresponding individual (this principle is often called Lamarckian inheritance). Based on this weight configuration the fitness of the individual, a weighted sum of the classification accuracy and the number of neurons, is calculated.

3.1.3. Experimental evaluation. We initialized the 25 individuals in the population of our EA with the expert-designed architecture proposed in [35]. This network has been tailored to the face detection task and has become the standard reference for neural network based face detection, see [51]. In the following, all results are given relative to the properties of the initial architecture.

Our EA successfully tackled the problem of reducing the number of hidden neurons of the face detection network without loss of detection accuracy [49]. The numbers of hidden neurons of the evolved networks are reduced by 27-35 %. This means, we could improve the speed of classification whether an image region corresponds to a face or not by approximately 30 %. By speeding up classification, the rate of complete scans of video-stream images of face recognition systems can

---

[1] A method that automatically adjusts the length of the learning period similar to the evolutionary strategy adaptation described in Section 2.1 has been proposed in [20].

be increased leading to a more accurate recognition. A generalization performance test on an external data set, which is independent from all data used for optimization, demonstrates that most of our considerably smaller networks perform at least as good as the expert-designed architecture. Some of the evolved classifiers even show an improvement of the classification accuracy by more than 10 %.

3.2. Adapting the weights of networks for reinforcement learning tasks

Reinforcement (RL, [47]) learning is an important, biologically plausible learning paradigm. In RL the feedback about the performance of an adaptive system may be sparse, unspecific, and delayed. Evolutionary algorithms have proven to be powerful and competitive approaches compared to standard RL methods [28]. The recent success of evolved NNs in game playing [7] underlines the potential of the combination of NNs and evolutionary computation for RL. In the following, we describe an application of the CMA-ES to the adaptation of the weights of a NN for solving a RL task. In this scenario, no gradient information is available to adapt the NN parameters.

3.2.1. Reinforcement learning. In the standard RL scenario, an agent interacts with its environment at discrete time steps t. It perceives the environment to be in state $s_t \in S$ and chooses a behavior a_t from the set of actions A according to its policy $\pi : S \to A$. After the execution of action a_t, the environment makes a possibly stochastic transition to a perceived state s_{t+1} and thereby emits a possibly stochastic numerical reward $r_{t+1} \in \mathbb{R}$. The objective of the agent is to adapt its policy such that the *expected discounted cumulative future reward* $R_t = \sum_{t'=t+1}^{\infty} \gamma^{t'-t-1} r_{t'}$ with discount rate $\gamma \in]0, 1]$ is maximized.

Two different (model-free) approaches to solve RL problems can be distinguished. The most common methods such as temporal-difference learning algorithms adapt value functions [47]. Usually, they learn a state-value function $V : S \to \mathbb{R}$ or a state-action-value function $Q : S \times A \to \mathbb{R}$ for judging states or state-action pairs, respectively. The policy π is then defined on top of this function. The second approach is to search directly in the space of policies [28]. However, the gradient $\partial R_t / \partial \pi_t$ can usually not be computed (this problem can be circumvented be *actor-critic* architectures, see [47]). When S or A is too large or generalization from experiences to new states and actions is desired, function approximators like NNs are used to model Q, V, or π.

The potential advantages of direct search methods like EAs compared to standard RL methods are that they

1. allow for direct search in the policy space,
2. are often easier to apply and are more robust with respect to the tuning of the meta-parameters (learning rates, etc.),
3. can be applied if the function approximators are non-differentiable, and
4. can also optimize the underlying structure of the function approximators.

In the following, we describe an application of the CMA-ES to the adaptation of the weights of an NN that directly represents a policy π.

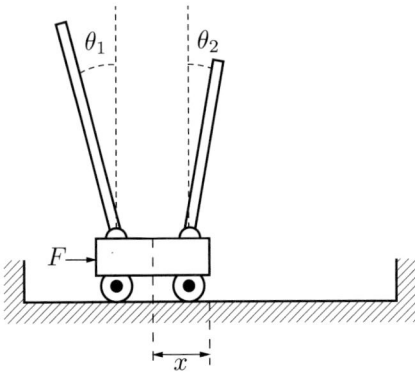

FIGURE 4. Double pole balancing problem. The parameters x, θ_1, and θ_2 are the offset of the cart from the center of the track and the angles from the vertical of the long and short pole, respectively.

3.2.2. Evolving networks for pole balancing. Pole balancing problems (also known as inverted pendulum problems) are standard benchmark tasks for EAs that adapt NNs for control, see [50] for early and [46] for recent references. In our example, the task is to balance two poles hinged on a wheeled cart, which can move on a finite length track, by exerting forces either left or right on the cart. The movements of the cart and the poles are constrained within the vertical plane. A balancing attempt fails if either the angle from the vertical of any pole exceeds a certain threshold or the cart leaves the track. Figure 4 illustrates the task (the corresponding equations of motion are given, e.g., in [21, 50]). The problem of designing a controller for the cart can be viewed as a RL task, where the actions are the applied forces and the perceived state corresponds to the information about the system provided to the controller.

We consider the *double pole without velocities* scenario, where the controller gets only x, θ_1, and θ_2 as inputs and the output is a force between $-10\,N$ and $10\,N$. The environment is only partially observable as information about the velocity of the cart and the angle velocities are needed for a successful control strategy. In order to distinguish between system states, e.g., whether a pole is moving up or down, the controller needs the capability to exploit history information. This can be achieved by recurrent neural networks (RNNs).

In our example, we use the following simple RNN architecture. Let $z_i(t)$ for $0 < i \leq n_{\text{in}}$ be the activation of the n_{in} input units at time step t of a network with a total of n_{neurons} neurons. The activation of the other neurons is given by

$$z_i(t) = g\left(\sum_{j=1}^{n_{\text{in}}} w_{ij} z_j(t) + \sum_{j=n_{\text{in}}+1}^{n_{\text{neurons}}} w_{ij} z_j(t-1) \right) \qquad (14)$$

for $n_{\text{in}} < i \leq n_{\text{neurons}}$, where the w_{ij} are the weights and g is a non-linear transfer function, see Section 3.1.1. All RNNs in this study have a single output neuron and the input signals provided to the networks are appropriately scaled.

The weights of the RNNs are adapted by the CMA-ES [21]. We use the same fitness function as in [17], see also [16, 21, 46]. This quality measure consists of two additive terms. The first addend is proportional to the number of time steps the controller manages to balance the poles starting from a fixed initial position (i.e., we do not test for generalization). The second term penalizes oscillations in order to exclude the control strategy to balance the poles by just moving the cart quickly back and forth from the set of solutions. A trial is stopped and regarded as successful when the fittest individual in the population balances the poles for 10^5 time steps.

method	evaluations	population size
ESP	6213	100
NEAT	6326	100
CMA-ES, $n_{\text{hidden}} = 3$	3521	13
CMA-ES, $n_{\text{hidden}} = 5$	4856	13
CMA-ES, $n_{\text{hidden}} = 7$	5029	16

TABLE 1. Number of balancing attempts needed to find an appropriate control strategy averaged over 75 (ESP and NEAT) and 50 (CMA-ES) trials, respectively. Additionally, the population sizes used in the experiments are given.

The results of our experiments using RNN architectures with different numbers of neurons are shown in table 1. They are compared to the best performing methods for evolving RNNs for this task so far, the *Enforced Sub-Population* (ESP, [16]) and the *NeuroEvolution of Augmenting Topologies* (NEAT, [46]) algorithm. In contrast to our approach, NEAT does not only adapt the weights of NNs, but also their structure – this weakens the comparison. Note that the ESP and NEAT results are better than those reported in [16, 21, 46], because they refer to new trials with optimized settings of the algorithms, in particular with reduced population sizes (Kenneth O. Stanley, private communication). However, even for $n_{\text{hidden}} = 7$ the CMA-ES performs statistically significantly better (t-test, $p < .05$) than the ESP method. The smaller the network size, the better are the results obtained by the CMA-ES.

Our experiments show that that standard RNN architectures combined with the CMA-ES are sufficient for achieving good results on pole balancing tasks when the architecture of the NNs are fixed. The efficient adaptation of the search strategy by the CMA-ES allows for fast optimization of the network weights by detecting correlations between object variables without requiring large population sizes. Still, the network structure matters, which becomes obvious from the differences between architectures with different numbers of hidden neurons. Hence, methods are required that evolve both the structure and the weights of NNs for RL tasks – although for nearly all initializations our approach outperformed the existing algorithms that additionally adapt the topology (more results and details are given in [21]).

3.3. Evolutionary tuning of support vector machines

Support vector machines (SVMs, e.g., [9, 37, 48]), which can be viewed as special regularization networks (at least there is a very close relation, see [13]), have become a standard method in machine learning. The main idea of SVMs for binary classification is to map the input vectors to a feature space and to classify the transformed data by a linear function that promises good generalization. The transformation is done implicitly by a kernel, which computes an inner product in the feature space. Constructing the linear decision function is a convex optimization problem that can be solved using quadratic programming. Thus, the problem of designing an appropriate model is elegantly separated into two stages: First, the kernel function (and its parameters) has to be chosen. In the general case of non-separable data, one also has to select a regularization parameter, which controls the trade-off between minimizing the training error and the complexity of the decision function. Second, the corresponding convex optimization problem has to be solved.

The first stage can be identified with choosing the structure of the adaptive systems, and the second stage with adapting its parameters by learning. The latter can be solved efficiently by "classical" algorithms, the former is a difficult multimodal optimization task requiring the use of heuristics. Often a parameterized family of kernel functions is considered and the kernel adaptation reduces to finding an appropriate parameter vector for the given problem. These parameters together with the regularization parameter are called the hyperparameters of the SVM.

In practice the hyperparameters are usually determined by grid search. That is, the hyperparameters are varied with a fixed step-size through a wide range of values and the performance of every combination is assessed using some performance measure. Because of the computational complexity, grid search is only suitable for the adjustment of very few parameters. Perhaps the most elaborate alternative techniques for choosing multiple hyperparameters are gradient descent methods [6, 8, 15]. However, these approaches have some severe drawbacks, for example: The kernel function has to be differentiable, which excludes for example string kernels. The score function for assessing the performance of the hyperparameters (or at least an accurate approximation of this function) also has to be differentiable with respect to kernel and regularization parameters, which excludes reasonable measures such as the number of support vectors.

In the following, we present the application of the CMA-ES for optimizing SVM hyperparameters – a new approach that does not suffer from the limitations described above [14]. Beforehand, we give a concise formal description of SVMs.

3.3.1. Support vector machines. We consider L_1-norm soft margin SVMs for the discrimination of two classes. Let $(\boldsymbol{x}_i, t_i), 1 \leq i \leq \ell$, be the training examples, where $t_i \in \{-1, 1\}$ is the label associated with input pattern $\boldsymbol{x}_i \in \mathbb{R}^{n_{\text{in}}}$. The mapping $\phi : \mathbb{R}^{n_{\text{in}}} \to F$ of the input vectors to the feature space F is implicitly done by a kernel $K : \mathbb{R}^{n_{\text{in}}} \times \mathbb{R}^{n_{\text{in}}} \to \mathbb{R}$. The kernel computes an inner product in the feature space, that is $K(\boldsymbol{x}_i, \boldsymbol{x}_j) = \langle \phi(\boldsymbol{x}_i), \phi(\boldsymbol{x}_j) \rangle$. The linear function for classification in the feature space is chosen according to a bound on the generalization error. This bound takes a target margin and the margin slack vector into account (cf. [9, 37]). The latter corresponds to the amounts by which individual training patterns fail to meet the target margin. This leads to the SVM decision function

$$h(\boldsymbol{x}) = \text{sign}\left(\sum_{i=1}^{\ell} t_i \alpha_i^* K(\boldsymbol{x}_i, \boldsymbol{x}) + b\right) , \tag{15}$$

where we define $\text{sign}(x) = -1$ if $x < 0$ and $\text{sign}(x) = 1$ otherwise. The coefficients α_i^* are the solution of the following quadratic optimization problem:

$$\text{maximize} \quad W(\boldsymbol{\alpha}) = \sum_{i=1}^{\ell} \alpha_i - \frac{1}{2} \sum_{i,j=1}^{\ell} t_i t_j \alpha_i, \alpha_j K(\boldsymbol{x}_i, \boldsymbol{x}_j) \tag{16}$$

$$\text{subject to} \quad \sum_{i=1}^{\ell} \alpha_i t_i = 0$$

$$0 \leq \alpha_i \leq C, \quad i = 1, \ldots, \ell.$$

The optimal value for b can then be computed based on the solution $\boldsymbol{\alpha}^*$. The vectors $\boldsymbol{x}_i$ with $\alpha_i > 0$ are called support vectors. The regularization parameter C controls the trade-off between maximizing the target margin and minimizing the L_1-norm of the margin slack vector of the training data.

3.3.2. Evolving SVM hyperparameters. We consider general Gaussian kernels

$$K_{\boldsymbol{A}}(\boldsymbol{x}, \boldsymbol{x}') = e^{-(\boldsymbol{x}-\boldsymbol{x}')^T \boldsymbol{A}(\boldsymbol{x}-\boldsymbol{x}')} , \tag{17}$$

where $\boldsymbol{x}, \boldsymbol{x}' \in \mathbb{R}^{n_{\text{in}}}$ and $\boldsymbol{A}$ is a symmetric positive definite $n_{\text{in}} \times n_{\text{in}}$ matrix. We allow arbitrary symmetric positive definite matrices $\boldsymbol{A}$; this means the input space can be scaled and rotated. The individuals in the ES encode C and the kernel parameters. When encoding $\boldsymbol{A}$, we have to ensure that after variation the genotype still corresponds to a feasible (i.e., symmetric, positive definite) matrix. We make use of the fact that for any symmetric and positive definite $n \times n$ matrix $\boldsymbol{A}$ there exists an orthogonal $n \times n$ matrix $\boldsymbol{T}$ and a diagonal $n \times n$ matrix $\boldsymbol{D}$ with positive entries such that $\boldsymbol{A} = \boldsymbol{T}^T \boldsymbol{D} \boldsymbol{T}$ and

$$\boldsymbol{T} = \prod_{i=1}^{n-1} \prod_{j=i+1}^{n} \boldsymbol{R}(\alpha_{i,j}) , \tag{18}$$

data		accuracy on test set	# SV
Breast-Cancer	grid-search	74.51	113.52
	evolutionary tuned	$75.38 \pm 0.42^\star$	$112.70 \pm 0.68^\star$
Diabetes	grid-search	76.67	247.83
	evolutionary tuned	76.73 ± 0.32	$235.73 \pm 3.43^\star$
Heart	grid-search	84.79	106.33
	evolutionary tuned	$85.14 \pm 0.33^\star$	$75.51 \pm 1.5^\star$
Thyroid	grid-search	95.83	16.36
	evolutionary tuned	$96.01 \pm 0.05^\star$	$15.42 \pm 0.18^\star$

TABLE 2. Results averaged over 20 trials $\pm$ standard deviations. The first column specifies the medical benchmark. The second column indicates whether the results refer to the initial grid search values or to the evolutionary optimized kernels. The percentages of correctly classified patterns on the test sets (averaged over 100 different partitions into training and test data) are given as well as the average numbers of support vectors (# SV). Results that are statistically significantly better compared to grid-search are marked with $\star$ (two-sided t-test, $p < .05$).

as proven in [36]. The $n \times n$ matrices $\boldsymbol{R}(\alpha_{i,j})$ are elementary rotation matrices. These are equal to the unit matrix except for $[\boldsymbol{R}(\alpha_{i,j})]_{ii} = [\boldsymbol{R}(\alpha_{i,j})]_{jj} = \cos \alpha_{ij}$ and $[\boldsymbol{R}(\alpha_{i,j})]_{ji} = -[\boldsymbol{R}(\alpha_{i,j})]_{ij} = \sin \alpha_{ij}$.

When using the evolution strategy, each genotype encodes the $n_{\text{in}} + (n_{\text{in}}^2 - n_{\text{in}})/2 + 1$ parameters

$$(C', d_1, \ldots, d_{n_{\text{in}}}, \alpha_{1,2}, \alpha_{1,3}, \ldots, \alpha_{1,n_{\text{in}}}, \alpha_{2,3}, \alpha_{2,4}, \ldots, \alpha_{2,n_{\text{in}}}, \ldots, \alpha_{n_{\text{in}}-1,n_{\text{in}}}) \ . \quad (19)$$

We encode $\boldsymbol{A}$ according to (18) and set $\boldsymbol{D} = \text{diag}(|d_1|, \ldots, |d_n|)$ and $C = |C'|$.

We evaluated our approach on common medical benchmark problems [14], namely *Breast-Cancer*, *Diabetes*, *Heart*, and *Thyroid* preprocessed and partitioned as proposed in [31]. These are binary classification problems where the task is to predict whether a patient suffers from a certain disease or not. There are 100 partitions of each dataset into disjoint training and test sets. In [27], appropriate SVM hyperparameters C and a for Gaussian kernels with $\boldsymbol{A} = a\boldsymbol{I}$ are determined using a two-stage grid search. For each hyperparameter combination, five SVMs are constructed using the training sets of the first five data partitions and the average of the classification rates on the corresponding five test sets determines the score value of this parameter vector. The hyperparameter vector with the best score is selected and its performance is finally measured by calculating the score function using all 100 partitions.

We initialized our populations with the values found in [27] and used the same score function to determine the fitness. The results are shown in table 2. Except for one case, the scaled and rotated kernels led to significantly ($p < .05$) better results. There is another remarkable advantage of the scaled and rotated kernels: The number of support vectors – and therefore the execution time and storage complexity of the classifier – decreases. For a detailed description and additional results see [14].

4. Conclusions

Finding an appropriate neural system for a given task usually requires the solution of difficult optimization problems. These include adapting the hyperparameters of support vector machines or finding the right topology of a multi-layer perceptron network. Evolutionary algorithms (EAs) are particularly well suited for such kinds of tasks, especially when higher order optimization methods cannot be applied. We demonstrated three successful applications of evolutionary computation to the optimization of neural systems, where the EAs made use of deterministic strategy adaptation to improve the search performance.

Of course, it is appealing to use evolutionary methods for the design of neural networks, because this resembles the natural evolution of nervous systems. However, as our examples have shown, evolutionary methods should be regarded as the state-of-the-art choice for many neural network optimization tasks not because of the biological metaphor but because they are highly competitive approaches often superior to alternative methods.

References

[1] M.A. Arbib, editor. *The Handbook of Brain Theory and Neural Networks*. MIT Press, 2 edition, 2002.

[2] M.A. Arbib. Towards a neurally-inspired computer architecture. *Natural Computing*, 2(1):1–46,, 2003.

[3] H.-G. Beyer and H.-P. Schwefel. Evolution strategies: A comprehensive introduction. *Natural Computing*, 1(1):3–52, 2002.

[4] H.-G. Beyer, H.-P. Schwefel, and I. Wegener. How to analyse evolutionary algorithms. *Theoretical Computer Science*, 287:101–130, 2002.

[5] C.M. Bishop. *Neural Networks for Pattern Recognition*. Oxford University Press, 1995.

[6] O. Chapelle, V. Vapnik, O. Bousquet, and S. Mukherjee. Choosing multiple parameters for support vector machines. *Machine Learning*, 46(1):131–159, 2002.

[7] K. Chellapilla and D.B. Fogel. Evolution, neural networks, games, and intelligence. *Proceedings of the IEEE*, 87(9):1471–1496, 1999.

[8] K.-M. Chung, W.-C. Kao, C.-L. Sun, and C.-J. Lin. Radius margin bounds for support vector machines with RBF kernel. *Neural Computation*, 15(11):2643–2681, 2003.

[9] N. Cristianini and J. Shawe-Taylor. *An Introduction to Support Vector Machines and other kernel-based learning methods*. Cambridge University Press, 2000.

[10] L. Davis. Adapting operator probabilities in genetic algorithms. In J.D. Schaffer, editor, *Proceedings of the Third International Conference on Genetic Algorithms, ICGA'89*, pages 61–69, Fairfax, VA, USA, 1989. Morgan Kaufmann.

[11] A.E. Eiben, R. Hinterding, and Z. Michalewicz. Parameter control in evolutionary algorithms. *IEEE Transactions on Evolutionary Computation*, 3(2):124–141, 1999.

[12] A.E. Eiben and G. Rudolph. Theory of evolutionary algorithms: A bird's eye view. *Theoretical Computer Science*, 229(1):3–9, 1999.

[13] T. Evgeniou, M. Pontil, and T. Poggio. Regularization networks and support vector machines. *Advances in Computational Mathematics*, 13:1–50, 2000.

[14] F. Friedrichs and C. Igel. Evolutionary tuning of multiple SVM parameters. In M. Verleysen, editor, *12th European Symposium on Artificial Neural Networks (ESANN 2004)*, pages 519–524. Evere, Belgium: d-side publications, 2004.

[15] C. Gold and P. Sollich. Model selection for support vector machine classification. *Neurocomputing*, 55(1-2):221–249, 2003.

[16] F.J. Gomez and R. Miikulainen. Solving non-markovian tasks with neuroevolution. In T. Dean, editor, *Proceeding of the Sixteenth International Joint Conference on Artificial Intelligence (IJCAI)*, pages 1356–1361, Stockholm, Sweden, 1999. Morgan Kaufmann.

[17] F. Gruau, D. Whitley, and L. Pyeatt. A comparison between cellular encoding and direct encoding for genetic neural networks. In J.R. Koza, D.E. Goldberg, D.B. Fogel, and R.L. Riolo, editors, *Genetic Programming 1996: Proceedings of the First Annual Conference*, pages 81–89, Stanford University, CA, USA, 1996. MIT Press.

[18] N. Hansen and A. Ostermeier. Convergence properties of evolution strategies with the derandomized covariance matrix adaptation: The $(\mu/\mu, \lambda)$-CMA-ES. In *5th European Congress on Intelligent Techniques and Soft Computing (EUFIT'97)*, pages 650–654. Aachen, Germany: Verlag Mainz, Wissenschaftsverlag, 1997.

[19] N. Hansen and A. Ostermeier. Completely derandomized self-adaptation in evolution strategies. *Evolutionary Computation*, 9(2):159–195, 2001.

[20] M. Hüsken and C. Igel. Balancing learning and evolution. In W.B. Langdon, E. Cantu-Paz, K. Mathias, R. Roy, D. Davis, R. Poli, K. Balakrishnan, V. Honavar, G. Rudolph, J. Wegener, L. Bull, M.A. Potter, A.C. Schultz, J.F. Miller, E. Burke, and N. Jonoska, editors, *Proceedings of the Genetic and Evolutionary Computation Conference (GECCO 2002)*, pages 391–398. Morgan Kaufmann, 2002.

[21] C. Igel. Neuroevolution for reinforcement learning using evolution strategies. In R. Sarker, R. Reynolds, H. Abbass, K.C. Tan, B. McKay, D. Essam, and T. Gedeon, editors, *Congress on Evolutionary Computation (CEC 2003)*, volume 4, pages 2588–2595. IEEE Press, 2003.

[22] C. Igel, W. Erlhagen, and D. Jancke. Optimization of dynamic neural fields. *Neurocomputing*, 36(1-4):225–233, 2001.

[23] C. Igel and M. Hüsken. Empirical evaluation of the improved Rprop learning algorithm. *Neurocomputing*, 50(C):105–123, 2003.

[24] C. Igel and M. Kreutz. Operator adaptation in evolutionary computation and its application to structure optimization of neural networks. *Neurocomputing*, 55(1-2):347–361, 2003.

[25] C. Igel and M. Toussaint. Neutrality and self-adaptation. *Natural Computing*, 2(2):117–132, 2003.

[26] M. Mandischer. A comparison of evolution strategies and backpropagation for neural network training. *Neurocomputing*, 42(1–4):87–117, 2002.

[27] P. Meinicke, T. Twellmann, and H. Ritter. Discriminative densities from maximum contrast estimation. In S. Becker, S. Thrun, and K. Obermayer, editors, *Advances in Neural Information Processing Systems 15*, Cambridge, MA, 2002. MIT Press.

[28] D.E. Moriarty, A.C. Schultz, and J.J. Grefenstette. Evolutionary Algorithms for Reinforcement Learning. *Journal of Artificial Intelligence Research*, 11:199–229, 1999.

[29] S. Nolfi. Evolution and learning in neural networks. In M.A. Arbib, editor, *The Handbook of Brain Theory and Neural Networks*, pages 415–418. MIT Press, 2 edition, 2002.

[30] M. Patel, V. Honavar, and K. Balakrishnan, editors. *Advances in the Evolutionary Synthesis of Intelligent Agents*. MIT Press, 2001.

[31] G. Rätsch, T. Onoda, and K.-R. Müller. Soft margins for adaboost. *Machine Learning*, 42(3):287–32, 2001.

[32] I. Rechenberg. *Evolutionsstrategie '94*. Werkstatt Bionik und Evolutionstechnik. Frommann-Holzboog, Stuttgart, 1994.

[33] R.D. Reed and R.J. Marks II. *Neural Smithing*. MIT Press, 1999.

[34] M. Riedmiller. Advanced supervised learning in multi-layer perceptrons – From backpropagation to adaptive learning algorithms. *Computer Standards and Interfaces*, 16(5):265–278, 1994.

[35] H.A. Rowley, S. Baluja, and T. Kanade. Neural network-based face detection. *IEEE Transactions on Pattern Analysis and Machine Intelligenc*, 20(1):23–38, 1998.

[36] G. Rudolph. On correlated mutations in evolution strategies. In R. Männer and B. Manderick, editors, *Parallel Problem Solving from Nature 2 (PPSN II)*, pages 105–114. Elsevier, 1992.

[37] B. Schölkopf and A. J. Smola. *Learning with Kernels: Support Vector Machines, Regularization, Optimization, and Beyond*. MIT Press, 2002.

[38] H.-P. Schwefel. *Evolution and Optimum Seeking*. Sixth-Generation Computer Technology Series. John Wiley & Sons, 1995.

[39] B.A. Sendhoff. *Evolution of Structures – Optimization of Artificial Neural Structures for Information Processing*. Shaker Verlag, Aachen, 1998.

[40] H.T. Siegelmann and E.D. Sontag. On the computational power of neural nets. *Journal of Computer and System Sciences*, 50(1):132–150, 1995.

[41] J. Šíma. Training a single sigmoidal neuron is hard. *Neural Computation*, 14:2709–2728, 2002.

[42] J. Šíma and P. Orponen. General-purpose computation with neural networks: A survey of complexity theoretix results. *Neural Computation*, 15(12):2727–2778, 2003.

[43] J.E. Smith and T.C. Fogarty. Operator and parameter adaptation in genetic algorithms. *Soft Computing*, 1(2):81–87, 1997.

[44] E.D. Sontag. Recurrent neural networks: Some systems-theoretic aspects. In M. Karny, K. Warwick, and V. Kurkova, editors, *Dealing with Complexity: A Neural Network Approach*, pages 1–12. Springer-Verlag, 1997.

[45] P. Stagge. *Strukturoptimierung rückgekoppelter neuronaler Netze. Konzepte neuronaler Informationsverarbeitung*. ibidem-Verlag, Stuttgart, 2001.

[46] K.O. Stanley and R. Miikkulainen. Evolving neural networks through augmenting topologies. *Evolutionary Computation*, 10(2):99–127, 2002.

[47] R.S. Sutton and A.G. Barto. *Reinforcement Learning: An Introduction*. MIT Press, 1998.

[48] V.N. Vapnik. *The Nature of Statistical Learning Theory*. Springer-Verlag, 1995.

[49] S. Wiegand, C. Igel, and U. Handmann. Evolutionary optimization of neural networks for face detection. In M. Verleysen, editor, *12th European Symposium on Artificial Neural Networks (ESANN 2004)*, pages 39–144. Evere, Belgium: d-side publications, 2004.

[50] A. Wieland. Evolving controls for unstable systems. In *Proceedings of the International Joint Conference on Neural Networks*, volume II, pages 667–673. IEEE Press, 1991.

[51] M.-H. Yang, D.J. Kriegman, and N. Ahuja. Detecting faces in images: A survey. *IEEE Transactions on Pattern Analysis and Machine Intelligence*, 24(1):34–58, 2002.

[52] X. Yao. Evolving artificial neural networks. *Proceedings of the IEEE*, 87(9):1423–1447, 1999.

[53] G. P. Zhang. Neural Networks for Classification: A Survey. *IEEE Transactions on System, Man, and Cybernetics – Part C*, 30(4), 2000.

[54] A.A. Zhigljavsky. *Theory of global random search*. Kluwer Academic Publishers, 1991.

Christian Igel, Stefan Wiegand and Frauke Friedrichs
Institut für Neuroinformatik
Ruhr-Universität Bochum
D 44780 Bochum, Germany
e-mail: `christian.igel@neuroinformatik.rub.de`

"Bochumer Research Group" with H.N. Mhaskar

Trends and Applications in Constructive Approximation
(Eds.) M.G. de Bruin, D.H. Mache & J. Szabados
International Series of Numerical Mathematics, Vol. 151, 125–134
© 2005 Birkhäuser Verlag Basel/Switzerland

Finding Relevant Input Arguments with Radial Basis Function Networks

Detlef H. Mache and Jennifer Meyer

Abstract. In this paper we will give new aspects of the problem of *finding relevant input arguments*. This topic is of great interest in several scientific fields, such as complexity reduction or in applications in the areas of medicine, biology or technical fields.

Approximation theorists know well the problem of *the curse of dimension*, which causes problems for applications using approximation methods.

Here we give an approach which makes use of the scattered data interpolation abilities of Radial Basis Function Networks to handle this problem.

1. Introduction

Our aim is to demonstrate that neural structures are good tools to handle complexity reduction problems. In fact the algorithm that will be introduced in this paper is based on a neural structure (a Radial Basis Function Network) and is able to detect relevant input arguments behind which a very complex and highly nonlinear structure is found.

To provide a first insight into how complexity problems are present in nature we take a look at the human brain: There is an unlimited number of information in the world between which the brain has to select the important ones and for that the brain has to find methods to present it compressed. This is a very complex task and how the brain functions in detail is in a lot of aspects unknown, but some concepts are still clear. Thus mathematicians, computerscientists and engineers are working on artificial intelligence strategies in the hope that evolution has found good ideas and so rebuilding that ideas might be powerful. What we know is that the brain consists of over 10^{11} little subunits, that work as tiny processors, called neurons and even 10^{14} synaptic connections between that neurons and that all in a brain of only about 1.35 kg. There is no computer that can simulate the possibilities that a human brain offers. But we still establish algorithms that make use of learning-strategies and structures we know from the brain.

In this article we thought about how useful computational neural networks are to do some special kind of complexity reduction.

2. About relevance

To achieve the competence to talk about complexity reduction by means of finding relevant input arguments in a complex dataset, we first of all have to discuss what is meant with the expression *relevance*. A lot of different definitions of relevance can be found in literature each of it serves its special use and aim. For our purpose we use a definition of CARUANA and FREITAG in [1] which describes the word relevance with the expression **incremetally useful** under certain conditions:

Definition 2.1. *Let L be a learning algorithm and $X = X_1 \times \cdots \times X_d$, $d \in \mathbb{N}$ be the set of attributes that is potentially available to L and $S \in X^N$, $N \in \mathbb{N}$ is a given dataset. The attribute X_i, $i \in \{1, \ldots, d\}$, is* **incrementally useful to L in respect to X**, *if the result of L with regard to $X \setminus X_i$ on S is worse than the result of L on S with regard to X.*

Our learning algorithm is a special computational neural network, called Radial Basis Function Network, that is a three-layered, feedforward network, that uses positive definite functions or conditionally positive definite functions as kernel functions (see for this [4]). Let us formulate the following definition of **RBF-Networks** by using Gaussian kernel functions:

Definition 2.2. *A **Radial Basis Function Network** (**RBF-Network**) is a function of the following form:*

$$\phi \ : \quad \mathbb{R}^d \to \mathbb{R},$$

$$\phi(x) \ = \ \sum_{i=1}^{p} w_i e^{-q\|x - x_i\|_2^2}, \quad q > 0,$$

where $x_i \in \mathbb{R}^d$, with $i = 1, \ldots, p \in \mathbb{N}$, are the interpolation points and the centres of the Gaussian Basis Functions and $w_i \in \mathbb{R}, i \in 1, \ldots, p$, are the so-called weights. $d \in \mathbb{N}$ denotes the input dimension of the RBF-Network.

In Fig. 1 the neural structure of an RBF-Network as a function approximator – as it will be used in this paper – is schematically shown.

MICCHELLI and others started around 1986 a function theory on the solvability of the interpolation problem with RBF-Networks, see for this [4].

We will now come to formulate the details of our algorithm that serves as an approach to detect the relevant input arguments.

3. Neural cross-validation with RBF-networks for finding relevant input arguments

The fundametal question is, if a neural network can be used to analyse the structure of a dataset in that way, to find *those* input arguments, that are relevant. If the human brain is so effective in its function to receive signals from the world and to compute them and to create actions, why should we not use some structures of the

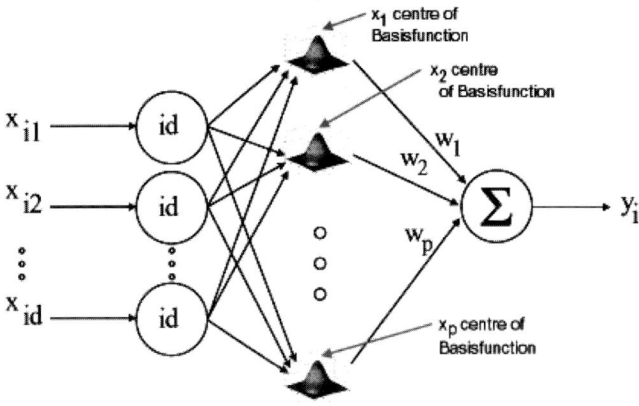

FIGURE 1. The neural structure of an RBF-Network

brain to analyse for example complex datasets and try to reduce their complexity by finding the irrelevant features?

The idea is now, to check how useful it is to use RBF-Networks for complexity reduction prolems. The choice of RBF-Networks emerged because of our need for a scattered data function approximation method, a field in which RBF-Networks established during the last few decades. The approach was a so-called cross-validated use of RBF-Networks. Therefore we will first of all present the following definition of a k-fold cross-validation that follows [5].

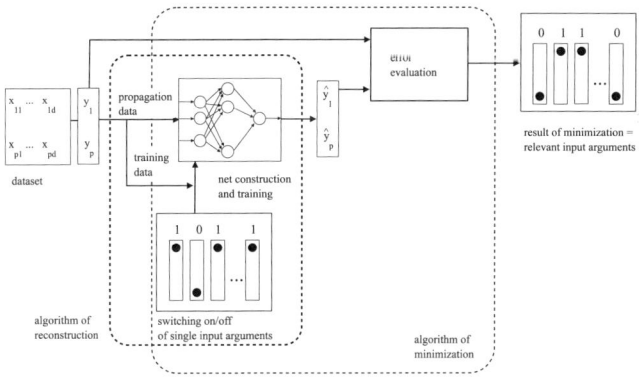

FIGURE 2. Schematic representation of the error minimization strategy

Definition 3.1. *A* **k-fold cross-validation**, *$k \in \mathbb{N}$, is the following proceeding: Let E be a set of examples and L be a learning process. Partition E into k parts of about equal size $E[1], \ldots, E[k]$. Hypothesise the results $h_1, \ldots, h_k$ on the following*

basis: h_i is for all $i = 1, \ldots, k$ defined as the result of L on the set of examples $E \setminus E[i]$. The error of L on E can be estimated as

$$\text{error}_{CV(E[1],\ldots,E[k])}(L, E) := \frac{\sum_{i=1}^{k} \text{error}_{E[i]}(h_i)}{k}.$$

For understanding how RBF-Networks are used in our context to detect the relevant input arguments, we will first of all introduce one main idea behind our algorithm. Fig. 2 depicts the framework of our approach:

The available dataset consists of pairs of the form $(x_i \in \mathbb{R}^d, y_i \in \mathbb{R})$, with $i = 1, \ldots, p$. We take this dataset and divide it into two parts, one for learning and the other for propagation. The one for learning trains 2^d neural networks. Each of the 2^d networks considers one pattern of input dimension-combination, which is in the figure indicated by means of showing a **1**, if the dimension is regarded and showing a **0** if the dimension is not regarded. As a result we have 2^d neural networks, which all represent one input pattern. We propagate each such network with the retained propagation data and build for them a reconstructed output vector $\hat{y} = (\hat{y}_1, \ldots, \hat{y}_p)^t$. For the case, that this output pattern differs from the actual output vector $y = y_1, \ldots, y_p$ (which is the default case) we measure an error value, for example the mean squared error (mse) between the two vectors,

$$e_{mse}(y_i, \hat{y}_i) = \frac{1}{p} \sum_{i=1}^{p} (y_i - \hat{y}_i)^2. \tag{1}$$

As a result we receive 2^d error values and we simply take the best value, that is the smallest one. The corresponding network now presents our model and the corresponding input combination is consequently defined as the best one.

Of cause we have to think about the exponential time, but instead using every 2^d networks it is also possible to use some evolutionary strategies or greedy ideas with time $O(n)$, see for this [2]. We now did a kind of a k-fold cross-validation strategy which is described in the following algorithm, in which some of the yet described ideas are sketched in a very short form:

1. Divide the given dataset S into two disjoint parts of equal size $S = SCV \cup SP$. The elements of the dataset can be randomly chosen. The part SCV is that one to be taken for training with the cross-validation method and the part SP denotes the part that is retained for later propagation and error evaluation.
2. Then do a k-fold cross-validation on the part SCV, but in each of the k learning-steps take every 2^d networks with each possible input argument-combination. Choose in each step the best network and the according best combination of input arguments with the belonging error-value.
3. Now we have k best input argument combinations. Take the data SP and propagate these best networks with SP and choose that network and its corresponding input combination with the smallest error.

We choose the mean-squared-error (1) as the error-measurement to take more emphasis on larger errors which is not the case if the mean absolute error (mae)

$$e_{mae}(y, \hat{y}) = \frac{1}{p} \sum_{i=1}^{p} |y_i - \hat{y}_i|$$

is considered. The whole algorithm is – for giving some overview – schematically summarized in Fig. 3.

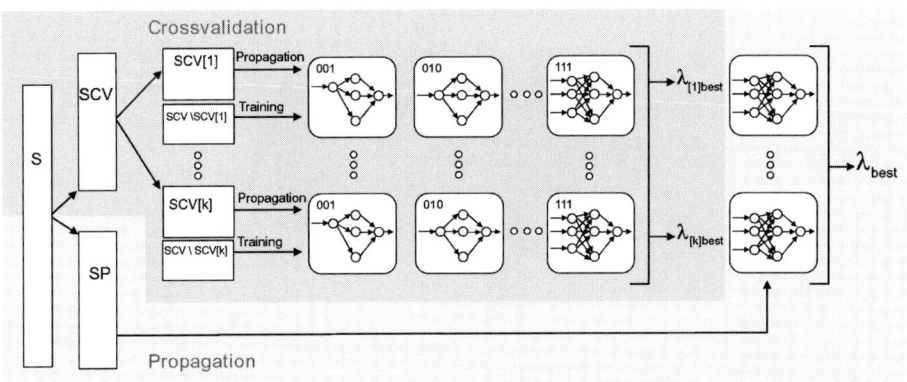

FIGURE 3. Schematic representation of the algorithm NCVRelevance

4. Some examples

In this section we will demonstrate how the algorithm works on some examples of data. We are going to construct artificial data with the help of common functions like ROSENBROCK's bananafunction (visualized in Fig. 4) and the so-called RAS-

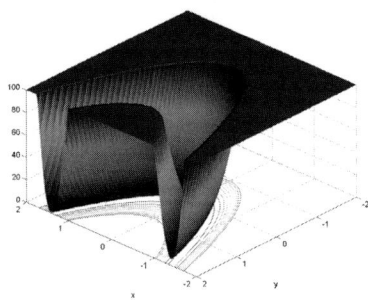

FIGURE 4. ROSENBROCK's function $f(x_1, x_2) = 100(x_2 - x_1^2)^2 + (1 - x_1)^2$

TRIGIN's function (see Fig. 5). Both functions are commonly used and famous for testing optimization strategies.

The generation of the dataset of 400 data points proceeds as following:

We build over 400 times two input arguments randomly between defined bounds (in the case of normalized data between -1 and 1). Then we construct two further 400 randomly distributed input data in the defined bounds and com-

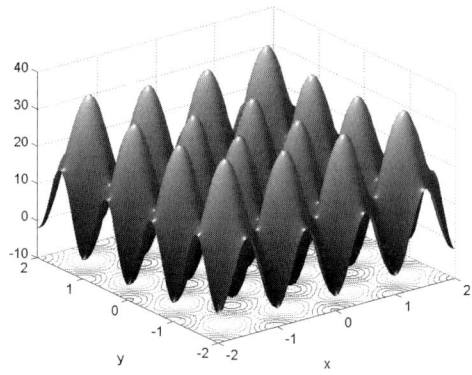

FIGURE 5. RASTRIGIN's function $f(x_1, x_2) = 10 + (x_1^2 - 10\cos(2\pi x_1)) + (x_2^2 - 10\cos(2\pi x_2))$

pute the belonging output of each function to these 400 pairs. We receive for each function data of over 400 data points of each four input data arguments and the according output data argument that stands in functional correlation with two of the input data. The question was now, if our algorithm is able to detect in both cases the two relevant input data arguments. In the case, where we use ROSEN-BROCK's banana function for the output data, we put the output of the function in functional dependence of the first and forth input argument. In the case, where we use RASTRINGIN's function we take the output data of the function in dependence of the first and third input argument.

Our algorithm now easily finds the relevant arguments as shown in Fig. 6 and Fig. 7. Some graphical tools for giving visual outputs such as linear ranking (see for this [7]) methods for stressing differences between the relevance of input arguments form the basis of the graphic 6. There are some types of functions generating datasets, that make it hart to algorithms to detect the right input arguments. Experiments with common algorithms (Rinciple Component Analysis (PCA, first described in [6], entropy-approaches (Shannon1948), etc.) on such data showed that these functions show in common a very rough structure. In Fig. 8 we present such a function that shows a very oscillating behaviour. Our algorithm NCVREL-EVANCE (Neural Cross-Validation on Relevance) has no problems to detect the right arguments as shown in Fig. 9. We tested further the algorithm [2] of KIENDL, MACHE, MEYER and SCHAUTEN on this data, the result is shown in Fig. 10, where

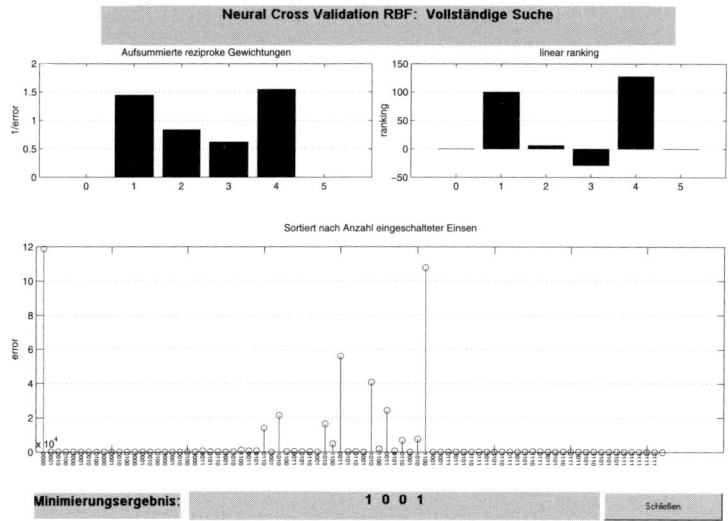

FIGURE 6. The visual output of the algorithm with $k = 5$ on data, that have been generated with ROSENBROCK's function. The first input argument corresponds to x_1, the forth to x_2. The second and third input argument are in no functional correlation to the output argument.

we can see that only one argument is detected as to be relevant, the second one is not found: The algorithm [2] has in fact problems to detect *that* argument of the function, that produces the oscillating structure.

5. Conclusions

In this article we showed some examples for demonstrating how the algorithm works and for giving an impression of the established software NCVRELEVANCE. We tested it in fact on numerous data, among that industrial, medical and further artificial data. But in that more real datasets the correct answer which input arguments are relevant is often not clear at all or only to some extend, so that our result cannot be proven to be correct. One main interesting fact we found out is that our algorithm is able to detect input arguments behind which a very rough structure is found, like it is the case in the example of data generated by the function in Fig. 8. We suppose that the reason for that are the high nonlinear structure of our used basis functions and the good abilities of RBF-Networks as scattered data approximators. In addition we found out that our algorithm often better detects irrelevant input arguments than other tested algorithms such as that of KIENDL, MACHE, MEYER and SCHAUTEN described in [2]. We tested that

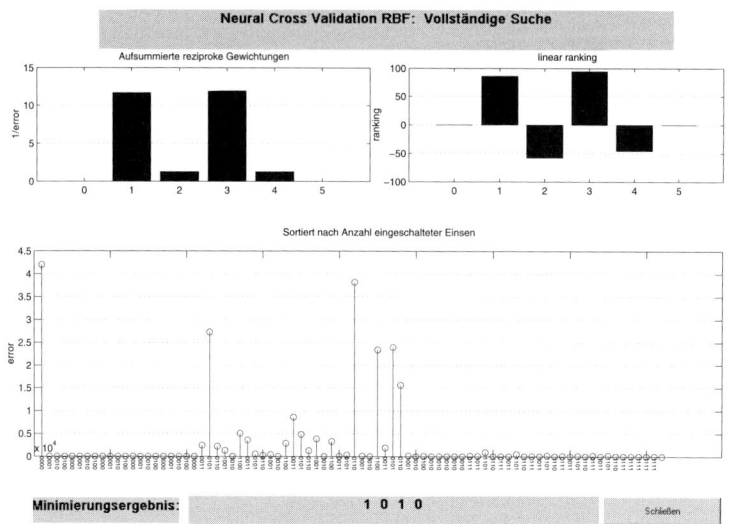

FIGURE 7. The visual output of the algorithm with $k = 5$ on data, that has been generated with RASTRIGIN's function. The first input argument corresponds to x_1, the third to x_2. The second and forth input argument are in no functional correlation to the output argument.

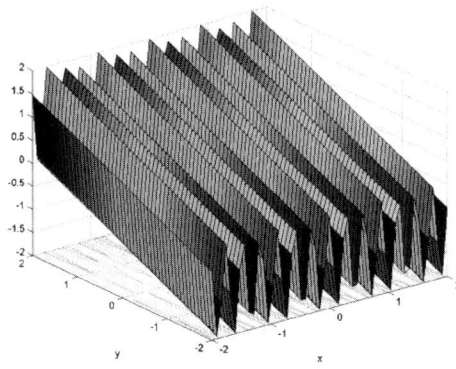

FIGURE 8. $f(x_1, x_2) = cos(100x_1) + 0.5x_2$

hypotheses on especially artificially constructed datasets with functions of difficult structures, such as functions that are very few smooth or oscillate very strong etc.

In conclusion we can summarize that, as expected, neural structures are good tools for complexity reduction, like detecting relevant input arguments and further research on a skilled use of neural structures in this issue is sincerely desirable.

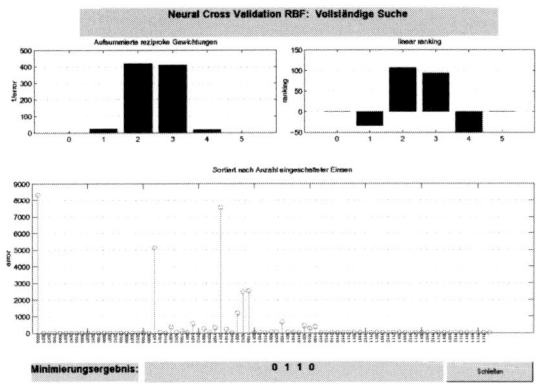

FIGURE 9. The visual output of the algorithm with $k = 5$ on data, that has been generated with the function $f(x_1, x_2) = cos(100x_1) + 0.5x_2$. The second input argument corresponds to x_1, the third to x_2. The first and fourth input argument are in no functional correlation to the output argument.

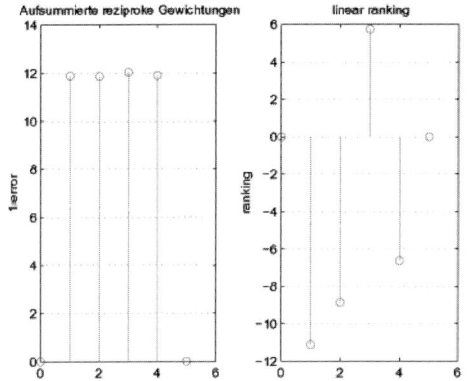

FIGURE 10. Visual output of the algorithm [2]. The second output corresponds to x_1 and is not detected to be relevant.

References

[1] Caruana, R.A., Freitag, D. (1994): *How useful is relevance?* In: Working Notes of the AAAI Fall Symposium on Relevance, pp. 25–29, New Orleans, LA: AAAI Press.

[2] Kiendl, H., Mache, D.H., Meyer, J., Schauten, D. (2003): *Rekonstruktionsbasierte Selektion relevanter Einflussgrößen*. In: *Reihe Computational Intelligence* des Sonderforschungsbereichs 531, *Design und Management komplexer technischer Prozesse und Systeme mit Methoden der Computational Intelligence*, Nr. CI-154/03.

[3] Meyer, J. (2003): *Multivariate Scattered Data Interpolation mit Radialen Basisfunktions Netzen als Ansatz zum Erkennen relevanter Eingabegrößen.* Diploma thesis at the Institute for Applied Mathematics, Approximationstheorie at the University of Dortmund, supervised by Prof. Dr. D.H. Mache.

[4] Micchelli, C.A. (1986): *Interpolation of Scattered Data: Distances Matrices and Conditionally Positive Definite Functions.* In: *Constructive Approximation* 2, pp. 11–22.

[5] Wrobel, S., Morik, K., Joachims, T. (2000): *Maschinelles Lernen und Data Mining.* In: Görz, G., Rollinger, C.-R., Schneeberger, J. (Hrsg.) *Handbuch der künstlichen Intelligenz.* 3., vollständig überarbeitete Auflage. Oldenbourg-Verlag: München, Wien.

[6] Pearson, K. (1901): *On lines and placcs of closest fit to systems of points in space* phil. Mag., 2: 559–572.

[7] Pohlheim, H.: (2000): *Evolutionäre Algorithmen.* Springer-Verlag Berlin, Heidelberg, New York.

[8] Shannon, C.E. (1948): *A Mathematical Theory of Communication.* In: The Bell System Technical Journal 27, 379–423 and 623–656.

Detlef H. Mache
University of Applied Sciences
TFH Georg Agricola Bochum
Applied Mathematics (FB 3), Constructive Approximation
Herner Str. 45
D-44787 Bochum, Germany
e-mail: `Mache@math.uni-dortmund.de`

Jennifer Meyer
Graduate Student at the International
Graduate School of Neuroscience (IGSN)
Ruhr-Universität Bochum
Institute for Neuroinformatics
Chair of Theoretical Biology
D-44780 Bochum, Germany
e-mail: `Jennifer.Meyer@neuroinformatik.ruhr-uni-bochum.de`

Trends and Applications in Constructive Approximation
(Eds.) M.G. de Bruin, D.H. Mache & J. Szabados
International Series of Numerical Mathematics, Vol. 151, 135–163
© 2005 Birkhäuser Verlag Basel/Switzerland

Subdivision Schemes and Non Nested Grids

Marie-Laurence Mazure

Abstract. Spline subdivision suggests the necessity of possibly using non nested grids to analyse the convergence of subdivision schemes, and also of changing grids to prove the smoothness of the limit curves. Inspired by this example, we introduce both non nested binary grids and equivalent grids. We give a sufficient condition for convergence, and we show how to use it through changes of grids to guarantee smoothness.

1. Introduction

As is well known, curve subdivision schemes are systematic procedures to build curves as limits of sequences of polygonal lines in $\mathbb{R}^d$, for some $d \geq 1$. At each level, each vertex of the polygonal line is calculated from the vertices of the polygonal line of the previous level. In some schemes this is done in a purely geometrical way, *c.g.*, in the very first subdivision schemes considered (see [13]), which we refer to as the *de Rham schemes*, and which we shall describe below. Once and for all, a positive number γ is selected. At each level, place two consecutive points of the next level on each segment of the polygonal line, so that they divide the segment with ratios $1 : \gamma : 1$. Denoting by $f_{j,k}$, $k \in \mathbb{Z}$, the vertices of level $j \geq 0$, this can be explicitly done through the following equalities:

$$f_{j+1,2k} = \frac{1+\gamma}{2+\gamma}\, f_{j,k-1} + \frac{1}{2+\gamma}\, f_{j,k}, \quad f_{j+1,2k+1} = \frac{1}{2+\gamma}\, f_{j,k-1} + \frac{1+\gamma}{2+\gamma}\, f_{j,k}. \quad (1)$$

On the other hand, some schemes do involve a grid in their definition, *e.g.*, all *Lagrange interpolatory subdivision schemes*. For all $j \geq 0$, and all $k \in \mathbb{Z}$, set $x_{j,k} := 2^{-j}k$. Given some fixed positive integer N, the vertices of level $(j+1)$ are now defined from those of level j by (see [6,9])

$$f_{j+1,2k} := f_{j,k}, \quad f_{j+1,2k+1} := P_{j,k}(x_{j+1,2k+1}), \qquad (2)$$

where $P_{j,k}$ denotes the polynomial function of degree at most 2N-1 with values in $\mathbb{R}^d$ which satisfies the Lagrange interpolating conditions $P_{j,k}(x_{j,r}) := f_{j,r}$ for $k-N+1 \leq r \leq k+N$. Both examples (1) and (2) are classical and they will serve as references in our introduction.

Now, whether or not a grid is involved in the definition of a given subdivision scheme, the simplest idea to analyse its convergence is to introduce parameterisations of the polygonal lines. This is generally done by reference to a regular grid, that is, by studying the convergence of the sequence of piecewise affine functions parameterising the polygonal line of level j, with knots at the points $2^{-j}k$, $k \in \mathbb{Z}$. This choice seems especially logical for all subdivision schemes which, like our two reference schemes, are regular both in space (uniform schemes: the procedure does not depend on k) and in time (stationary schemes: the procedure is the same at any level). Nonetheless, it is quite usual to also resort to regular grids for nonstationary/nonuniform schemes (see, for instance, [8]). Still, the necessity of considering irregular grids recently came out, in particular in relation with the construction of second generation wavelets. In [10], this was our original motivation for studying non regular Lagrange interpolatory schemes. Such schemes are defined exactly as in (2), but using now non regularly spaced points $x_{j,k}$ (see also [4]), assumed to satisfy $x_{j,k} < x_{j,k+1}$, along with the nestedness condition: $x_{j+1,2k} = x_{j,k}$ for all $j \geq 0$ and all $k \in \mathbb{Z}$. This in turn justified our interest in convergence of general subdivision schemes in connection with irregular grids. In particular, inspired by a well-known necessary and sufficient condition for stationary and uniform schemes associated with the regular grid [7], we established in [10] an interesting sufficient condition of convergence in terms of the difference schemes. We also showed how to guarantee smoothness of the limit curves by applying the sufficient condition to the derived schemes. Most logically, the grid which intervenes for the study of Lagrange interpolatory schemes is the grid composed of the points $x_{j,k}$ involved in its definition.

Let us now come back to the de Rham scheme (1) associated with the particular value $\gamma = 2$: two consecutive vertices of level $j + 1$ are then located at one fourth and three fourths respectively on a given segment of level j. This is also known as the Chaikin algorithm [1]. It is well known that each step describes insertion of knots for quadratic C^1 polynomial spline curves with regularly spaced knots at each level. The limit curve is then the quadratic spline the poles of which are the initial vertices [14]. Hence, although not visible at first sight, the Chaikin algorithm does implicitly involve a regular grid. This can be seen as one more justification for analysing its convergence using the points $2^{-j}k$. Knot insertion for splines is not limited to the case of regularly spaces knots, so let us see if this is confirmed by the extension to the nonregular case. At a given level j, we start with a strictly increasing sequence of non regularly spaced knots $x_{j,k}$, $k \in \mathbb{Z}$, and for each k, we insert a new knot $x_{j+1,2k+1}$, while keeping all old knots through the nestedness condition $x_{j+1,2k} = x_{j,k}$. The poles (see section 4) $f_{j+1,k}$, $k \in \mathbb{Z}$, of level $j + 1$ are then obtained from those of level j by means of the following equalities:

$$\begin{cases} f_{j+1,2k} = \dfrac{x_{j+1,2k+1} - x_{j,k+2}}{x_{j,k} - x_{j,k+2}}\, f_{j,k-1} + \dfrac{x_{j+1,2k+1} - x_{j,k}}{x_{j,k+2} - x_{j,k}}\, f_{j,k}, \\[2em] f_{j+1,2k+1} = \dfrac{x_{j+1,2k+3} - x_{j,k+2}}{x_{j,k} - x_{j,k+2}}\, f_{j,k-1} + \dfrac{x_{j+1,2k+3} - x_{j,k}}{x_{j,k+2} - x_{j,k}}\, f_{j,k}. \end{cases} \tag{3}$$

At first sight, the nested grid composed of the knots $x_{j,k}$ may seem the natural grid to associate with the subdivision scheme (3). Still, it is not so. This becomes totally clear when trying to use derived schemes to verify the smoothness of the limit curves, but it was already suggested by the fact that the pole $f_{j,k}$ is not linked to one of the knots of level j, but to a couple of consecutive knots. Both reasons clearly indicate that the proper grid is that composed of all points $(x_{j,k+1} + x_{j,k+2})/2$. It is easy to check that this is not a nested grid. In the Chaikin algorithm we should thus use the points $2^{-j}(k + \frac{3}{2})$, but the regularity makes it possible to replace it by the usual nested regular grid as well.

Therefore, spline subdivision does highlight the necessity of not limiting ourselves to nested grids, and it was the actual motivation of the present work. As a matter of fact, there is a crucial difference between Lagrange interpolatory schemes and de Rham schemes: the former are interpolatory, in the sense that all vertices of level j are maintained at level $j + 1$ by the left equality in (2), while the latter are not. Being nested is for grids the exact analogue of being interpolary for subdivision schemes. Therefore, while it is most logical to use nested grids for the study of interpolatory schemes, it is no longer so for non interpolatory ones. On the other hand, considering spline subdivision of higher degrees suggests another crucial idea, namely the fact that we have to reserve the right to change grid as we want to prove more smoothess of the limit curves.

The need of choosing a grid adapted to the subdivision scheme when leaving the interpolatory framework was already pointed out in the most interesting paper [5], in particular in order to construct derived schemes. In the latter paper too non nested grids are considered, which in some sense become nested at infinite, but both the definitions and the purposes are different from ours.

The paper is organized as follows. The second section contains all necessary definitions and technical results about grids and convergence of sequences of polygonal lines. The two main ideas are the use of non nested binary grids, supposed to meet as weak as possible requirements, and the possibility of changing grids to analyse the smoothness of the limit curves. In the third section we recall the basic tools concerning subdivision schemes and we extend to non nested grids the sufficient condition for convergence established for nested ones in [10]. We also show how to use the results of the first section to guarantee smoothness of limit curves produced by subdivision schemes. Finally, the fourth section illustrates all results of the previous one through the example of spline subdivision.

2. Preliminaries

This section contains preliminaries about curves obtained as limits of sequences of polygonal lines. A natural way to define such limits is to involve parameterisations relative to grids. Such grids are generally assumed to be nested. However, motivated by the example of spline subdivision, in the present paper we relax the assumption of nestedness.

2.1. Convergence of polygonal lines

In this subsection we are concerned with the convergence of sequences of polygonal lines. Let L_j, $j \geq 0$, be such a sequence in the space $\mathbb{R}^d$, $d \geq 1$. At each level j, we assume that the polygonal line L_j is given by its vertices $f_{j,k} \in \mathbb{R}^d$, $k \in \mathbb{Z}$. The simplest idea to define the convergence is to introduce parameterisations and to consider the convergence of the corresponding sequence of parameterisations.

With this in mind, for each $j \geq 0$, we choose a *grid of level j*, that is, a bi-infinite sequence $\mathcal{X}_j := (x_{j,k})_{k \in \mathbb{Z}}$, of real numbers, meeting the following two basic requirements:

(G$_1$) $x_{j,k} < x_{j,k+1}$ *for all $k \in \mathbb{Z}$*;

(G$_2$) $\lim_{k \to -\infty} x_{j,k} = -\infty$, $\lim_{k \to +\infty} x_{j,k} = +\infty$.

Without any more requirements, we shall refer to the sequence $\mathcal{X} := (\mathcal{X}_j)_{j \geq 0}$ as the chosen *grid*. The two properties (G$_1$) and (G$_2$) are exactly what is needed to ensure that, for any nonnegative integer j, the function $F_j : \mathbb{R} \to \mathbb{R}^d$ such that

$$F_j(x_{j,k}) = f_{j,k}, \quad F_j \text{ is affine on } [x_{j,k}, x_{j,k+1}], \quad k \in \mathbb{Z}, \tag{4}$$

is well defined. However, through elementary examples, it is easy to get convinced that defining the convergence of the sequence L_j, $j \geq 0$, through the convergence of the sequence of functions F_j, $j \geq 0$, (pointwise for instance) would be nonsense without some additional requirement on how consecutive levels are linked. This can be done for instance by considering *binary* grids, that is grids $\mathcal{X}$ in which any two points $x_{j+1,2k}$ and $x_{j+1,2k+1}$ of level $j + 1$ are "affiliated" in some sense to the point $x_{j,k}$ of level j. This affiliation is generally meant as the fact that that $x_{j+1,2k} = x_{j,k}$ for all $k \in \mathbb{Z}$, and therefore, for all $k \in \mathbb{Z}$, $x_{j+1,2k+1}$ is located in the interval $]x_{j,k}, x_{j,k+1}[$. In such a case we shall say that the grid is a *nested binary grid*. In the present paper, the expression binary grid is to be understood with the following larger meaning:

(G$_3$)   *the grid $\mathcal{X}$ is said to be binary if there exist two integers $N_1, N_2 \in \mathbb{Z}$ such that*:

$$x_{j+1,2k+N_1} \leq x_{j,k} \leq x_{j+1,2k+N_2}, \quad k \in \mathbb{Z}, \ j \geq 0. \tag{5}$$

The particular case of nested binary grid thus corresponds to $N_1 = N_2 = 0$. Of course, when the grid is binary, it is sufficient to require (G$_2$) to be satisfied by the grid of level 0. The latter definition of binary grids is inspired by the example of spline subdivision as we shall see in Section 4. Clearly, the definition could easily be adapted to *ternary* grids, or more generally to *p-ary* grids.

Definition 2.1. *We say that the sequence of polygonal lines L_j, $j \geq 0$, converges if there exists a binary grid $\mathcal{X}$ such that the corresponding sequence of parameterisations F_j, $j \geq 0$, converges uniformly on any compact subset of $\mathbb{R}$. If so, we shall say that the sequence of polygonal lines converges relative to the grid $\mathcal{X}$.*

Given another binary grid $\widehat{\mathcal{X}} = \left(\widehat{\mathcal{X}}_j\right)_{j \geq 0}$, with $\widehat{\mathcal{X}}_j := \left(\widehat{x}_{j,k}\right)_{k \in \mathbb{Z}}$, consider the piecewise affine functions $\varphi_j : \mathbb{R} \to \mathbb{R}$, $j \geq 0$, defined by:

$$\varphi_j(x_{j,k}) = \widehat{x}_{j,k}, \quad \varphi_j \text{ is affine on } [x_{j,k}, x_{j,k+1}], \quad k \in \mathbb{Z}. \tag{6}$$

Due to (G_1), each function φ_j is strictly increasing on $\mathbb{R}$. The convergence of the same sequence of polygonal lines L_j, $j \geq 0$, will now be studied by considering the convergence of the piecewise affine functions $\widehat{F}_j$ defined as in (4) but using the new grid $\widehat{\mathcal{X}}$, so that

$$F_j = \widehat{F}_j \circ \varphi_j, \quad j \geq 0. \tag{7}$$

The following statement makes Definition 2.1 consistent.

Proposition 2.2. *If a sequence of polygonal lines in $\mathbb{R}^d$ converges relative to two binary grids $\mathcal{X}$ and $\widehat{\mathcal{X}}$, then the corresponding limit curves $\mathcal{C}$ and $\widehat{\mathcal{C}}$ are identical.*

Proof. Denote by $\widehat{N}_1, \widehat{N}_2$ the two integers provided by (G_3) concerning the grid $\widehat{\mathcal{X}}$. Then, applying (5) repeatedly gives, for any $k \in \mathbb{Z}$:

$$\begin{aligned} x_{j,2^j k + (2^j-1)N_1} &\leq x_{0,k} \leq x_{j,2^j k + (2^j-1)N_2}, \\ \widehat{x}_{j,2^j k + (2^j-1)\widehat{N}_1} &\leq \widehat{x}_{0,k} \leq \widehat{x}_{j,2^j k + (2^j-1)\widehat{N}_2}. \end{aligned} \tag{8}$$

Let t_0 be a real number. Due to (G_1) and (G_2) we can consider the two integers ℓ_0 and r_0 uniquely defined by $x_{0,\ell_0} := \text{Max}\{x_{0,k}, \ k \in \mathbb{Z} \text{ and } x_{0,k} \leq t_0\}$ and $x_{0,r_0} := \text{Min}\{x_{0,k}, \ k \in \mathbb{Z} \text{ and } x_{0,k} \geq t_0\}$. From the left part of (8) we can deduce that, for all $j \geq 0$,

$$x_{j,2^j \ell_0 + (2^j-1)N_1} \leq t_0 \leq x_{j,2^j r_0 + (2^j-1)N_2},$$

whence, using definition (6):

$$\widehat{x}_{j,2^j \ell_0 + (2^j-1)N_1} \leq \varphi_j(t_0) \leq \widehat{x}_{j,2^j r_0 + (2^j-1)N_2}.$$

Using now the right part of (8), this ensures that:

$$\widehat{x}_{0,k_1} \leq \varphi_j(t_0) \leq \widehat{x}_{0,k_2}, \quad j \geq 0,$$

where k_1, k_2 are any two integers chosen so that

$$k_1 \leq \ell_0 + (1 - 2^{-j})(N_1 - \widehat{N}_2), \quad k_2 \geq r_0 + (1 - 2^{-j})(N_2 - \widehat{N}_1) \quad \text{for all } j \geq 0.$$

In the compact interval $[\widehat{x}_{0,k_1}, \widehat{x}_{0,k_2}]$ we can find a convergent subsequence $\varphi_{j_r}(t_0)$, $r \geq 0$. Set $\widehat{t}_0 := \lim_{r \to +\infty} \varphi_{j_r}(t_0)$. On account of (7), the uniform convergence of the sequence $\widehat{F}_j$, $j \geq 0$, to a function $\widehat{F}$, guarantees that $F(t_0) = \widehat{F}(\widehat{t}_0)$. Therefore, the two sets $\mathcal{C} := \{F(t), \ t \in \mathbb{R}\}$ and $\widehat{\mathcal{C}} := \{\widehat{F}(t), \ t \in \mathbb{R}\}$ are identical, *i.e.*, the functions F and $\widehat{F}$ are two different parameterisations of the same curve. $\square$

Remark 2.3. *Due to the two assumptions (G_1) and (G_2), for any given $t_0 \in \mathbb{R}$, and any $j \geq 0$, we can choose an integer $k_j(t_0)$ such that*

$$|x_{j,k_j(t_0)} - t_0| = \underset{k \in \mathbb{Z}}{\text{Min}} \, |x_{j,k} - t_0|. \tag{9}$$

Indeed, generalising what we did in the proof of proposition 2.2, we can consider the two integers $\ell_j(t_0)$, $r_j(t_0)$ $(0 \leq r_j(t_0) - \ell_j(t_0) \leq 1)$, defined by $x_{j,\ell_j(t_0)} :=$ Max$\{x_{j,k}, \ k \in \mathbb{Z}$ and $x_{j,k} \leq t_0\}$, $x_{j,r_j(t_0)} :=$ Min$\{x_{j,k}, \ k \in \mathbb{Z}$ and $x_{j,k} \geq t_0\}$. For any $j \geq 0$, $k_j(t_0)$ is equal to either $\ell_j(t_0)$ or $r_j(t_0)$. Arguments similar to those used in the proof of Proposition 2.2 lead to

$$x_{0,k_1} \leq x_{j,2^j \ell_0(t_0)+(2^j-1)N_1} \leq t_0 < x_{j,2^j(r_0(t_0))+(2^j-1)N_2} \leq x_{0,k_2}$$

for all $j \geq 0$, with $k_1 := \ell_0(t_0) - N_2 + N_1$, $k_2 := r_0(t_0) + N_2 - N_1$. Accordingly, for any $t \in \mathbb{R}$, the sequence $x_{j,k_j(t)}$, $j \geq 0$, is bounded, and therefore admits a subsequence which converges, but a priori nothing guarantees that it converges to t. It is natural to introduce a kind of "density assumption" for the grids, the weakest possible will be the following one:

(G$_4$)   *for each real number t, the (bounded) sequence $x_{j,k_j(t)}$, $j \geq 0$, admits a subsequence converging to t.*

Note that, if the binary grid $\mathcal{X}$ is nested, then (G$_4$) *holds iff* $\lim_{j \to +\infty} x_{j,k_j(t)} = t$ *for all $t \in \mathbb{R}$.*

2.2. Equivalent grids

Later on we shall be interested not only in getting the same limit curve relative to different grids, but even the same limit function. This is why we introduce the following definition.

Definition 2.4. *We say that two grids $\mathcal{X}$ and $\widehat{\mathcal{X}}$ are equivalent if the sequence of functions φ_j, $j \geq 0$, defined in (6) is pointwise convergent to $Id_\mathbb{R}$.*

Let us now compare convergence relative to equivalent grids.

Proposition 2.5. *Let $\mathcal{X}$ and $\widehat{\mathcal{X}}$ be two equivalent binary grids. Then, any sequence L_j, $j \geq 0$, of polygonal lines converges relative to $\mathcal{X}$ iff it converges relative to $\widehat{\mathcal{X}}$, with the same limit function.*

Proof. Each function φ_j defined in (6) is an increasing bijection of $\mathbb{R}$. Accordingly, if the sequence φ_j, $j \geq 0$ is pointwise convergent to $Id_\mathbb{R}$,

(a) so is the sequence φ_j^{-1}, $j \geq 0$;

(b) the convergence is uniform on compact sets of $\mathbb{R}$.

If $\widehat{F}_j$ converges to F uniformly on compact sets of $\mathbb{R}$, due to (7) and (b), so does F_j. On account of (a), the proof is complete. $\square$

Remark 2.6.

1. *Two nested binary grids are equivalent iff they are identical.*

2. *If a binary grid $\mathcal{X}$ satisfies* (G$_4$), *then any binary grid $\widehat{\mathcal{X}}$ which is equivalent to $\mathcal{X}$ also satisfies* (G$_4$).

3. *In the proof of Proposition 2.5, we already observed the reflexivity of the relation "being equivalent". It is also transitive, hence it is a relation of equivalence in the set of all binary grids.*

4. *It is possible to introduce a weaker relation of equivalence between binary grids by requiring more generally that the sequence φ_j, $j \geq 0$, converges to a continuous (or, more generally C^p for some $p \geq 0$) bijection $\varphi : \mathbb{R} \to \mathbb{R}$. This would maintain the result of Proposition 2.5, but now the limit functions F and $\widehat{F}$ would satisfy $F = \widehat{F} \circ \varphi$, i.e., φ would be a C^0 (resp. C^p) change of parameterisation.*

Let us now present an interesting example of equivalent binary grids which will intervene in Section 4. We first need to introduce the following definition, in which we adopt the terminology of [4].

Definition 2.7. *We say that a binary grid $\mathcal{X}$ is homogenous, if*

$$\eta := \operatorname*{Sup}_{j \geq 0,\ k \in \mathbb{Z}} \operatorname{Max}\left(\frac{d_{j,k+1}}{d_{j,k}}, \frac{d_{j,k}}{d_{j,k+1}}\right) < +\infty, \tag{10}$$

where

$$d_{j,k} := x_{j,k} - x_{j,k-1}, \quad k \in \mathbb{Z},\ j \geq 0. \tag{11}$$

Proposition 2.8. *Let $\mathcal{X}$ be a homogenous nested binary grid. Then, for any given $p_1 < p_2$ in $\mathbb{Z}$, the grid $\widehat{X}$ defined by*

$$\widehat{x}_{j,k} := \frac{x_{j,k+p_1+1} + \cdots + x_{j,k+p_2}}{p_2 - p_1}, \quad k \in \mathbb{Z}, j \geq 0, \tag{12}$$

is a homogenous binary grid which is equivalent to $\mathcal{X}$ and which satisfies (G_4).

Although Proposition 2.8 could be proved more directly, we will obtain it by means of the following lemma which we shall need in Section 4 for other purposes.

Lemma 2.9. *Suppose that the binary grid $\mathcal{X}$ is nested and homogenous. Then, for any $n \geq 1$, there exists a positive number $A_n < 1$ such that, for any $k, \ell \in \mathbb{Z}$,*

$$2\ell \leq k \leq 2\ell + n \quad \Rightarrow \quad x_{j+1,k+n} - x_{j+1,k} \leq A_n\,(x_{j,\ell+n} - x_{j,\ell}). \tag{13}$$

We can choose for instance $A_n := \dfrac{\eta^{\lceil \frac{n+1}{2} \rceil}}{1+\eta^{\lceil \frac{n+1}{2} \rceil}}$, where η is defined by (10).

Proof. Although the proof will not be done by induction, we start by proving (13) for $n = 1$. Dividing by $(1 + \eta)$ both hand sides of each inequality $d_{j+1,2\ell+1} \leq \eta d_{j+1,2\ell+2}$ and $d_{j+1,2\ell+2} \leq \eta d_{j+1,2\ell+1}$, yields, with $A_1 := \eta/(1+\eta) \in [1/2, 1[$:

$$(1 - A_1)\,d_{j+1,2\ell+1} \leq A_1\,d_{j+1,2\ell+2}, \quad (1 - A_1)\,d_{j+1,2\ell+2} \leq A_1\,d_{j+1,2\ell+1}.$$

The grid $\mathcal{X}$ being nested, we have $d_{j,\ell+1} = d_{j+1,2\ell+1} + d_{j+1,2\ell+2}$. Whence $d_{j+1,2\ell+1} \leq A_1\,d_{j,\ell+1}$ and $d_{j+1,2\ell+2} \leq A_1\,d_{j,\ell+1}$. This is the announced result for $n = 1$.

Let us now consider the case $n = 2m \geq 2$. Let $k, \ell \in \mathbb{Z}$ be two integers such that $2\ell \leq k \leq 2\ell + n$. Suppose first that $k = 2p$, so that $\ell \leq p \leq \ell + m$. Then, again due to the grid $\mathcal{X}$ being nested, we have:

$$x_{j+1,k+n} - x_{j+1,k} = x_{j,p+m} - x_{j,p} = \sum_{r=p+1}^{p+m} d_{j,r}, \qquad (14)$$

$$x_{j,\ell+n} - x_{j,\ell} = \sum_{r=\ell+1}^{p} d_{j,r} + \sum_{r=p+1}^{p+m} d_{j,r} + \sum_{r=p+m+1}^{\ell+n} d_{j,r}. \qquad (15)$$

Using (10) repeatedly leads to:

$$d_{j,r} \geq \frac{d_{j,r+p-\ell}}{\eta^{p-\ell}} \text{ for } \ell+1 \leq r \leq p, \ d_{j,r} \geq \frac{d_{j,r+p-m-\ell}}{\eta^{\ell+m-p}} \text{ for } p+m+1 \leq r \leq \ell+n.$$

Hence (15) gives:

$$x_{j,\ell+n} - x_{j,\ell} \geq \sum_{r=p+1}^{2p-\ell} d_{j,r}\left(1 + \frac{1}{\eta^{p-\ell}}\right) + \sum_{r=2p-\ell+1}^{p+m} d_{j,r}\left(1 + \frac{1}{\eta^{\ell+m-p}}\right).$$

Therefore, due to (14), the announced inequality $x_{j+1,k+n} - x_{j+1,k} \leq A_n(x_{j,\ell+n} - x_{j,\ell})$ will be satisfied for any A_n such that

$$A_n \geq \text{Max}\left(\frac{\eta^{p-\ell}}{\eta^{p-\ell}+1}, \frac{\eta^{\ell+m-p}}{\eta^{\ell+m-p}+1}\right). \qquad (16)$$

Suppose now that $k = 2p+1$, so that $\ell \leq p \leq \ell+m-1$. Then, taking the result proved for $n = 1$ into account, we can write:

$$\begin{aligned}
x_{j+1,k+n} - x_{j+1,k} &= d_{j+1,2p+2} + d_{j+1,2p+2m+1} + \sum_{r=p+2}^{p+m} d_{j,r} \\
&\leq A_1(d_{j,p+1} + d_{j,p+m+1}) + \sum_{r=p+2}^{p+m} d_{j,r}. \qquad (17)
\end{aligned}$$

On the other hand,

$$x_{j,\ell+n} - x_{j,\ell} = d_{j,p+1} + d_{j,p+m+1} + \sum_{r=\ell+1}^{p} d_{j,r} + \sum_{r=p+2}^{p+m} d_{j,r} + \sum_{r=p+m+2}^{\ell+n} d_{j,r}. \quad (18)$$

Using (10) repeatedly as previously, equality (18) leads to:

$$x_{j,\ell+n} - x_{j,\ell} \geq d_{j,p+1} + d_{j,p+m+1} + \sum_{r=p+2}^{2p-\ell+1} d_{j,r}\left(1 + \frac{1}{\eta^{p-\ell+1}}\right) + \sum_{r=2p-\ell+2}^{p+m} d_{j,r}\left(1 + \frac{1}{\eta^{\ell+m-p}}\right).$$

By comparison with (17), the inequality $x_{j+1,k+n} - x_{j+1,k} \leq A_n(x_{j,\ell+n} - x_{j,\ell})$ will now be satisfied for any A_n such that

$$A_n \geq \text{Max}\left(A_1, \frac{\eta^{p-\ell+1}}{\eta^{p-\ell+1}+1}, \frac{\eta^{\ell+m-p}}{\eta^{\ell+m-p}+1}\right). \qquad (19)$$

Gathering all cases (16) and (19), the expected result is obtained by choosing A_n so that:

$$1 > A_n \geq \frac{\eta^m}{\eta^m + 1} = \frac{\eta^{\lceil \frac{n+1}{2} \rceil}}{1 + \eta^{\lceil \frac{n+1}{2} \rceil}}.$$

The case $n = 2m + 1$ can be proved in a similar way.   □

Lemma 2.10. *Suppose that the binary grid $\mathcal{X}$ is nested and homogenous. Then, for any given $n \in \mathbb{Z}$, we have, with the notations introduced in (9):*

$$\lim_{j \to +\infty} x_{j,k_j(t)+n} = t \quad \text{for all } t \in \mathbb{R}. \tag{20}$$

Proof. Consider any sequence of integers m_j, $j \geq 0$, such that $2m_j \leq m_{j+1} \leq 2m_j + 1$. Then, for any integer $n \geq 1$, iteration of (13) yields:

$$x_{j,m_j+n} - x_{j,m_j} \leq A_n(x_{j-1,m_{j-1}+n} - x_{j-1,m_{j-1}}) \leq \cdots \leq A_n{}^j(x_{0,m_0+n} - x_{0,m_0}).$$

Hence

$$\lim_{j \to +\infty} (x_{j,m_j+n} - x_{j,m_j}) = 0. \tag{21}$$

For a given $t \in \mathbb{R}$, using the notations introduced in Remark 2.3, it is straightforward to derive, for all $j \geq 0$:

$$x_{j,\ell_j(t)} = x_{j+1,2\ell_j(t)} \leq x_{j+1,\ell_{j+1}(t)} \leq t < x_{j+1,\ell_{j+1}(t)+1} \leq x_{j+1,2\ell_j(t)+2} = x_{j,\ell_j(t)+1}.$$

Hence

$$2\ell_j(t) \leq \ell_{j+1}(t) \leq 2\ell_j(t) + 1, \quad j \geq 0.$$

Applied to the sequence $\ell_j(t)$, $j \geq 0$, (21) enables us to conclude that:

$$\lim_{j \to +\infty} (x_{j,\ell_j(t)+n} - x_{j,\ell_j(t)}) = 0.$$

On the other hand the two sequences $x_{j,\ell_j(t)}$ and $x_{j,\ell_j(t)+1}$, $j \geq 0$, are respectively nondecreasing, and nonincreasing. It follows that:

$$\lim_{j \to +\infty} x_{j,\ell_j(t)} = \lim_{j \to +\infty} x_{j,\ell_j(t)+1} = \cdots = \lim_{j \to +\infty} x_{j,\ell_j(t)+n} = t, \quad t \in \mathbb{R}. \tag{22}$$

From (13), it also results that

$$2\ell - n \leq k \leq 2\ell \quad \Rightarrow \quad x_{j+1,k} - x_{j+1,k-n} \leq A_n (x_{j,\ell} - x_{j,\ell-n}).$$

Taking into account the fact that, for all $j \geq 0$, $2r_j(t) - 1 \leq r_{j+1}(t) \leq 2r_j(t)$ would similarily lead to

$$\lim_{j \to +\infty} x_{j,r_j(t)} = \lim_{j \to +\infty} x_{j,r_j(t)-1} = \cdots = \lim_{j \to +\infty} x_{j,r_j(t)-n} = t, \quad t \in \mathbb{R}. \tag{23}$$

The announced relation (20) follows easily from (22) and (23).   □

Proof of Proposition 2.8. All grids (12) do satisfy (G_1) and (G_2), but none of them is nested. They are examples of binary grids since they do satisfy (G_3) too: indeed, for all $j \geq 0$ and all $k \in \mathbb{Z}$, we have

$$\widehat{x}_{j+1,2k+\widehat{N}_1} \leq \widehat{x}_{j,k} \leq \widehat{x}_{j+1,2k+\widehat{N}_2}, \quad \widehat{N}_1 := p_1 + 1, \widehat{N}_2 := p_2, \tag{24}$$

with strict inequalities as soon as $p_2 - p_1 \geq 2$.

On the other hand, let t be a given real number. From (20) we can derive that, for all $n \in \mathbb{Z}$,

$$\lim_{j \to +\infty} \widehat{x}_{j,k_j(t)+n} = t. \tag{25}$$

Transforming the double inequality $x_{j,k_j(t)-1} \le t \le x_{j,k_j(t)+1}$ via the increasing function φ_j defined in (6) gives:

$$\widehat{x}_{j,k_j(t)-1} \le \varphi_j(t) \le \widehat{x}_{j,k_j(t)+1}.$$

Relation (25) proves that $\lim_{j \to +\infty} \varphi_j(t) = t$. $\square$

2.3. On the regularity of the limits

We now address the problem of how to guarantee the regularity of the limit curve of a convergent sequence of polygonal lines. This will be done by studying the regularity of the limit function involved. For this purpose we shall associate with the initial sequence a new one, obtained by means of divided differences of order one.

For a given $j \ge 0$, we consider the divided differences of the sequence F_j based on consecutive points of the grids at level j, *i.e.*,

$$f_{j,k}^{[1]} := [x_{j,k}, x_{j,k-1}]F_j := \frac{f_{j,k} - f_{j,k-1}}{x_{j,k} - x_{j,k-1}}, \quad k \in \mathbb{Z}. \tag{26}$$

We denote by $L_j^{[1]}$ the polygonal line with vertices $f_{j,k}^{[1]}$, and we call the sequence $L_j^{[1]}$, $j \ge 0$, *the derived sequence* of L_j, $j \ge 0$, w.r. to $\mathcal{X}$. Similarly to (4), we shall denote by $F_j^{[1]} : \mathbb{R} \to \mathbb{R}$ the function defined by

$$F_j^{[1]}(x_{j,k}) = f_{j,k}^{[1]}, \quad F_j^{[1]} \text{ is affine on } [x_{j,k}, x_{j,k+1}], \quad k \in \mathbb{Z}. \tag{27}$$

The following proposition extends [4,Lemma 5].

Proposition 2.11. *Let $\mathcal{X}$ be a binary grid satisfying* (G_4). *Assume that the sequence L_j, $j \ge 0$, and its derived sequence w.r. to $\mathcal{X}$ both converge relative to $\mathcal{X}$, and let F and $F^{[1]}$ be the corresponding limit functions. Then, the function F is continuously differentiable and we have:*

$$F' = F^{[1]}. \tag{28}$$

Proof. Given $y \in \mathbb{R}$, let us introduce, for $y_1 \ne y_2$, the quantity:

$$H(y_1, y_2) := \frac{F(y_1) - F(y_2)}{y_1 - y_2} - F^{[1]}(y). \tag{29}$$

Given any $j \ge 0$, and any $k_1, k_2 \in \mathbb{Z}$, with $x_{j,k_1} \ne x_{j,k_2}$ (that is, $k_1 \ne k_2$), we can write:

$$|H(y_1, y_2)| \le |H_1(y_1, y_2)| + |H_2(y_1, y_2)| + |H_3(y_1, y_2)|. \tag{30}$$

where:

$$H_1(y_1, y_2) : = \frac{F(y_1) - f_{j,k_1}}{y_1 - y_2} - \frac{F(y_2) - f_{j,k_2}}{y_1 - y_2},$$

$$H_2(y_1, y_2) : = (f_{j,k_1} - f_{j,k_2}) \left(\frac{1}{y_1 - y_2} - \frac{1}{x_{j,k_1} - x_{j,k_2}} \right), \tag{31}$$

$$H_3(y_1, y_2) : = \frac{f_{j,k_1} - f_{j,k_2}}{x_{j,k_1} - x_{j,k_2}} - F^{[1]}(y).$$

In order to prove that $\lim_{y_1, y_2 \to y} H(y_1, y_2) = 0$, we consider a given $\varepsilon > 0$. It is sufficient to show that, for any $y_1 \neq y_2$ sufficiently close to y, a convenient choice of the integers j, k_1, k_2 will enable us to bound each of the three quantities $|H_i(y_1, y_2)|$, $i = 1, 2, 3$, by ε.

Let us first transform the quantity $H_3(y_1, y_2)$. Being interested only by its absolute value, without loss of generality we can suppose that $k_1 < k_2$. We can then write, taking account of (26):

$$x_{j,k_2} - x_{j,k_1} = \sum_{k=k_1+1}^{k_2} (x_{j,k} - x_{j,k-1}), \tag{32}$$

$$f_{j,k_2} - f_{j,k_1} = \sum_{k=k_1+1}^{k_2} (f_{j,k} - f_{j,k-1}) = \sum_{k=k_1+1}^{k_2} f_{j,k}^{[1]} (x_{j,k} - x_{j,k-1}).$$

Accordingly:

$$|H_3(y_1, y_2)| = \frac{1}{x_{j,k_2} - x_{j,k_1}} \left| \sum_{k=k_1+1}^{k_2} (f_{j,k}^{[1]} - F^{[1]}(y)) (x_{j,k} - x_{j,k-1}) \right|.$$

Equality (32) makes it clear that, in order to ensure $|H_3(y_1, y_2)| < \varepsilon$, it is sufficient to choose the integers j, k_1, and k_2 so that

$$|f_{j,k}^{[1]} - F^{[1]}(y)| \leq \varepsilon \quad \text{for } k_1 + 1 \leq k \leq k_2. \tag{33}$$

Now, using both the uniform continuity of $F^{[1]}$ on any interval $[y - \alpha, y + \alpha]$, $\alpha > 0$ and the uniform convergence on $[y - \alpha, y + \alpha]$ of the sequence $F_j^{[1]}$, $j \geq 0$, to $F^{[1]}$, we can derive the existence of a real number $\delta > 0$ and of an integer $J(\varepsilon) \geq 0$, so that :

$$|x - y| \leq \delta \text{ and } j \geq J(\varepsilon) \quad \Rightarrow \quad |F^{[1]}(y) - F_j^{[1]}(x)| \leq \varepsilon. \tag{34}$$

Once and for all, we choose two points y_1, y_2 such that

$$y_1 \neq y_2, \quad |y_1 - y| \leq \frac{\delta}{2}, \quad |y_2 - y| \leq \frac{\delta}{2}. \tag{35}$$

As observed in Remark 2.3, we know that the two sequences $x_{j,k_j}(y_1)$, $x_{j,k_j}(y_2)$, $j \geq 0$, are bounded. This, along with the fact that the grid $\mathcal{X}$ satisfies (G$_4$), ensures the existence of a strictly increasing sequence of nonnegative integers j_r, $r \geq 0$, and two sequences k_r^1, k_r^2, in $\mathbb{Z}$, such that

$$y_1 = \lim_{r \to +\infty} x_{j_r, k_r^1}, \quad y_2 = \lim_{r \to +\infty} x_{j_r, k_r^2}. \tag{36}$$

Denote by $R(\varepsilon)$ an integer such that:

$$r \geq R(\varepsilon) \quad \Rightarrow \quad \begin{cases} j_r \geq J(\varepsilon) \\ |y_i - x_{j_r, k_r^i}| \leq \frac{\delta}{2}, & i = 1, 2. \end{cases} \tag{37}$$

Choose $r \geq R(\varepsilon)$. Due to (35) and (37), any real number x located between x_{j_r, k_r^1} and x_{j_r, k_r^2} satisfies $|x - y| \leq \delta$. Hence, due to (34), it satisfies $|F^{[1]}(y) - F_{j_r}^{[1]}(x)| \leq \varepsilon$. In particular, for any k between k_r^1 and k_r^2, the point $x_{j_r, k}$ is located between x_{j_r, k_r^1} and x_{j_r, k_r^2}. Since $F_{j_r}^{[1]}(x_{j_r, k}) = f_{j_r, k}^{[1]}$, we can thus state

$$|F^{[1]}(y) - f_{j_r, k}^{[1]}| \leq \varepsilon \text{ for any } k \text{ between } k_r^1 \text{ and } k_r^2. \tag{38}$$

According to (33), we have thus proved that

$$j := j_r, k_1 := k_r^1, k_2 := k_r^2, \ r \geq R(\varepsilon) \quad \Rightarrow \quad |H_3(y_1, y_2)| \leq \varepsilon. \tag{39}$$

On the other hand, the uniform convergence of F_j and the uniform continuity of F on a compact containing y_1, y_2 guarantee that

$$F(y_i) = \lim_{r \to +\infty} F_{j_r}(x_{j_r, k_r^i}) = \lim_{r \to +\infty} f_{j_r, k_r^i}, \quad i = 1, 2. \tag{40}$$

In particular, relation (40) enables us to bound $|f_{j_r, k_r^1} - f_{j_r, k_r^2}|$ independently of r. From (36), we know that

$$\lim_{r \to +\infty} \left(x_{j_r, k_r^1} - x_{j_r, k_r^2} \right) = y_1 - y_2 \neq 0.$$

It is thus possible to choose $R'(\varepsilon) \geq 0$ so that,

$$j := j_r, k_1 := k_r^1, k_2 := k_r^2, \ r \geq R'(\varepsilon) \quad \Rightarrow \quad |H_i(y_1, y_2)| \leq \varepsilon \text{ for } i = 1, 2. \tag{41}$$

Choose $j := j_r$, with $r \geq \max(R(\varepsilon), R'(\varepsilon))$. On account of (39) and (41), the corresponding inequality (30) ensures $|H(y_1, y_2)| \leq 3\varepsilon$. $\qquad\square$

Remark 2.12. *The proof of (28) does not require the grid to be binary, it only uses condition* (G_4). *Even* (G_4) *is only a sufficient condition as shown by the following trivial example. Choose* $f_j := \mathcal{X}_j$ *for all* $j \geq 0$. *Then, obviously,* $f_j^{[1]} = \mathbb{1}$, *where* $\mathbb{1}$ *denotes the bi-infinite sequence all components of which are equal to 1. With this choice, (28) is satisfied with no condition on the grid other than the basic requirements* (G_1) *and* (G_2), *since* $F(t) = F_j(t) = t$ *and* $F^{[1]}(t) = F_j^{[1]}(t) = 1$ *for all* t.

Corollary 2.13. *Let* L_j, $j \geq 0$, *be a sequence of polygonal lines, and let* $\mathcal{X}^0$, $\mathcal{X}^1, \ldots, \mathcal{X}^P$ *be a sequence of equivalent binary grids. For* $1 \leq p \leq P$, *we define recursively the sequence* $L_j^{[p]}$, $j \geq 0$, *as the derived sequence of* $L_j^{[p-1]}$, $j \geq 0$, *w.r. to* $\mathcal{X}^p$, *with* $L_j^{[0]} := L_j$. *Suppose that, for* $0 \leq p \leq P$, *the sequence* $L_j^{[p]}$, $j \geq 0$, converges relative to $\mathcal{X}^p$, with limit function $F_j^{[p]}$. Then, if the grid $\mathcal{X}^0$ satisfies (G_4), the function $F := F^{[0]}$ is C^P on $\mathbb{R}$, with

$$F^{(p)} = F^{[p]}, \quad 0 \le p \le P. \tag{42}$$

Proof. Since $\mathcal{X}^0$ satisfies (G_4), so does any of the grids $\mathcal{X}^1, \ldots, \mathcal{X}^P$ (see Remark 2.6, 2). For any $j \ge 0$, let $f_j = f_j^{[0]}$ be the sequences of vertices of the polygonal line L_j. For $1 \le p \le P$, and $j \ge 0$, the vertices $f_j^{[p]}$ of the polygonal line $L_j^{[p]}$ are defined by:

$$f_{j,k}^{[p]} := [x_{j,k}^p, x_{j,k-1}^p] F_j^{[p-1]} := \frac{f_{j,k}^{[p-1]} - f_{j,k-1}^{[p-1]}}{x_{j,k}^p - x_{j,k-1}^p}, \tag{43}$$

where, for $0 \le p \le P$, and $j \ge 0$, $F_j^{[p]}$ denotes the following function:

$$F_j^{[p]}(x_{j,k}^p) = f_{j,k}^{[p]}, \quad F_j^{[p]} \text{ is affine on } [x_{j,k}^p, x_{j,k+1}^p], \quad k \in \mathbb{Z}. \tag{44}$$

The assumption ensures that, for $0 \le p \le P$, the sequence $F_j^{[p]}$, $j \ge 0$, converges to $F^{[p]}$ uniformly on compact sets. Let us fix $p \le P - 1$. The two grids $\mathcal{X}^p$ and $\mathcal{X}^{p+1}$ being equivalent the sequence $L_j^{[p]}$, $j \ge 0$, is convergent relative to $\mathcal{X}^{p+1}$ too, with limit function $F^{[p]}$, which means that the sequence $\widehat{F}_j^{[p]}$, $j \ge 0$, defined by:

$$\widehat{F}_j^{[p]}(x_{j,k}^{p+1}) = f_{j,k}^{[p]}, \quad \widehat{F}_j^{[p]} \text{ is affine on } [x_{j,k}^{p+1}, x_{j,k+1}^{p+1}], \quad k \in \mathbb{Z}. \tag{45}$$

converges to $F^{[p]}$ uniformly on any compact of $\mathbb{R}$. By application of Proposition 2.11 to the sequences of polygonal lines $L_j^{[p]}$, $L_j^{[p+1]}$, $j \ge 0$, and to the grid $\mathcal{X}^{[p+1]}$ (that is, to the two sequences $\widehat{F}_j^{[p]}$, $F_j^{[p+1]}$ defined in (45) and (44), respectively), it follows that $F^{[p]}$ is C^1 on $\mathbb{R}$, with $F^{[p]'} = F^{[p+1]}$. $\square$

Remark 2.14. 1. *We have worked with backward divided differences, but all previous results would be valid with forward divided differences as well. Our choice will become clear in Section 4.*

2. *Even though we were interested in differentiability of order P, we limited ourselves to considering everywhere divided differences of order 1. This was made possible by allowing changes of grids. Indeed, up to multiplication by a constant, divided differences of any order p can be viewed as divided differences of order one w.r. to a new grid. Starting from an initial binary grid $\mathcal{X}$, let us define the grids $\mathcal{X}^1, \ldots, \mathcal{X}^P$ as follows:*

$$x_{j,k}^p := \frac{x_{j,k-p+1} + \cdots + x_{j,k}}{p}, \quad j \ge 0, \ k \in \mathbb{Z}, \ 1 \le p \le P. \tag{46}$$

Then, up to multiplication by $p!$, the quantity $f_{j,k}^{[p]}$ defined in (43), is nothing but the divided difference of order p of the function F_j, based on the $p + 1$ points $x_{j,k}, x_{j,k-1}, \ldots, x_{j,k-p}$, i.e.,

$$f_{j,k}^{[p]} = p! \, [x_{j,k}, \ldots, x_{j,k-p}] F_j.$$

*Had we worked with $\widehat{f}_{j,k}^{[p]} := [x_{j,k}, \dots, x_{j,k-p}]F_j$ instead of $f_{j,k}^{[p]}$ to define the polyg-
onal lines $L_j^{[p]}$, formula (42) would then become $F^{(p)} = p!\, F^{[p]}$. Actually, our definition of $f_{j,k}^{[p]}$ by means of the grids introduced in (46) might be the proper way to
define divided differences of any order. This way, for instance, the pth order divided difference of a sufficiently differentiable function F would be equal to $F^{(p)}(\xi)$
for some point ξ located in any interval containing the $p+1$ real numbers on which
the divided difference is based (see [5] for similar observations).*

3. Non nested grids for subdivision schemes

A subdivision scheme is a systematic procedure to produce sequences of polygonal
lines from given initial ones. We present here the tools about linear binary subdivision schemes which will be useful to state a sufficient condition for convergence
of the sequences of polygonal lines such a scheme produces.

3.1. Subdivision schemes, difference and derived schemes

A linear subdivision scheme is defined as an infinite sequence $\mathcal{S} = \{S_j,\, j \geq 0\}$ of
matrices $S_j := (S_{j,k,\ell})_{k,\ell \in \mathbb{Z}}$. By means of such a sequence, it is possible to construct
sequences of polygonal lines L_j, $j \geq 0$, in $\mathbb{R}^d$, $d \geq 1$, in the following way. For any
$j \geq 0$, we denote by $f_j := (f_{j,k})_{k \in \mathbb{Z}}$ a bi-infinite sequence of vertices in $\mathbb{R}^d$, so that
f_j actually represents a matrix with d columns and a bi-infinite number of rows.
The vertices of level $j + 1$ are calculated from those of level j as follows:

$$f_{j+1} := S_j f_j, \quad j \geq 0. \tag{47}$$

For this to be valid without any restriction on f_j, it is assumed that each row
of S_j contains only a finite number of nonzero elements. For binary schemes, the
elements $f_{j+1,2k}$ and $f_{j+1,2k+1}$ are affiliated to $f_{j,k}$, generally by assuming that the
nonzero elements of the row of index p to have indices centered around $p/2$. The
number of nonzero elements is generally assumed to be bounded independently of
the row, which then amounts to require that the number of nonzero elements of the
column of any index $\ell \in \mathbb{Z}$ to be bounded independently of ℓ, the corresponding
indices being centered around 2ℓ. When the bound is also independent of the
level, the scheme is said to be local. For the sake of uniformity of the notations
throughout the paper, here local schemes will rather be defined as follows.

(SS$_1$)   *We say that the scheme $\mathcal{S}$ is local if there exists two integers M_1, M_2
such that for any $j \geq 0$, and any $k, \ell \in \mathbb{Z}$,*

$$S_{j,k,\ell} \neq 0 \;\Rightarrow\; 2\ell + M_1 \leq k \leq 2\ell + M_2. \tag{48}$$

This will provide a better localisation of the nonzero elements.

With any local subdivision scheme $\mathcal{S}$, it is classical to associate its *difference
scheme* $\mathcal{D} = \{D_j,\, j \geq 0\}$, defined by

$$D_{j,k,\ell} := \sum_{i \geq \ell}(S_{j,k,i} - S_{j,k-1,i}), \quad j \geq 0, \quad k, \ell \in \mathbb{Z}. \tag{49}$$

Each matrix D_j is well defined and it satisfies the so-called *commutation formula*:

$$\Delta S_j = D_j \Delta, \quad j \geq 0, \tag{50}$$

where the matrix $\Delta := (\Delta_{k,\ell})_{k,\ell \in \mathbb{Z}}$ is given by:

$$\Delta_{k,\ell} := \delta_{k,\ell} - \delta_{k-1,\ell}. \tag{51}$$

In order to make the equality

$$\Delta(S_j w) = D_j(\Delta w), \quad j \geq 0, \tag{52}$$

valid for any sequence $w = (w_k)_{k \in \mathbb{Z}}$ in $\mathbb{R}^d$, we have to make sure that each row of D_j contains only a finite number of nonzero elements. This is not guaranteed, although $D_{j,k,\ell} = 0$ for $k < 2\ell + M_1$. This justifies the introduction of the following property:

(SS$_2$)   *We say that S reproduces constants if $f_{0,k} = a$ for all $k \in \mathbb{Z}$ implies $f_{j,k} = a$ for all $k \in \mathbb{Z}$ and for $j \geq 0$.*

Equivalently, the scheme S reproduces constants iff, for all $j \geq 0$, $S_j \mathbb{1} = \mathbb{1}$, where $\mathbb{1}$ denotes the bi-infinite sequence all components of which are equal to 1, *i.e.*, iff all matrices S_j satisfy

$$\sum_{\ell \in \mathbb{Z}} S_{j,k,\ell} = 1, \quad j \geq 0, \quad k \in \mathbb{Z}. \tag{53}$$

From now on, we assume the local scheme S to reproduce constants. Then, for $k \geq 2\ell + M_2$, $\sum_{i \geq \ell} S_{j,k,i} = \sum_{i \geq \ell} S_{j,k-1,i} = 1$. Hence, the difference scheme is local, with, for any $j \geq 0$,

$$D_{j,k,\ell} \neq 0 \;\Rightarrow\; 2\ell + M_1 \leq k \leq 2\ell + M_2 - 1. \tag{54}$$

On account of (54), for any initial f_0, we can apply (52) with $w = f_j$. The difference scheme thus enables us to calculate the differences of order 1 (*i.e.*, $\Delta f_j = (f_{j,k} - f_{j,k-1})_{k \in \mathbb{Z}}$) as

$$\Delta f_{j+1} = D_j \Delta f_j, \quad j \geq 0. \tag{55}$$

Choose a binary grid $\mathcal{X}$. At each level j we can now consider the function F_j and its first order divided differences w.r. to $\mathcal{X}$ defined as in (4) and (26), that is, with the notation introduced in (11),

$$f_{j,k}^{[1]} = (\Delta f_j)_k / d_{j,k}, \quad k \in \mathbb{Z}.$$

They can thus be calculated by:

$$f_{j+1}^{[1]} = S_j^{[1]} f_j^{[1]}, \quad j \geq 0, \tag{56}$$

where the subdivision scheme $S^{[1]}$ is defined by:

$$S_{j,k,\ell}^{[1]} := \frac{d_{j,\ell}}{d_{j+1,k}} D_{j,k,\ell}, \quad j \geq 0, \; k, \ell \in \mathbb{Z}. \tag{57}$$

The scheme $\mathcal{S}^{[1]}$ is called the *derived scheme of $\mathcal{S}$ w.r. to the grid $\mathcal{X}$*. Like the difference scheme, it is local, with

$$S^{[1]}_{j,k,\ell} \neq 0 \; \Rightarrow \; 2\ell + M_1 \leq k \leq 2\ell + M_2 - 1. \tag{58}$$

It is often usual to introduce the following additional requirement on subdivision schemes:

(SS$_3$)   *We say that the scheme $\mathcal{S}$ is bounded if*

$$\|\mathcal{S}\|_\infty := \sup_{j \geq 0} \|S_j\|_\infty < +\infty, \tag{59}$$

where

$$\|S_j\|_\infty := \sup_{k \in \mathbb{Z}} \sum_{\ell \in \mathbb{Z}} |S_{j,k,\ell}|, \quad j \geq 0.$$

The latter requirement (SS$_3$) guarantees in particular that, if we start with a bounded f_0 (in the sense that $\|f_0\|_\infty := \sup_{k \in \mathbb{Z}} \|f_{0,k}\| < +\infty$, where $\|.\|$ denotes a given norm in $\mathbb{R}^d$), then each f_j will also be bounded for each $j \geq 0$.

If, in addition to be local and to reproduce constants, the scheme $\mathcal{S}$ is also assumed to be bounded, then its difference scheme is bounded too, but this may not be true for the derived scheme $\mathcal{S}$.

Proposition 3.1. *With respect to any homogenous binary grid, the derived scheme of a local and bounded subdivision scheme which reproduces constants is not only local but also bounded.*

On account of (57) and (58), Proposition 3.1 is an immediate consequence of the following lemma.

Lemma 3.2. *If the binary grid $\mathcal{X}$ is homogenous, then, for any K_1, $K_2 \in \mathbb{Z}$, $K_1 \leq K_2$,*

$$\theta := \sup_{\substack{j \geq 0, \ k,\ell \in \mathbb{Z} \\ 2\ell + K_1 \leq k \leq 2\ell + K_2}} \frac{d_{j,\ell}}{d_{j+1,k}} < +\infty. \tag{60}$$

Proof. The inequalities $x_{j+1,2\ell-2+N_1} \leq x_{j,\ell-1} < x_{j,\ell} \leq x_{j+1,2\ell+N_2}$ lead to

$$d_{j,\ell} \leq \sum_{i=N_1-1}^{N_2} d_{j+1,2\ell+i}, \quad j \geq 0 \,, \ell \in \mathbb{Z}.$$

On the other hand, if the grid satisfies (10), then, for any integers $L_1 \leq L_2$ and any $j \geq 0$

$$k + L_1 \leq r \leq k + L_2 \; \Rightarrow \; d_{j,r} \leq \eta^{\mathrm{Max}(|L_1|,|L_2|)} \, d_{j,k}.$$

The expected result follows easily. $\square$

3.2. Convergence and differentiability

Definition 3.3. *We say that the subdivision scheme converges relative to a grid $\mathcal{X}$ if, for any initial sequence of vertices f_0, the sequence of polygonal lines L_j with vertices f_j obtained by* (47) *converges relative to $\mathcal{X}$.*

From now on, if there is no ambiguity about the binary grid $\mathcal{X}$ relative to which we are considering the convergence of a given local subdivision scheme $\mathcal{S}$, we shall denote by $\mathcal{S}_j f_0$ the parameterisation (relative to $\mathcal{X}$) of the polygonal line L_j with vertices f_j provided at level $j \geq 0$ by the scheme $\mathcal{S}$ (that is, $f_j = \mathcal{S}_{j-1} \dots \mathcal{S}_0 f_0$). Moreover, in case the scheme converges (relative to $\mathcal{X}$), $\mathcal{S} f_0$ will denote the limit of the sequence $\mathcal{S}_j f_0$, $j \geq 0$.

Suppose now that the scheme $\mathcal{S}$ (also named $\mathcal{S}^{[0]}$) is local and reproduces constants. Let $\mathcal{S}^{[1]}$ be its derived scheme w.r. to a binary grid $\mathcal{X}^1$. It is local. Suppose that $\mathcal{S}^{[1]}$ too reproduces constants. Then its derived scheme w.r. to a binary grid $\mathcal{X}^2$ is local. Continuing the same way, if the scheme $\mathcal{S}^{[p-1]}$ is local and reproduces constants, then its derived scheme w.r. to a binary grid $\mathcal{X}^p$ is local. If we are able to define in this way

$$\mathcal{S}^{[p]} := \text{ the derived scheme of } \mathcal{S}^{[p-1]} \text{ w.r. to a binary } \mathcal{X}^p, \quad 1 \leq p \leq P, \quad (61)$$

then we shall say that the subdivision scheme $\mathcal{S}$ is *of order greater than or equal to P relative to the sequence* $\mathcal{X}^1, \dots, \mathcal{X}^P$ *of binary grids*. Note that this is independent of the last grid $\mathcal{X}^P$. In particular being of order greater than or equal to 1 does not depend on any grid since it just means reproducing constants.

With the notations introduced above, from a given sequence $f_{j,k}^{[0]} := f_j = \mathcal{S}_{j-1} \dots \mathcal{S}_0 f_0$, $j \geq 0$, provided by the scheme $\mathcal{S}$, for $1 \leq p \leq P$, we can then define recursively the sequences $f_j^{[p]}$, $j > 0$, by

$$f_{j,k}^{[p]} := [x_{j,k}^p, x_{j,k-1}^p] \mathcal{S}_j^{[p-1]} f_0^{[p-1]} = \frac{f_{j,k}^{[p-1]} - f_{j,k-1}^{[p-1]}}{x_{j,k}^p - x_{j,k-1}^p}, \quad j \geq 0, \ k \in \mathbb{Z}. \quad (62)$$

As a direct application of Corollary 2.13, we can state the following result.

Theorem 3.4. *Let $\mathcal{S}$ be a local subdivision scheme which converges relative to a binary grid $\mathcal{X}$ supposed to satisfy* (G_4), *and let $\mathcal{X}, \mathcal{X}^1, \dots, \mathcal{X}^P$ be any sequence of binary grids equivalent to $\mathcal{X}$. Assume that $\mathcal{S}$ is of order greater than or equal to $P \geq 1$ relative to the sequence $\mathcal{X}^1, \dots, \mathcal{X}^{P-1}$ of binary grids and denote by $\mathcal{S}^{[1]}, \dots, \mathcal{S}^{[P]}$ the corresponding local schemes defined according to* (61). *If, for $1 \leq p \leq P$, the scheme $\mathcal{S}^{[p]}$ converges relative to $\mathcal{X}^p$, then, for all initial f_0, the function $\mathcal{S} f_0$ is C^P on $\mathbb{R}$, with*

$$\left(\mathcal{S} f_0 \right)^{(p)} = \mathcal{S}^{[p]} f_0^{[p]}, \quad 1 \leq p \leq P. \quad (63)$$

Remark 3.5. *To be in a position to apply Theorem 3.4 to a given subdivision scheme $\mathcal{S}$, we first need to ensure existence of the derived schemes $\mathcal{S}^{[p]}$, $1 \leq p \leq P$. Supposing that $\mathcal{S}^{[p-1]}$ exists and reproduces constants for some p, $1 \leq p \leq P-2$, the problem consists in finding a convenient grid $\mathcal{X}^p$ so that the scheme $\mathcal{S}^{[p]}$ defined* *by (61) reproduces constants too, i.e., so that it satisfies $S_j^{[p]}\mathbb{1} = \mathbb{1}$ for all $j \geq 0$. It is sufficient to define at each level*

$$\mathcal{X}_{j+1}^{[p]} := S_j^{[p-1]}\mathcal{X}_j^{[p]}, \tag{64}$$

provided that (64) yields a grid, that is, provided that the sequence $\mathcal{X} := (\mathcal{X}_j)_{j \geq 0}$ obtained through (64) satisfies (G$_1$) and (G$_2$). We shall see an illustration of (64) in the fourth section.

3.3. A sufficient condition for convergence

In [10] we gave a sufficient condition on to ensure the convergence of a subdivision scheme $\mathcal{S}$ relative to any nested binary grid. Here, we shall adapt the proof to the more general situation of binary grids which are not necessarily nested.

Theorem 3.6. *Assume that the local and bounded subdivision scheme $\mathcal{S}$ reproduces constants and that its difference scheme $\mathcal{D}$ satisfies the following property:*

There exist two integers J, $K \geq 0$, there exists a number $\mu \in]0,1[$ such that

$$\|D_{j+K} \ldots D_{j+1} D_j\|_\infty \leq \mu \ \text{for all} \ j \geq J. \tag{*}$$

Then the scheme $\mathcal{S}$ converges relative to any binary grid $\mathcal{X}$. More precisely, for any bounded f_0 the sequence $\mathcal{S}_j f_0$, $j \geq 0$, converges to $\mathcal{S} f_0$ uniformly on the whole of $\mathbb{R}$, and there exists a positive constant C such that

$$\|\mathcal{S} f_0 - \mathcal{S}_j f_0\|_\infty \leq C \, \widehat{\mu}^j \, \|\Delta f_J\|_\infty, \quad j \geq J, \tag{65}$$

where $\widehat{\mu} := \mu^{1/(K+1)}$.

As in [10], the following lemma is crucial. Since neither the result nor the proof involve any grid, we refer to the proof given in [10].

Lemma 3.7. *If the difference scheme of the subdivision scheme $\mathcal{S}$ satisfies (*), then, there exists a positive constant H such that, for any initial vector f_0,*

$$\|\Delta f_j\|_\infty \leq H \, \widehat{\mu}^j \, \|\Delta f_J\|_\infty, \quad j \geq J, \tag{66}$$

with $H := \widehat{\mu}^{-(J+K)} \underset{0 \leq s \leq K-1}{\text{Max}} \|D_{J+s-1} D_{J+s-2} \ldots D_J\|_\infty$ and $\widehat{\mu} := \mu^{1/(K+1)}$.

Proof of Theorem 3.6. Assume the local subdivision scheme $\mathcal{S}$ to reproduce constants and its difference scheme to satisfy (*). Given a fixed binary grid $\mathcal{X}$ satisfying (5), we shall prove the convergence of $\mathcal{S}$ relative to $\mathcal{X}$. Let $f_0 = (f_{0,k})_{k \in \mathbb{Z}}$ denote a fixed element of $\mathbb{R}^{\mathbb{Z}}$. For any nonnegative integer j, $f_j = (f_{j,k})_{k \in \mathbb{Z}}$ is obtained through (47) and F_j denotes the corresponding piecewise affine interpolating function as introduced in (4). This function can be written as follows:

$$F_j(x) = \sum_{\ell \in \mathbb{Z}} f_{j,\ell} \, \Lambda_{j,\ell}(x), \quad x \in \mathbb{R}, \tag{67}$$

where, for any $\ell \in \mathbb{Z}$, $\Lambda_{j,\ell} : \mathbb{R} \to \mathbb{R}$ denotes the function which is affine on each $[x_{j,k}, x_{j,k+1}]$ and which satisfies the interpolation conditions $\Lambda_{j,\ell}(x_{j,k}) = \delta_{k,\ell}$, $k, \ell \in \mathbb{Z}$. For any $j \geq 0$, let us calculate the difference $F_{j+1} - F_j$. For any $x \in \mathbb{R}$,

$$F_{j+1}(x) - F_j(x) = \sum_{k \in \mathbb{Z}} f_{j+1,k} \, \Lambda_{j+1,k}(x) - \sum_{\ell \in \mathbb{Z}} f_{j,\ell} \, \Lambda_{j,\ell}(x). \qquad (68)$$

Using (47) and the fact that all sums are actually finite ones, one can write the latter equality as follows:

$$F_{j+1}(x) - F_j(x) = \sum_{\ell \in \mathbb{Z}} \left[\sum_{k \in \mathbb{Z}} S_{j,k,\ell} \, \Lambda_{j+1,k}(x) - \Lambda_{j,\ell}(x) \right] f_{j,\ell}. \qquad (69)$$

For a given $j \geq 0$, we denote by $Y_j = (y_{j,r})_{r \in \mathbb{Z}}$ the strictly increasing bi-infinite sequence composed of all distinct elements of the set $\{x_{j,p}, \ p \in \mathbb{Z}\} \cup \{x_{j+1,p}, \ p \in \mathbb{Z}\}$. Associated with Y_j, we introduce the functions $M_{j,r}$, $r \in \mathbb{Z}$, affine on each interval $[y_{j,r}, y_{j,r+1}]$, and such that: $M_{j,s}(y_{j,r}) = \delta_{r,s}$, for all $r, s \in \mathbb{Z}$. For any $k, \ell \in \mathbb{Z}$, we have:

$$\Lambda_{j+1,k}(x) = \sum_{r \in \mathbb{Z}} A_{j,r,k} \, M_{j,r}(x), \quad \Lambda_{j,\ell}(x) = \sum_{r \in \mathbb{Z}} B_{j,r,\ell} \, M_{j,r}(x), \quad x \in \mathbb{R}, \qquad (70)$$

with

$$A_{j,r,k} := \Lambda_{j+1,k}(y_{j,r}), \quad B_{j,r,\ell} := \Lambda_{j,\ell}(y_{j,r}).$$

For a fixed $r \in \mathbb{Z}$, let us examine the row of index r of the corresponding matrices $A_j := (A_{j,r,k})_{r,k \in \mathbb{Z}}$ and $B_j := (B_{j,r,\ell})_{r,\ell \in \mathbb{Z}}$. From the definition of the functions $\Lambda_{j,k}$ we know that:

$$A_{j,r,k} \neq 0 \quad \Leftrightarrow \quad x_{j+1,k-1} < y_{j,r} < x_{j+1,k+1},$$
$$B_{j,r,\ell} \neq 0 \quad \Leftrightarrow \quad x_{j,\ell-1} < y_{j,r} < x_{j,\ell+1}.$$

Suppose first that $y_{j,r} = x_{j,s}$. Then, by comparison of the latter relations with (5), we obtain:

$$A_{j,r,k} \neq 0 \quad \Rightarrow \quad 2s + N_1 \leq k \leq 2s + N_2, \quad B_{j,r,\ell} = \delta_{\ell,s}$$

Similarly, if now $y_{j,r} = x_{j+1,s}$, then

$$A_{j,r,k} = \delta_{k,s}, \quad B_{j,r,\ell} \neq 0 \quad \Rightarrow \quad \left[\frac{s - N_2}{2} \right] \leq \ell \leq \left[\frac{s - N_1 + 1}{2} \right].$$

In particular the number of nonzero elements of each row of either A_j or B_j is finite and bounded above independently of the row and independently of j. On the other hand, for any $x \in \mathbb{R}$, we have $1 = \sum_{k \in \mathbb{Z}} \Lambda_{j+1,k}(x) = \sum_{\ell \in \mathbb{Z}} \Lambda_{j,\ell}(x)$. Hence,

$$\sum_{k \in \mathbb{Z}} A_{j,r,k} = 1, \quad \sum_{\ell \in \mathbb{Z}} B_{j,r,\ell} = 1, \quad r \in \mathbb{Z}.$$

Since $A_{j,r,k}$ and $B_{j,r,k}$ are nonnegative, this implies that $\|A_j\|_\infty = \|B_j\|_\infty = 1$ for all $j \geq 0$. Combining (70) and (69) leads to:

$$F_{j+1}(x) - F_j(x) = \sum_{r \in \mathbb{Z}} \sum_{\ell \in \mathbb{Z}} [(A_j S_j)_{r,\ell} - B_{j,r,\ell}] f_{j,\ell} \, M_{j,r}(x), \quad x \in \mathbb{R}.$$

In other words, the quantity $(F_{j+1} - F_j)(y_{j,r})$ is the component of index r of the vector $(A_j S_j - B_j) f_j$. Hence, on account of the piecewise affinity of $F_{j+1} - F_j$, we have

$$\|F_{j+1} - F_j\|_\infty = \mathrm{Sup}_{r\in\mathbb{Z}} |(F_{j+1} - F_j)(y_{j,r})| = \|(A_j S_j - B_j) f_j\|_\infty. \qquad (71)$$

The matrix $U_j := (U_{j,k,\ell})_{k,\ell\in\mathbb{Z}}$, where:

$$U_{j,k,\ell} := \sum_{r\leq\ell} B_{j,k,r} - (A_j S_j)_{k,r}, \qquad (72)$$

is well defined, and it satisfies:

$$A_j S_j - B_j = U_j \Delta, \quad j \geq 0. \qquad (73)$$

The number of nonzero elements of each row of either matrix $A_j S_j$ or B_j is bounded above independently of the row and of j. Moreover all elements of any row of either matrix $A_j S_j$ or B_j add to unity. From (72) it follows that each row of U_j has only a finite number of nonzero elements independent of the row, and this number itself is bounded above independently of j. The previous considerations also make the equality $(U_j \Delta)w = U_j(\Delta w)$ valid without restriction on $w = (w_k)_{k\in\mathbb{Z}}$.

From now on, we assume the subdivision scheme S to be bounded. Then, from (73) one can also conclude that $\|U_j\|_\infty$ is bounded independently of j, say $\|U_j\|_\infty \leq M < +\infty$ for all $j \geq 0$. Accordingly, relations (71) and (73) enable us to write:

$$\|F_{j+1} - F_j\|_\infty = \|U_j \Delta f_j\|_\infty \leq M \|\Delta f_j\|_\infty, \quad j \geq 0.$$

Using Lemma 3.6, we thus have:

$$\|F_{j+1} - F_j\|_\infty \leq \Gamma \, \widehat{\mu}^j \, \|\Delta f_J\|_\infty, \quad j \geq J,$$

with $\widehat{\mu} := \mu^{1/(K+1)}$ and where $\Gamma := HM$ is independent of j and f_0. It follows that

$$\|F_{j+q} - F_j\|_\infty \leq C \, \widehat{\mu}^j \, \|\Delta f_J\|_\infty, \quad j \geq J, \quad q \geq 0,$$

with $C := \Gamma/(1 - \widehat{\mu})$. Provided that f_0 is assumed to be bounded, the uniform convergence on $\mathbb{R}$ of the sequence F_j, $j \geq 0$, is proved, along with (65). □

Theorem 3.8. *Let S $(=S^{[0]})$ be a local and bounded subdivision scheme which is of order greater than or equal to $P + 1 \geq 1$ relative to a sequence $\mathcal{X}^1, \ldots, \mathcal{X}^P$ of equivalent homogenous binary grids satisfying (G_4). Denote by $S^{[1]}, \ldots, S^{[P]}$ the corresponding local schemes defined according to (61). Suppose that, for $0 \leq p \leq P$, the difference scheme $\mathcal{D}^{[p+1]}$ of the scheme $S^{[p]}$ satisfies $(*)$. Then, relative to any binary grid equivalent to $\mathcal{X}^1, \ldots, \mathcal{X}^P$, the scheme S produces C^P functions, with, for all initial f_0,*

$$\left(S f_0\right)^{(p)} = S^{[p]} f_0^{[p]}, \quad 1 \leq p \leq P. \qquad (74)$$

Proof. Proposition 3.1 guarantees the boundedness of all schemes $S^{[p]}$, $0 \leq p \leq P$. Since their difference schemes all satisfy condition $(*)$, they all converge according to Theorem 3.6. The announced result thus follows from Theorem 3.4. □

4. Spline subdivision

This section is devoted to spline subdivision schemes. In order to obtain explicit expressions for such schemes in an elegant way, it is natural to use blossoms. For the sake of simplicity, we shall focus on splines with simple knots only.

4.1. Blossoms and poles

Given a polynomial function P of degree less than or equal to n, its *blossom* [12] is the unique function p of n variables which is symmetric, n-affine (*i.e.*, affine in each variable), and which gives P by restriction to the diagonal of $\mathbb{R}^n$, that is,

$$p(x^{[n]}) = P(x), \quad x \in \mathbb{R}. \tag{75}$$

In (75) as in the rest of the paper, the notation $x^{[k]}$ means x repeated k times. For example, for $n = 3$, $p(x_1, x_2, x_3) = a_0 + a_1(x_1 + x_2 + x_3)/3 + a_2(x_1 x_2 + x_2 x_3 + x_3 x_1)/3 + a_3 x_1 x_2 x_3$ is the blossom of $P(x) = a_0 + a_1 x + a_2 x^2 + a_3 x^3$.

All results recalled in this section and the next two ones are direct consequences of the three properties of blossoms: symmetry, n-affinity, diagonal property. For details, we refer the reader to [12,11]. As an instance, by differentiation of (75), the symmetry and the n-affinity of p yield a straightforward calculation of the derivative of P as follows:

$$P'(x) = \frac{n}{h} \left[p(x^{[n-1]}, u+h) - p(x^{[n-1]}, u) \right], \tag{76}$$

where u, h denote any real numbers with $h \neq 0$. As a polynomial function of degree less than or equal to $(n-1)$, P' has a blossom $p^{\{1\}}$, which is the unique function of $(n-1)$ variables which is symmetric, $(n-1)$-affine, and which gives P' by restriction to the diagonal of $\mathbb{R}^{n-1}$. Hence, (76) readily leads to:

$$p^{\{1\}}(x_1, \ldots, x_{n-1}) = \frac{n}{h} \left[p(x_1, \ldots, x_{n-1}, u+h) - p(x_1, \ldots, x_{n-1}, u) \right]. \tag{77}$$

Again this equality is valid for any real numbers u, h, with $h \neq 0$. From (77) and from the properties of blossoms, it is classical to derive the following characterization of contact of order $r \leq n$ between two polynomial function P, Q of degree less than or equal to n, with values in $\mathbb{R}^d$, through their blossoms p and q:

$$\left. \begin{array}{c} P^{(i)}(a) = Q^{(i)}(a) \\ \text{for } 0 \leq i \leq r \end{array} \right\} \Longleftrightarrow \left\{ \begin{array}{c} p(a^{[n-r]}, x_1, \ldots, x_r) = q(a^{[n-r]}, x_1, \ldots, x_r) \\ \text{for all } x_1, \ldots, x_r \in \mathbb{R}. \end{array} \right. \tag{78}$$

Given a bi-infinite sequence of knots $\mathcal{K} := \{t_k, \ k \in \mathbb{Z}\}$, with $t_k < t_{k+1}$ for all k and $|t_k| \to +\infty$ when $|k| \to +\infty$, for each $p \geq 1$, we denote by $\mathcal{A}_p$ the set of all p-tuples which are admissible relative to $\mathcal{K}$, that is, all p-tuples $(\zeta_1, \ldots, \zeta_p)$ such that each t_ℓ satisfying $\text{Min}(\zeta_1, \ldots, \zeta_p) < t_\ell < \text{Max}(\zeta_1, \ldots, \zeta_p)$ appears at least once in the sequence denote $\zeta_1, \ldots, \zeta_p$. Moreover, given $(\zeta_1, \ldots, \zeta_p) \in \mathcal{A}_p$, we denote by $\mathcal{J}_p(\zeta_1, \ldots, \zeta_p)$ the set of all (consecutive) integers i such that the interval $[t_i, t_{i+1}]$ contains at least one of the points $\zeta_1, \ldots, \zeta_p$.

From now on, we suppose that $S : \mathbb{R} \to \mathbb{R}^d$ is a given polynomial spline relative to the knot vector $\mathcal{K}$, meaning that S is C^{n-1} on $\mathbb{R}$ and that there exist polynomial functions $G_q : \mathbb{R} \to \mathbb{R}^d$, $q \in \mathbb{Z}$, of degree less than or equal to n, such that

$$S(x) = G_q(x), \quad x \in [t_q, t_{q+1}]. \tag{79}$$

The blossom s of S is now the function of n variables defined on the set $\mathcal{A}_n$ by

$$s(\zeta_1, \ldots, \zeta_n) := g_i(\zeta_1, \ldots, \zeta_n), \quad i \in \mathcal{J}_n(\zeta_1, \ldots, \zeta_n), \tag{80}$$

where, for each $q \in \mathbb{Z}$, g_q is the blossom of the polynomial function G_q. The equality (80) is made meaningful by the fact that S is C^{n-1} and by (78). The *poles* of the spline S are the points

$$P_k := s(t_{k+1}, \ldots, t_{k+n}), \quad k \in \mathbb{Z}. \tag{81}$$

All values of the blossom s of S can be calculated as convex combinations of the poles. Indeed, given any admissible n-tuple $(\zeta_1, \ldots, \zeta_n)$, and any integer $i \in \mathcal{J}_n(\zeta_1, \ldots, \zeta_n)$, as the result of the symmetry and the n-affinity of blossoms, the point $s(\zeta_1, \ldots, \zeta_n) = g_i(\zeta_1, \ldots, \zeta_n)$ will be obtained as an affine combination of the $(n+1)$ poles which can be labelled using the blossom g_i, namely the poles $P_{i-n}, \ldots, P_i$. This affine combination is actually a convex one, since the admissibility of $(\zeta_1, \ldots, \zeta_n)$ guarantees that $t_{i-n+1} \leq \zeta_1, \ldots, \zeta_n \leq t_{i+n}$. The particular case $\zeta_1 = \cdots = \zeta_n = x \in [t_i, t_{i+1}]$ gives the de Boor algorithm and, as is well known, it readily leads to existence of a B-spline basis. We will not insist on this well-known fact, being only interested here by insertion of knots.

4.2. Insertion of knots

With this aim in view let us clarify the calculations in the following case. For any $k \in \mathbb{Z}$, we choose a new knot u_k in $]t_k, t_{k+1}[$. We can now consider S as a polynomial spline relative to the knot sequence $\widehat{\mathcal{K}}$ with elements $\ldots, t_k, u_k, t_{k+1}, \ldots$. We shall recall here how to calculate its new poles.

1) Suppose first that $n = 2N$. The poles of S relative to $\widehat{\mathcal{K}}$ are the points $s(\mathcal{T}_k)$, $s(\mathcal{U}_k)$, $k \in \mathbb{Z}$, where

$$\mathcal{T}_k := (t_{k+1}, u_{k+1} \ldots, t_{k+N}, u_{k+N}), \quad \mathcal{U}_k := (u_k, t_{k+1}, \ldots, u_{k+N-1}, t_{k+N}).$$

For a given $k \in \mathbb{Z}$, both n-tuples $\mathcal{T}_k$ and $\mathcal{U}_k$ are admissible (relative to $\mathcal{K}$) and the integer k belongs to $\mathcal{J}_n(\mathcal{T}_k) = \mathcal{J}_n(\mathcal{U}_k)$. Now, denoting by $\mathcal{T}$ the N-tuple $(t_{k+1}, \ldots, t_{k+N})$, the value of s at any admissible n-tuple $(\mathcal{T}, y_1, \ldots, y_N)$ can always be obtained as a convex combination of the $N+1$ poles involving $\mathcal{T}$, namely the poles $P_{k-N}, \ldots, P_k$. Let us show how to calculate $s(\mathcal{T}_k)$.

Starting from the $(N+1)$ points $Q_{i,0} := P_i = g_k(t_{i+1}, \ldots, t_{i+n})$, $k - N \leq i \leq k$, the properties of blossoms enable us to calculate recursively the points

$$Q_{i,r} := g_k(t_{i+1}, t_{i+2}, \ldots, t_{i+n-r}, u_{k+1}, \ldots, u_{k+r}), \quad k - N + r \leq i \leq k, \tag{82}$$

for $0 \leq r \leq N$. Indeed, for $0 \leq r < N$, due to the symmetry and n-affinity of g_k, we have, for $k - N + r + 1 \leq i \leq k$:

$$Q_{i,r+1} = \left(\frac{t_{i+n-r} - u_{k+r+1}}{t_{i+n-r} - t_i}\right) Q_{i,r-1} + \left(\frac{u_{k+r+1} - t_i}{t_{i+n-r} - t_i}\right) Q_{i+1,r-1}. \tag{83}$$

In (83), the two coefficients are positive. Indeed, for $k - N + r + 1 \leq i \leq k$ and $0 \leq r \leq N - 1$, we have $t_i < t_{k+r+1} < u_{k+r+1} < t_{k+r+2} \leq t_{i+n-r}$.

The last step $r = N$ gives $Q_{k,N} = g_k(\mathcal{T}, u_{k+1}, \ldots, u_{k+N}) = s(\mathcal{T}_k)$ as a strictly convex combination (independent of S) of the $(N + 1)$ poles $P_{k-N}, \ldots, P_k$. The calculation of $s(\mathcal{U}_k)$ as a strictly convex combination of the same $(N + 1)$ poles $P_{k-N}, \ldots, P_k$ can be done in a similar way.

2) <u>Suppose now that $n = 2N + 1$</u>. We now have to consider separately

$$\mathcal{U}_k := (u_k, t_{k+1}, \ldots, u_{k+N-1}, t_{k+N}, u_{k+N})$$

and

$$\mathcal{T}_k := (t_{k+1}, u_{k+1}, \ldots, t_{k+N}, u_{k+N}, t_{k+N+1}).$$

Following the same procedure as in the case $n = 2N$, the point $s(\mathcal{U}_k) = g_k(\mathcal{U}_k)$ will be obtained as a strictly convex combination of the $(N+2)$ poles which involve the N-tuple $(t_{k+1}, \ldots, t_{k+N})$, i.e., $P_{k-N-1}, \ldots, P_k$. As for the point $s(\mathcal{T}_k) = g_k(\mathcal{T}_k)$, it will be obtained as a strictly convex combination of the $(N + 1)$ poles involving the $(N + 1)$-tuple $(t_{k+1}, \ldots, t_{k+N+1})$, which are $P_{k-N}, \ldots, P_k$.

4.3. Poles of the derivatives

In this subsection we shall examine the links between the poles of a spline and those of its first derivative. Suppose that $n \geq 2$. We can then differentiate the spline S given by (79). It is a spline related to the same knot sequence $\mathcal{K}$, with sections of degree less than or equal to $n - 1$. From the obvious equalities

$$S'(x) = G_q'(x), \quad x \in [t_q, t_{q+1}], \quad q \in \mathbb{Z},$$

we can deduce that the blossom $s^{\{1\}}$ of S' can be defined on the set $\mathcal{A}_{n-1}$ of all admissible $(n - 1)$-tuples as follows:

$$s^{\{1\}}(\zeta_1, \ldots, \zeta_{n-1}) := g_i^{\{1\}}(\zeta_1, \ldots, \zeta_{n-1}), \quad \text{for all } i \in \mathcal{J}_{n-1}(\zeta_1, \ldots, \zeta_{n-1}), \quad (84)$$

that is, on account of (77):

$$s^{\{1\}}(\zeta_1, \ldots, \zeta_{n-1}) = \frac{n}{h} \left[g_i(\zeta_1, \ldots, \zeta_{n-1}, u + h) - g_i(\zeta_1, \ldots, \zeta_{n-1}, u) \right], \quad (85)$$

for any $u, h \in \mathbb{R}, h \neq 0$. We define the domain of the admissible p-tuple $(\zeta_1, \ldots, \zeta_p)$ as the set $\mathcal{D}(\zeta_1, \ldots, \zeta_p)$ of all $x \in \mathbb{R}$ such that $(\zeta_1, \ldots, \zeta_{n-1}, x) \in \mathcal{A}_{p+1}$. In other words $\mathcal{D}(\zeta_1, \ldots, \zeta_p)$ is the union of all intervals $[t_i, t_{i+1}], i \in \mathcal{J}(\zeta_1, \ldots, \zeta_p)$. For any $x \in \mathcal{D}(\zeta_1, \ldots, \zeta_{n-1})$, we clearly have $\mathcal{D}(\zeta_1, \ldots, \zeta_{n-1}) \subset \mathcal{D}(\zeta_1, \ldots, \zeta_{n-1}, x)$. Hence, for any $(\zeta_1, \ldots, \zeta_{n-1}) \in \mathcal{A}_{n-1}$ and any $x, y \in \mathcal{D}(\zeta_1, \ldots, \zeta_{n-1}), x \neq y$, (85) gives:

$$s^{\{1\}}(\zeta_1, \ldots, \zeta_{n-1}) = \frac{n}{y - x} \left[s(\zeta_1, \ldots, \zeta_{n-1}, y) - s(\zeta_1, \ldots, \zeta_{n-1}, x) \right]. \quad (86)$$

According to (81), the poles of the spline S' are the points $P_k^{\{1\}}$ defined as

$$P_k^{\{1\}} := s^{\{1\}}(t_{k+1}, \ldots, t_{k+n-1}), \quad k \in \mathbb{Z}.$$

Both t_k and t_{k+n} belong to $\mathcal{D}(t_{k+1}, \ldots, t_{k+n-1})$. Therefore, applying equality (86) with $(\zeta_1, \ldots, \zeta_{n-1}) = (t_{k+1}, \ldots, t_{k+n-1})$, $x = t_k$, and $y = t_{k+n}$, we eventually obtain:

$$P_k^{\{1\}} = \frac{n}{t_{k+n} - t_k} \left[P_k - P_{k-1} \right], \quad k \in \mathbb{Z}. \tag{87}$$

So, if the C^{n-1} spline S is given by its poles P_k, $k \in \mathbb{Z}$, we can obtain the poles of all its derivatives $S^{(p)}$, $p \leq n - 1$, recursively by means of (87).

4.4. Spline subdivision schemes and their difference schemes

We now consider a nested binary grid $\mathcal{X} = \left(\mathcal{X}_j \right)_{j \geq 0}$, with $\mathcal{X}_j = \left(x_{j,k} \right)_{k \in \mathbb{Z}}$, that is, a grid satisfying (G$_1$) and (G$_2$) along with the nestedness property:

$$x_{j+1,2k} = x_{j,k}, \quad j \geq 0, \; k \in \mathbb{Z}.$$

For any given $j \geq 0$, we denote by $\mathcal{E}_j^n$ the spline space associated with the knot vector $\mathcal{X}_j$. Due to the nestedness of $\mathcal{X}$, we have

$$\mathcal{E}_j^n \subset \mathcal{E}_{j+1}^n, \quad j \geq 0,$$

the spline space $\mathcal{E}_{j+1}^n$ being obtained from $\mathcal{E}_j^n$ by insertion of the knots $x_{j+1,2k+1}$, $k \in \mathbb{Z}$. Let S be an element of the initial space $\mathcal{E}_0^n$, defined by its poles $f_{0,k}$, $k \in \mathbb{Z}$. For any $j \geq 0$, we denote by $f_{j,k}$, $k \in \mathbb{Z}$, the poles of S considered as an element of $\mathcal{E}_j^n$. Therefore:

$$f_{j,k} := s(x_{j,k+1}, \ldots, x_{j,k+n}), \quad k \in \mathbb{Z}. \tag{88}$$

From our previous considerations, we can assert that the poles of level $j+1$ can be obtained as convex combinations of poles of level j. More precisely, for any $k \in \mathbb{Z}$ and any $j \geq 0$, the pole $f_{j+1,2k}$ is a strictly convex combination of the points $f_{j,k-[\frac{n+1}{2}]}, \ldots, f_{j,k}$, and the pole $f_{j+1,2k+1}$ is a strictly convex combination of the points $f_{j,k-[\frac{n}{2}]}, \ldots, f_{j,k}$, i.e.,

$$f_{j+1,2k} = \sum_{\ell=k-[\frac{n+1}{2}]}^{k} S_{j,2k,\ell}^n \, f_{j,\ell}, \quad f_{j+1,2k+1} = \sum_{\ell=k-[\frac{n}{2}]}^{k} S_{j,2k+1,\ell}^n \, f_{j,\ell}, \tag{89}$$

with positive coefficients with sum equal to 1 in both equalities. The coefficients involved in (89) do not depend on the chosen spline $S \in \mathcal{E}_0^n$. We can write (89) as follows:

$$f_{j+1} = S_j^n f_j, \quad j \geq 0, \tag{90}$$

with

$$\sum_{\ell \in \mathbb{Z}} S_{j,p,\ell}^n = 1 \text{ for all } p \in \mathbb{Z}, \quad S_{j,p,\ell}^n \geq 0 \text{ for all } p, \ell \in \mathbb{Z}, \tag{91}$$

and

$$S_{j,p,\ell}^n \neq 0 \quad \Leftrightarrow \quad 2\ell \leq p \leq 2\ell + n + 1. \tag{92}$$

We shall refer to the corresponding subdivision scheme $\mathcal{S}^n := \{S_j^n, \, j \geq 0\}$ as the *spline subdivision scheme of degree n associated with the nested grid $\mathcal{X}$*. Note that, when addressing the problem of convergence of the scheme $\mathcal{S}^n$, the underlying natural grid to use is not the initial grid $\mathcal{X}$. In fact, taking account of the meaning

(88) of the vertices $f_{j,k}$ we are working with, it seems more consistent to use the grid $\widetilde{\mathcal{X}}^n = \left(\widetilde{\mathcal{X}}^n_j\right)_{j \geq 0}$, where $\widetilde{\mathcal{X}}^n_j := \{\widetilde{x}^n_{j,k}, \ j \geq 0, k \in \mathbb{Z}\}$ is defined by

$$\widetilde{x}^n_{j,k} := \frac{x_{j,k+1} + \cdots + x_{j,k+n}}{n}, \quad j \geq 0, \ k \in \mathbb{Z}, \tag{93}$$

that is, to parameterise the control polygon of level j by the function $\widetilde{F}_j$ defined as follows:

$$\widetilde{F}_j(\widetilde{x}^n_{j,k}) := s(x_{j,k+1}, \ldots, x_{j,k+n}), \quad \widetilde{F}_j \text{ is affine on } [\widetilde{x}^n_{j,k}, \widetilde{x}^n_{j,k+1}]. \tag{94}$$

The validity of this choice will be confirmed by the link between spline subdivision schemes of consecutive degrees. Indeed, if $n \geq 2$, the derivative S' is an element of $\mathcal{E}^{n-1}_0$. Due to the inclusions $\mathcal{E}^{n-1}_j \subset \mathcal{E}^{n-1}_{j+1}$, $j \geq 0$, it can also be considered as an element of any $\mathcal{E}^{n-1}_j$. As so, we denote by $f^{\{1\}}_{j,k}$, $k \in \mathbb{Z}$, its poles, that is:

$$f^{\{1\}}_{j,k} = s^{\{1\}}(x_{j,k+1}, \ldots, x_{j,k+n-1}), \quad k \in \mathbb{Z}. \tag{95}$$

Similarly to (90), we know that the poles of level $j + 1$ of S' can be derived from those of level j by means of the spline subdivision scheme $\mathcal{S}^{n-1}$ of degree $n-1$ associated with the grid $\mathcal{X}$, namely:

$$f^{\{1\}}_{j+1,k} = S^{n-1}_j f^{\{1\}}_{j,k}, \quad j \geq 0, \tag{96}$$

where

$$S^{n-1}_{j,p,\ell} \neq 0 \quad \Leftrightarrow \quad 2\ell \leq p \leq 2\ell + n. \tag{97}$$

Now, from (87), we can deduce the following expression of the poles of S':

$$f^{\{1\}}_{j,k} = \frac{n}{x_{j,k+n} - x_{j,k}} \left[f_{j,k} - f_{j,k-1} \right] = \frac{f_{j,k} - f_{j,k-1}}{\widetilde{x}^n_{j,k} - \widetilde{x}^n_{j,k-1}}, \quad k \in \mathbb{Z}, \ j \geq 0. \tag{98}$$

Therefore, $f^{\{1\}}_{j,k}$ is nothing but the divided difference of order one of the function $\widetilde{F}_j$ defined in (94) w.r. to the binary grid $\widetilde{\mathcal{X}}^n$ introduced in (93):

$$f^{\{1\}}_{j,k} = \widetilde{f}^{[1]}_{j,k} := [\widetilde{x}^n_{j,k}, \widetilde{x}^n_{j,k-1}] \widetilde{F}_j, \quad k \in \mathbb{Z}, \ j \geq 0. \tag{99}$$

We can thus state the following result.

Theorem 4.1. *Let $\mathcal{S}^n$ denote the spline subdivision scheme of degree $n \geq 1$ related to the nested binary grid $\mathcal{X}$. For any $n \geq 2$, the scheme $\mathcal{S}^{n-1}$ is the derived scheme of the scheme $\mathcal{S}^n$ w.r. to the (non nested) binary grid $\widetilde{\mathcal{X}}^n$ defined in (93).*

As observed in the proof of Proposition 2.8, the grid $\widetilde{\mathcal{X}}$ is a non nested binary grid. For any degree $n \geq 2$ its satisfies the following property (G_3) (see (24)):

$$\widetilde{x}_{j+1,2k+1} < \widetilde{x}_{j,k} < \widetilde{x}_{j+1,2k+n}, \quad k \in \mathbb{Z}. \tag{100}$$

Theorem 4.1 was our actual motivation for introducing non nested grids.

Remark 4.2.

1. *It is our labelling of the poles (88) which led us to use backward divided differences. In order to work with forward differences, we should label them as* $f_{j,k} := s(x_{j,k-n+1}, \ldots, x_{j,k})$.

2. *Denoting by* Ξ *the identity of* $\mathbb{R}$*, i.e.,* $\Xi(t) := t$ *for all* $t \in \mathbb{R}$*, and by* ξ_n *the blossom of* Ξ *viewed as an element of* $\mathcal{P}_n$*, i.e.,* $\xi_n(t_1, \ldots, t_n) = (t_1 + \cdots + t_n)/n$*, we have*

$$\widetilde{x}_{j,k}^n = \xi_n(x_{j,k+1}, \ldots, x_{j,k+n}), \quad k \in \mathbb{Z}, \ j \geq 0.$$

Hence, each $\widetilde{\mathcal{X}}_j^n$ *is the sequence of poles of the function* Ξ *considered as an element of* $\mathcal{E}_j^n$*. It can thus be calculated according to the subdivision scheme* $\mathcal{S}^n$:

$$\widetilde{\mathcal{X}}_{j+1}^n = \mathcal{S}_j^n \widetilde{\mathcal{X}}_j^n, \quad j \geq 0.$$

This is a particular case of relation (64).

3. *What about the derived scheme of the scheme* $\mathcal{S}^1$ *w.r. to the grid* $\mathcal{X}^1$*? According to (88), for* $n = 1$*, at level* j*, the poles of a spline* S *are defined as*

$$f_{j,k} := s(x_{j,k+1}) = S(x_{j,k+1}) = S(\widetilde{x}_{j,k}^1), \quad k \in \mathbb{Z}.$$

The spline subdivision scheme $\mathcal{S}^1$ *can thus be described by:*

$$f_{j+1,2k} = \frac{d_{j+1,2k+2}}{d_{j,k+1}} f_{j,k-1} + \frac{d_{j+1,2k+1}}{d_{j,k+1}} f_{j,k}$$

and $f_{j+1,2k+1} = f_{j,k}$*. Hence the difference scheme* $\mathcal{D}^1$ *of* $\mathcal{S}^1$ *is given by*

$$D_{j,2k,\ell}^1 = \begin{cases} \frac{d_{j+1,2k+1}}{d_{j,k+1}} & \text{if } \ell = k \\ 0 & \text{otherwise} \end{cases}, \quad D_{j,2k+1,\ell}^1 = \begin{cases} \frac{d_{j+1,2k+2}}{d_{j,k+1}} & \text{if } \ell = k \\ 0 & \text{otherwise} \end{cases}.$$

The latter formulae show that the derived scheme $\mathcal{S}^{1[1]}$ *w.r. to* $\mathcal{X}^1$ *satisfies*

$$S_{j,2k,\ell}^{1[1]} = S_{j,2k+1,\ell}^{1[1]} = \delta_{k,\ell}, \quad j \geq 0, \ k, \ell \in \mathbb{Z}.$$

Clearly, $\mathcal{S}^{1[1]}$ *reproduces constants. It can be considered as the spline subdivision scheme* $\mathcal{S}^0$ *of degree 0 which, at level* j*, associates with any piecewise constant function* $S \in \mathcal{E}_0^0$ *its values* $f_{j,k}$ *on all intervals* $[x_{j,k}, x_{j,k+1}[$.

4.5. Convergence of spline subdivision schemes

The aim of this subsection is to show that spline subdivision schemes provide a good illustration of the convergence results obtained in Section 3. Rates of convergence can be found in [2,3] for instance. As in the previous subsection, we denote by $\mathcal{S}^n$ the spline subdivision scheme of degree n associated with a given nested binary grid $\mathcal{X}$. We shall see that a weak assumption on the grid $\mathcal{X}$ ensures that the scheme $\mathcal{S}^n$ satisfies condition (∗).

Proposition 4.3. *Suppose that the nested binary grid $\mathcal{X}$ is homogenous, then, for all $n \geq 1$, the difference scheme $\mathcal{D}^n$ of the spline subdivision scheme $\mathcal{S}^n$ of degree n satisfies:*

$$\|\mathcal{D}^n\|_\infty < 1. \tag{101}$$

Proof. In accordance with our previous notations, we shall set:

$$\widetilde{d}_{j,k}^n := \widetilde{x}_{j,k}^n - \widetilde{x}_{j,k-1}^n = \frac{1}{n}\left(x_{j,k+n} - x_{j,k}\right). \tag{102}$$

We know that the subdivision scheme $\mathcal{S}^{n-1}$ is the derivative scheme of $\mathcal{S}^n$ w.r. to the grid $\widetilde{\mathcal{X}}^n$. Hence, the difference scheme $\mathcal{D}^n$ of $\mathcal{S}^n$ satisfies

$$D_{j,k,\ell}^n = \frac{\widetilde{d}_{j+1,k}^n}{\widetilde{d}_{j,\ell}^n}\, S_{j,k,\ell}^{n-1}, \quad j \geq 0,\ k,\ell \in \mathbb{Z}. \tag{103}$$

In (103) the nonzero elements are positive and they correspond to $2\ell \leq k \leq 2\ell+n$. Moreover, for any $k \in \mathbb{Z}$, $\sum_{\ell \in \mathbb{Z}} S_{j,k,\ell}^{n-1} = 1$. Therefore, due to (102) and (13), relation (103) proves that, for all $j \geq 0$, $\|D_j^n\|_\infty \leq A_n \|S_j^n\|_\infty = A_n < 1$. $\qquad\square$

Relation (101) is a particular case of condition (∗) with $J = K = 0$. Hence, as an immediate consequence of Theorem 3.6, we can state:

Corollary 4.4. *Let $\mathcal{S}^n$ be the spline subdivision scheme of degree $n \geq 2$ relative to a homogenous nested binary grid $\mathcal{X}$. Then, $\mathcal{S}^n$ converges relative to any binary grid $\widehat{\mathcal{X}}$, that is, for any initial f_0, the sequence $\mathcal{S}_j^n f_0$, $j \geq 0$, defined by*

$$\mathcal{S}_j^n f_0(\widehat{x}_{j,k}) = f_{j,k}, \quad \mathcal{S}_j^n f_0 \text{ affine on } [\widehat{x}_{j,k}, \widehat{x}_{j,k+1}], \quad k \in \mathbb{Z},$$

converges uniformly on compact sets of $\mathbb{R}$ to a function $\mathcal{S}^n f_0$. More precisely, for any initial bounded f_0, the convergence is uniform on $\mathbb{R}$, with

$$\|\mathcal{S}_j^n f_0 - \mathcal{S}^n f_0\|_\infty \leq C_n\, A_n{}^j.$$

Let us now illustrate Theorem 3.8 by the example of spline subdivision schemes associated with a homogenous nested binary grid $\mathcal{X}$. Below we recapitulate what we know:

1) On account of Theorem 4.1 and Remark 4.2, the scheme $\mathcal{S}^n$ is of exact order $n+1$ relative to the sequence of binary grids $\widetilde{\mathcal{X}}^n, \widetilde{\mathcal{X}}^{n-1}, \ldots, \widetilde{\mathcal{X}}^1$ introduced in (93), and relative to the latter sequence of grids, we have

$$\mathcal{S}^{n[i]} = \mathcal{S}^{n-i}, \quad 0 \leq i \leq n. \tag{104}$$

2) According to Proposition 4.3, because the grid $\mathcal{X}$ is homogenous, the difference schemes $\mathcal{D}^n, \mathcal{D}^{n-1}, \ldots, \mathcal{D}^1$ of , $\mathcal{S}^n, \mathcal{S}^{n-1}, \ldots, \mathcal{S}^1$ all satisfy (⋆).

3) From Proposition 2.8, we know that, because the grid $\mathcal{X}$ is homogenous, all grids $\mathcal{X}, \widetilde{\mathcal{X}}^n, \widetilde{\mathcal{X}}^{n-1}, \ldots, \widetilde{\mathcal{X}}^1$ are equivalent and satisfy (G$_4$). now that $n \geq 2$.

Let S be the element of $\mathcal{E}_0^n$ with poles given by f_0. For $i \leq n-1$, we denote by $f_j^{\{i\}} = \left(f_{j,k}^{\{i\}}\right)_{k \in \mathbb{Z}}$, the sequence of poles of the ith derivative of S viewed as an element of $\mathcal{E}_j^{n-i}$, that is,

$$f_{j,k}^{\{i\}} := s^{\{i\}}(x_{j,k+1}, \ldots, x_{j,k+n-i}), \quad k \in \mathbb{Z},$$

where $s^{\{i\}}$ denotes the blossom of $S^{(i)}$ (with $s^{\{0\}} := s$ and $f_{j,k}^{\{0\}} := f_{j,k}$). By application of (99) and Theorem 3.8, we can state that the limit function $\mathcal{S}^n f_0$ produced by the scheme $\mathcal{S}^n$ (relative to $\widetilde{\mathcal{X}}^n$ or to any grid equivalent to $\widetilde{\mathcal{X}}^n$) is C^{n-1} on $\mathbb{R}$, with

$$\left(\mathcal{S}^n f_0\right)^{(i)} = \mathcal{S}^{n-i} f_0^{\{i\}}, \quad 0 \leq i \leq n-1.$$

Now, when using the grid $\mathcal{X}^1$, it is obvious that $\mathcal{S}_j^1 f_0^{\{n-1\}} = S^{(n-1)}$ for all $j \geq 0$. This eventually guarantees that

$$\left(\mathcal{S}^n f_0\right)^{(n-1)} = S^{(n-1)}. \tag{105}$$

If we were working on a closed bounded interval with appropriate multiplicities at the end points, from (105) would immediately guarantee that $\mathcal{S}^n f_0 = S$. Let us conclude with:

Corollary 4.5. *The assumptions are the same as in Corollary 4.4. Then, as soon as the grid $\widehat{\mathcal{X}}$ satisfies (G_4) (for instance if $\widehat{\mathcal{X}}$ is equivalent to $\mathcal{X}$), for any initial f_0, the limit function $\mathcal{S}^n f_0$ is the spline $S \in \mathcal{E}_0^n$ with poles $f_{0,k}$.*

Proof. Given $t \in \mathbb{R}$, by (G_4), $t = \lim_{r \to +\infty} \widehat{x}_{j_r, k_{j_r}}(t)$. The convergence stated in Corollary 4.3 enables us to conclude that

$$\mathcal{S}^n f_0(t) = \lim_{r \to +\infty} f_{j_r, k_{j_r}}(t) = s(x_{j_r}, k_{j_r}(t)+1, \ldots, x_{j, k_{j_r}(t)+n}),$$

where s is the blossom of the spline S. The expected equality $\mathcal{S}^n f_0(t) = S(t)$ simply results from the diagonal property of s and from its continuity on the set of admissible n-tuples relative to the grid of level 0. □

References

[1] G.M. Chaikin: An algorithm for high speed curve convergence, *Computer Graphics and Image Processing*, **3** (1974), 346–349.

[2] E. Cohen and L.L. Schumaker: Rates of convergence of control polygons, *Computer Aided Geometric Design*, **2** (1985), 229–235.

[3] W. Dahmen: Subdivision algorithms converge quadratically, *Journal of Computational and Applied Mathematics*, **16** (1986), 145–158.

[4] I. Daubechies, I. Guskov and W. Sweldens: Regularity of irregular subdivision, *Constructive Approximation*, **15** (1999), 381–426.

[5] I. Daubechies, I. Guskov and W. Sweldens: Commutation for irregular subdivision, *Constructive Approximation*, **17** (2001), 479–514.

[6] G. Deslauriers and S. Dubuc; Interpolation dyadique, in *Fractals. Dimensions non entières et applications*, Paris, Masson, pp. 44–55, 1987.

[7] N. Dyn, J. Gregory, and D. Levin: Analysis of uniform binary subdivision schemes for curve design. *Constructive Approximation*, **7** (1991), 127–147.

[8] N. Dyn and D. Levin: Analysis of asymptotically equivalent binary subdivision schemes, *Journal of Mathematical Analysis and Applications*, **193** (1995), 594–621.

[9] N. Dyn, D. Levin, and J. Gregory: A 4-point interpolatory subdivision scheme for curve design, *Computer Aided Geometric Design*, **4** (1987), 57–268.

[10] V. Maxim and M.-L. Mazure: Subdivision schemes and irregular grids, *Numerical Algorithms*, **35** (2004), 1–28.

[11] M.-L. Mazure: Blossoming and CAGD algorithms, in *Shape preserving representations for Computer-Aided Design*, J.-M. Peña (ed.), Nova Science Pub., 1999, 99–117.

[12] L. Ramshaw: Blossoms are polar forms, *Computer Aided Geometric Design* **6** (1989), 323–358.

[13] G. de Rham: Sur une courbe plane, *Journal de Mathématiques Pures et Appliquées*, **35** (1956), 25–42.

[14] R. Riesenfeld: On Chaikin's algorithm, *Computer Graphics and Image Processing*, **4** (1975), 304–310.

Marie-Laurence Mazure
Université Joseph Fourier
Laboratoire de Modélisation et Calcul
BP 53
F-38041 Grenoble Cedex 9, France
e-mail: `mazure@imag.fr`

J.P. Berrut, H.N. Mhaskar, M. de Bruin and P. Vertesi

Trends and Applications in Constructive Approximation
(Eds.) M.G. de Bruin, D.H. Mache & J. Szabados
International Series of Numerical Mathematics, Vol. 151, 165–180
© 2005 Birkhäuser Verlag Basel/Switzerland

A Markov-Bernstein Inequality for Gaussian Networks

H.N. Mhaskar

Abstract. Let $s \geq 1$ be an integer. A Gaussian network is a function on $\mathbb{R}^s$ of the form $g(\mathbf{x}) = \sum_{k=1}^{N} a_k \exp(-\|\mathbf{x} - \mathbf{x}_k\|^2)$. The minimal separation among the centers, defined by $\min_{1 \leq j \neq k \leq N} \|\mathbf{x}_j - \mathbf{x}_k\|$, is an important characteristic of the network that determines the stability of interpolation by Gaussian networks, the degree of approximation by such networks, etc. We prove that if $g(\mathbf{x}) = \sum_{k=1}^{N} a_k \exp(-\|\mathbf{x} - \mathbf{x}_k\|^2)$, the minimal separation of g exceeds $1/m$, and $\log N = \mathcal{O}(m^2)$ then for any integer $r \geq 1$, any partial derivative $\mathcal{D}g$ of order r of g satisfies $\|\mathcal{D}g\|_{p,\mathbb{R}^s} \leq cm^r \|g\|_{p,\mathbb{R}^s}$.

1. Introduction

Let $s, N \geq 1$ be integers. A *Gaussian network* with N *neurons* is a function on the Euclidean space $\mathbb{R}^s$ of the form $\mathbf{x} \mapsto \sum_{k=1}^{N} a_k \exp(-\|\mathbf{x} - \mathbf{x}_k\|^2)$, where $\| \circ \|$ denotes the Euclidean norm on $\mathbb{R}^s$, the *centers* $\mathbf{x}_k$ are in $\mathbb{R}^s$, and $a_k \in \mathbb{R}$, $k = 1, \cdots, N$. These functions can be evaluated in hardware using parallel computation of the exponential terms, and are used extensively in many applications in pattern recognition, computer graphics, antenna array theory, probability density estimation, etc. A typical problem in all these applications is to approximate an unknown function (the *target function*) by such networks.

An important characteristic of Gaussian networks is the minimal separation among the centers, defined by $\min_{1 \leq j \neq k \leq N} \|\mathbf{x}_j - \mathbf{x}_k\|$. Many results in the theory of stability of interpolation by Gaussian networks, the degree of approximation by such networks, etc. depend upon the minimal separation. For example, Narcowich and Ward [9] have estimated the condition numbers of the interpolation matrices in the context of a general scattered data interpolation. Their estimates are in terms of the minimal separation between the interpolation points, independent of the number of points (and hence, of neurons) involved. In [7], we have argued that treating the minimal separation among the centers as the "cost of approximation" (rather than the more apparent cost in terms of the number of neurons) leads to matching direct and converse theorems in the theory of approximation by Gaussian networks. In particular, under certain conditions, if a function can be approximated by Gaussian networks at a polynomial rate, measured in terms of the minimal separation of the networks, then it can also be approximated at the same rate by the linear processes of weighted polynomial approximation.

The purpose of this paper is to prove a Markov-Bernstein inequality for Gaussian networks in terms of the minimal separation. We note that such an inequality was obtained by Erdélyi [3] in terms of the number of neurons. Also, if $r \geq 1$ is an integer, and $\mathcal{D}$ is a partial derivative operator of order r, our results in [7] immediately yield an inequality of the form $\|\mathcal{D}g\|_{p,\mathbb{R}^s} \leq c \exp(Am^2)\|g\|_{p,\mathbb{R}^s}$ for networks g where the minimal separation exceeds $1/m$. In this paper, we prove a substantially better inequality of the form $\|\mathcal{D}g\|_{p,\mathbb{R}^s} \leq cm^r\|g\|_{p,\mathbb{R}^s}$ for such networks, provided that the number of neurons is not too large (cf. Theorem 2.1 below). Our proof involves a good deal of book-keeping in estimating the degree of weighted polynomial approximation in a more careful way than what is available in the literature that we are aware of so far.

In Section 2, we formulate our main result (Theorem 2.1) regarding Gaussian networks. In Section 3, we discuss the background and prove the necessary new results on weighted polynomial approximation. In Section 4, we review some results regarding Gaussian networks, and prove Theorem 2.1.

2. Main result

Let $s \geq 1$ be an integer. The notation for the class of Gaussian networks will involve different bounds on the centers as well as the number of neurons involved. Thus, for $m, M, N > 0$, the symbol $\mathbb{G}_{N,M,m,s}$ denotes the class of functions of the form

$$\mathbf{x} \mapsto \sum_{1 \leq k \leq N, \ k \in \mathbb{Z}} a_k \exp(-\|\mathbf{x} - \mathbf{x}_k\|^2), \qquad \mathbf{x}, \mathbf{x}_k \in \mathbb{R}^s, \ a_k \in \mathbb{R}, \ 1 \leq k \leq N, \quad (1)$$

where $\max_{1 \leq k \leq N} \|\mathbf{x}_k\| \leq M$ and the *minimal separation*, $\min_{1 \leq k,j \leq N, k \neq j} \|\mathbf{x}_j - \mathbf{x}_k\| \geq m^{-1}$. Also, the union of the class of networks over a certain parameter will be denoted by writing the symbol ∞ in place of that parameter; for example, $\mathbb{G}_{N,\infty,m,s} := \cup_{M>0}\mathbb{G}_{N,M,m,s}$, etc. For $A, C, m > 0$, we write

$$\mathbb{B}(A,C;m,s) := \{g \in \mathbb{G}_{N,\infty,m,s} \ : \ N \leq C \exp(Am^2)\}. \tag{2}$$

We remark that if $A_1 \geq A$, $C_1 \geq C$, $m_1 \geq m$, then $\mathbb{B}(A,C;m,s) \subseteq \mathbb{B}(A_1,C_1;m_1,s)$.

If $1 \leq p \leq \infty$, $f : \mathbb{R}^s \to \mathbb{R}$ is a Lebesgue measurable function, and $S \subseteq \mathbb{R}^s$ is a Lebsegue measurable set having positive measure, we write

$$\|f\|_{p,S} := \begin{cases} \{\int_S |f(\mathbf{x})|^p d\mathbf{x}\}^{1/p}, & \text{if } 1 \leq p < \infty, \\ \text{ess sup}_{\mathbf{x} \in S}|f(\mathbf{x})|, & \text{if } p = \infty. \end{cases} \tag{3}$$

The set of all functions for which $\|f\|_{p,S} < \infty$ is denoted by $L^p(S)$, where, as usual, two functions that are equal almost everywhere on S are considered equal as elements of $L^p(S)$. Let D_j denote the operation of partial differentiation with respect the jth variable, and $\mathbf{D}^{\mathbf{k}} := \prod_{j=1}^{s} D_j^{k_j}$. For a suitably smooth function f, we write

$$\|f\|_{p,r,S} := \sum_{|\mathbf{k}| \leq r} \|\mathbf{D}^{\mathbf{k}} f\|_{p,S}. \tag{4}$$

In the sequel, we adopt the following convention regarding constants. The symbols $c, c_1, \cdots$ will denote positive constants depending only on A, C, s, p, r, and other similarly fixed parameters, but their values may be different at different occurrences, even within a single formula. Constants denoted by capital letters retain their values, subject to the choice of the parameters on which they depend.

Our main theorem in this paper is the following.

Theorem 2.1. *Let $s, r \geq 1$ be integers, $1 \leq p \leq \infty$, $m \geq 1$, $A, C > 0$. Then there exists a positive constant c depending only on A, C, p, r, and s, such that*

$$\|g\|_{p,r,\mathbb{R}^s} \leq cm^r \|g\|_{p,\mathbb{R}^s}, \qquad g \in \mathbb{B}(A, C; m, s). \tag{5}$$

The idea behind the proof of Theorem 2.1 is the following. In [7], we have established a connection between the ℓ^1 norm of the coefficients of a Gaussian network, and the norm of this network. Using the partial sums of the series in (43), we will approximate the basic Gaussian $\exp(-\| \circ -\mathbf{x}_k\|^2)$ by weighted polynomials, and hence, approximate the network by weighted polynomials. Unfortunately, this can be done adequately only if $\|\mathbf{x}_k\| \leq cm$. Therefore, we use a partition of unity so that the norms of different networks and their derivatives are essentially confined to cubes with side proportional to m. However, this involves estimating a norm of the form $\|\phi g\|_{p,r,\mathbb{R}^s}$ for a compactly supported ϕ. Here, we use Theorem 3.2 below to approximate ϕ by weighted polynomials, the partial sums of the series in (43) to approximate the part of g with centers in a cube of side proportional to m by weighted polynomials as well, and estimate the remaining part of g using the results in [7]. A repeated application of the Markov-Bernstein inequality (8) enables us to estimate the derivatives of the approximating weighted polynomials thus obtained in terms of their norms. A reverse process then takes us to (5).

3. Weighted polynomial approximation

In this section, we write $w(x) := \exp(-x^2)$. Our results here are in the univariate case, $s = 1$, but can be extended easily to the multivariate case by a simple tensor product argument. For $x \geq 0$, let Π_x denote the class of all univariate algebraic polynomials of degree at most x. First, we recall a few properties of polynomials. The following proposition will be used often, sometimes without an explicit reference.

Proposition 3.1. *Let* $m \geq 0$, $1 \leq p \leq \infty$, $\lambda > 0$, *and* $P \in \Pi_{m^2}$.

(a) (Infinite-finite range inequality) *For any* $\gamma > 0$, *there exists* $a > 1$, *depending only on* γ *and* λ *such that*

$$\|w^\lambda P\|_{p,\mathbb{R}\setminus[-am,am]} \leq c(\lambda,\gamma)\exp(-\gamma^2 m^2)\|w^\lambda P\|_{p,[-am,am]}. \tag{6}$$

In particular,

$$\|xw^\lambda P\|_{p,\mathbb{R}} \leq c(\lambda)m\|w^\lambda P\|_{p,\mathbb{R}}. \tag{7}$$

(b) (Markov-Bernstein inequality)

$$\|(w^\lambda P)'\|_{p,\mathbb{R}} \leq c(\lambda)m\|w^\lambda P\|_{p,\mathbb{R}}. \tag{8}$$

Proof. The inequality (6) follows from [6, Theorem 6.2.4, Lemma 7.2.2]. The inequality (7) is then clear. Part (b) follows from [6, Theorem 6.2.9, Theorem 3.4.2] and (7). □

In this section, we adopt the following notation. The space of all 2π-periodic continuous functions on $\mathbb{R}$, equipped with the norm $\|F\|^* := \|F\|_{\infty,[-\pi,\pi]}$, will be denoted by C^*, the class of all trigonometric polynomials of order at most n will be denoted by $\mathbf{H}_n$, and

$$E_n^*(F) := \inf_{T \in \mathbf{H}_n} \|F - T\|^*.$$

Our main theorem in this section is the following.

Theorem 3.2. *Let* $0 < a < 1$, $f : \mathbb{R} \to \mathbb{R}$ *be a continuous function, with* $f(x) = 0$ *if* $|x| \geq a$. *Let* $F(\theta) := f(3\cos(2\theta))$, $\theta \in \mathbb{R}$. *There exists a sequence of polynomials* $P_n \in \Pi_n$ *such that*

$$\max_{x \in \mathbb{R}} |f(x) - P_n(x)\exp(-nx^2)| \leq c_1 E_{c\sqrt{n}}^*(F) + \exp(-c_2\sqrt{n})\|f\|_{\infty,\mathbb{R}}, \tag{9}$$

where c, c_1, c_2 *are positive constants depending only on* a.

We note that the fact that the left-hand side of (9) tends to zero has been known for a very long time, and has been generalized a great deal (cf. [6, 1] and references therein). The novelty here is the rate of convergence. Our proof consists of a book keeping in the proof of Theorem 10.1.1 in [6], obtaining an intermediate polynomial involved in that proof using the following Theorem 3.3 of Gaier [4] instead of Jackson's theorem. The same proof can also be adapted at least to the more general situation discussed in [6, Section 10.1]. However, we do not wish to introduce here the additional notation that would be necessary to formulate this general version. We will give a simple proof of Theorem 3.3 which does not involve complex variable techniques, and yields estimates using the norm of the functions on the interval rather than unspecified constants depending upon the functions.

Theorem 3.3. *There exists a sequence of linear operators* $\mathcal{G}_n$ *on* $C[-3,3]$, *such that for each* $f \in C[-3,3]$, *and integer* $n \geq 1$, $\mathcal{G}_n(f) \in \Pi_n$, *and satisfies the following conditions: Writing* $F(t) := f(3\cos(2t))$ $(F \in C^*)$,

$$\|f - \mathcal{G}_n(f)\|_{\infty,[-3,3]} \leq c_1 E_{n/3}^*(F) + c_2\exp(-c_3 n)\|f\|_{\infty,[-3,3]}, \tag{10}$$

where c_1, c_2, c_3 are absolute positive constants. Further, if $a \in (0,3)$, $f(x) = 0$ for $|x| \geq a$, and $a < b < 3$, then

$$\|\mathcal{G}_n(f)\|_{\infty,[-3,-b]\cup[b,3]} \leq c_4 \exp(-cn)\|f\|_{\infty,[-3,3]},\tag{11}$$

where c_4, c are positive constants depending only on a and b.

Proof. We recall (cf. [2, Chapter 9, Theorem 3.1]) that the expression

$$v_\ell^*(t) := \frac{1}{\ell}\sum_{m=\ell}^{2\ell-1}\sum_{|k|\leq m} e^{ikt} = \frac{\cos \ell t - \cos(2\ell t)}{2\ell \sin^2(t/2)}$$

is an even, trigonometric polynomial of order at most $2\ell - 1$, the operator

$$V_\ell^*(F,x) := \frac{1}{2\pi}\int_{-\pi}^{\pi} F(t)v_\ell^*(x-t)dt, \qquad F \in C^*,\tag{12}$$

satisfies $\|V_\ell^*(F)\|^* \leq c\|F\|^*$, and

$$E_{2\ell}^*(F) \leq \|F - V_\ell^*(F)\|^* \leq cE_\ell^*(F), \qquad F \in C^*.\tag{13}$$

We write $V_0^*(F) := \frac{1}{2\pi}\int_{-\pi}^{\pi} F(t)dt$, $v_0^*(t) := 1$, and

$$\mathcal{G}_n^*(F) = 2^{-n}\sum_{\ell=0}^{n}\binom{n}{\ell}V_\ell^*(F).\tag{14}$$

Since [5, 8]

$$2^{-n}\sum_{\ell=0}^{\lfloor n/3 \rfloor-1}\binom{n}{\ell} \leq c_1 \exp(-cn),$$

we conclude that

$$\|F - \mathcal{G}_n^*(F)\|^* < c_1 \exp(-cn)\|F\|^* + c_2 E_{n/3}^*(F).\tag{15}$$

Next, we observe that for $\ell \geq 1$,

$$v_\ell^*(t) = \frac{1}{2\sin^2(t/2)}\int_t^{2t} \sin(\ell u)du.$$

Therefore,

$$2^{-n}\sum_{\ell=1}^{n}\binom{n}{\ell}v_\ell^*(t) = \frac{1}{2^{n+1}\sin^2(t/2)}\int_t^{2t}\left\{\sum_{\ell=0}^{n}\binom{n}{\ell}\sin(\ell u)\right\}du$$

$$= \frac{1}{2\sin^2(t/2)}\int_t^{2t}\sin(nu/2)\cos^n(u/2)du$$

$$= \frac{1}{\sin^2(t/2)}\int_{t/2}^{t}\sin(nu)\cos^n(u)du.$$

Thus,

$$\left|2^{-n}\sum_{\ell=1}^{n}\binom{n}{\ell}v_\ell^*(t)\right| \leq c_1(\delta)\exp(-c(\delta)n), \qquad t \in [\delta, \pi - \delta].$$

Now, let $F(t) = 0$ for $t \notin [\delta, \pi - \delta]$ (and hence, also for $t \notin [-\pi + \delta, -\delta]$). Then for $0 < \delta_1 < \delta$, and $|x| \leq \delta_1$,

$$
\begin{aligned}
|\mathcal{G}_n^*(F, x)| &\leq c \int_{\delta - \delta_1 \leq |t| \leq \pi - \delta + \delta_1} |F(x - t)| \left| \left\{ 2^{-n} \sum_{\ell=0}^{n} \binom{n}{\ell} v_\ell^*(t) \right\} \right| dt \\
&\leq c_1(\delta, \delta_1) \exp(-c(\delta, \delta_1)n) \|F\|^*.
\end{aligned}
\tag{16}
$$

Now, we observe that $F(t) = F(-t) = F(\pi - t)$ for $t \in \mathbb{R}$. Therefore, since each v_ℓ^* is an even function, it is not difficult to see that for $x \in \mathbb{R}$

$$
\begin{aligned}
2\pi V_\ell^*(F, \pi - x) &= \int_0^\pi (F(\pi - x - t) + F(\pi - x + t)) v_\ell^*(t) dt \\
&= \int_0^\pi (F(x + t) + F(x - t)) v_\ell^*(t) dt = 2\pi V_\ell^*(F, x).
\end{aligned}
$$

Hence, $\mathcal{G}_n^*(F, x)$ is a linear combination of $\cos 2kx$, $k = 0, \cdots, n$. Thus, the operator defined by $\mathcal{G}_n(f, 3\cos 2\theta) := \mathcal{G}_n^*(F, \theta)$ satisfies $\mathcal{G}_n(f) \in \Pi_n$. The estimates (15) and (16) lead to (10) and (11) respectively. $\qquad\square$

We resume our proof of Theorem 3.2 as in [6, Section 10.1], sketching only enough details to make the paper self-sufficient, and to point out the necessary differences.

Let T_n be the extremal polynomial satisfying

$$
\|w\mathsf{T}_n\|_{\infty, \mathbb{R}} = \inf_{P \in \Pi_{n-1}} \|((\circ)^n - P)w\|_{\infty, \mathbb{R}}
$$

Let

$$
\xi_n = \max\{\xi \in \mathbb{R} : w(\xi)\mathsf{T}_n(\xi) = \|w\mathsf{T}_n\|_{\infty, \mathbb{R}}\},
$$

and $Z_n = \xi_n/\sqrt{n}$. It is known [6, Chapter 6] that $Z_n \in (-1, 1)$, $\lim_{n \to \infty} Z_n = 1$, and that T_n has n zeros in $[-\sqrt{n}, \sqrt{n}]$. In this section only, we will write $T_n(x) := n^{-n/2}\mathsf{T}_n(\sqrt{n}x)$, $\mathcal{E}_n := \mathsf{T}_n(Z_n)w(\xi_n)$ (our notation here being different from that in [6]).

For $\epsilon > 0$ and integer $n \geq 1$, let

$$
\begin{aligned}
\Gamma_{n,1,\epsilon} &:= \{Z_n + (1 - Z_n)\frac{y^2}{\epsilon^2} + iy : 0 \leq y \leq \epsilon\}, \\
\Gamma_{n,2,\epsilon} &:= \{x + i\epsilon : -1 \leq x \leq 1\} \\
\Gamma_{n,\epsilon} &:= \Gamma_{n,1,\epsilon} \cup (-\Gamma_{n,1,\epsilon}) \cup \overline{\Gamma_{n,1,\epsilon}} \cup (-\overline{\Gamma_{n,1,\epsilon}}) \cup \Gamma_{n,2,\epsilon} \cup \overline{\Gamma_{n,2,\epsilon}}, \tag{17}
\end{aligned}
$$

where, for a set $S \subseteq \mathbb{C}$, $\overline{S} := \{\overline{z} : z \in S\}$ and $-S := \{-z \; z \in S\}$.

We recall the following facts from [6, Chapter 10.1] (cf. the proof of Lemma 10.1.3 there).

Lemma 3.4. *There exists an $\epsilon_0 > 0$ such that the following statements hold for each ϵ, $0 < \epsilon \le \epsilon_0$, where the constants may depend upon ϵ.*

$$\oint_{\Gamma_{n,2,\epsilon} \cup \overline{\Gamma}_{n,2,\epsilon}} \mathcal{E}_n |T_n(z) \exp(-n|z|^2)|^{-1} |dz| \le c_1 \exp(-cn), \tag{18}$$

$$\oint_{\Gamma_{n,1,\epsilon} \cup (-\Gamma_{n,1,\epsilon}) \cup \overline{\Gamma}_{n,1,\epsilon} \cup (-\overline{\Gamma}_{n,1,\epsilon})} \mathcal{E}_n |T_n(z) \exp(-n|z|^2)|^{-1} |dz| \le cn^{-1/2}, \tag{19}$$

and

$$\left| \frac{Z_n^2 - \xi^2}{x - \xi} \right| \le c, \qquad x \in \mathbb{R}, \ \xi \in \Gamma_{n,\epsilon}. \tag{20}$$

Proof. The estimates (18) and (19) are the estimates (10.1.24) and (10.1.25) respectively in [6, P. 260], the estimate (20) is not difficult, and is shown in [6] near the end of page 261. $\qquad\square$

Proof of Theorem 3.2. We follow the proof of [6, Proposition 10.1.2]. Let $a < b < 1$, and n be chosen large enough so that $Z_n > (1 + b)/2$. In this proof only, let $f_{1,n}(x) := f(x)/(Z_n^2 - x^2)$. Then for each $n \ge c$, each $f_{1,n}$ is continuous on $\mathbb{R}$ and $f_{1,n}(x) = 0$ if $|x| \ge a$. Moreover, $\|f_{1,n}\|_{\infty,\mathbb{R}} \le c\|f\|_{\infty,\mathbb{R}}$. We estimate $\|f_{1,n} - \mathcal{G}_m(f_{1,n})\|_{\infty,\mathbb{R}}$. We observe that for each $N \ge 1$, $R(x) := \sum_{k=0}^{N/2} Z_n^{-2k-2} x^{2k} \in \Pi_N$, and

$$\left| \frac{1}{Z_n^2 - x^2} - R(x) \right| \le c_1 \exp(-c_2 N), \qquad x \in [-b, b].$$

For $x \in [-3, -b] \cup [b, 3]$, $|R(x)| \le c(3/b)^N$. In this proof only, let the constant denoted by c in (11) be denoted by γ. We find an integer $\beta > (1/\gamma) \log(4/b)$. Then (11) implies that for $x \in [-3, -b] \cup [b, 3]$

$$|f_{1,n}(x) - \mathcal{G}_{\beta N}(f, x) R(x)| = |\mathcal{G}_{\beta N}(f, x) R(x)| \le c(3/4)^N \|f\|_{\infty,\mathbb{R}}. \tag{21}$$

For $|x| \le b$,

$$\left| \frac{f(x)}{Z_n^2 - x^2} - \mathcal{G}_{\beta N}(f, x) R(x) \right| \le c_1 E_{cN}^*(F) + c_2 \exp(-c_3 N) \|f\|_{\infty,\mathbb{R}}.$$

Writing $F_{1,n}(t) := f_{1,n}(3 \cos(2t))$, this estimate and (21) imply that for any integer $N \ge 1$,

$$\begin{aligned} E_{2(\beta+1)N}^*(F_{1,n}) &\le \|F_{1,n}(3\cos(2\circ)) - \mathcal{G}_{\beta N}(f, 3\cos(2\circ)) R(3\cos(2\circ))\|^* \\ &\le c_1 E_{cN}^*(F) + c_2 \exp(-c_3 N) \|f\|_{\infty,\mathbb{R}}, \end{aligned}$$

or equivalently,

$$E_m^*(F_{1,n}) \le c_1 E_{cm}^*(F) + c_2 \exp(-c_3 m) \|f\|_{\infty,\mathbb{R}}, \qquad m \ge 1.$$

Using Theorem 3.3 again with $f_{1,n}$ in place of f, we conclude that

$$\|f_{1,n} - \mathcal{G}_m(f_{1,n})\|_{\infty,\mathbb{R}} \le c_1 E_{cm}^*(F) + c_2 \exp(-c_3 m) \|f\|_{\infty,\mathbb{R}}, \qquad m \ge 1, \ n \ge c, \tag{22}$$

and

$$\|\mathcal{G}_m(f_{1,n})\|_{\infty,[-3,-b]\cup[b,3]} \le c_1 \exp(-cm) \|f\|_{\infty,\mathbb{R}}. \tag{23}$$

In view of Bernstein's inequality [2, Theorem 2.2, Chapter 4], we may then choose an $\epsilon > 0$ such that

$$|\mathcal{G}_{\lfloor\sqrt{n}\rfloor}(f_{1,n},z)| \leq c_1 \exp(-c\sqrt{n})\|f\|_{\infty,\mathbb{R}}, \quad z \in \Gamma_{n,1,\epsilon} \cup (-\Gamma_{n,1,\epsilon}) \cup \overline{\Gamma_{n,1,\epsilon}} \cup (-\overline{\Gamma_{n,1,\epsilon}}),$$

(24)

and in addition, the conclusions of Lemma 3.4 hold. In this proof only, let

$$
\begin{aligned}
g(z) &:= \mathcal{G}_{\lfloor\sqrt{n}\rfloor}(f_{1,n},z), \\
g_n(z) &:= \exp(nz^2)(Z_n^2 - z^2)g(z), \quad z \in \mathbb{C}, \\
h_n(x) &:= \begin{cases} g(x)(Z_n^2 - x^2), & \text{if } x \in (-Z_n, Z_n), \\ 0, & \text{if } x \in \mathbb{R} \setminus (-Z_n, Z_n), \end{cases}
\end{aligned}
$$

(25)

and L_n be polynomial interpolating g_n at the zeros of T_n. As in [6, p. 261], we obtain for $x \in \mathbb{R}$, $x \neq \pm Z_n$,

$$h_n(x)\exp(nx^2) - L_n(x) = \frac{T_n(x)}{2\pi i}\oint_{\Gamma_{n,\epsilon}} \frac{g_n(\xi)}{T_n(\xi)(\xi - x)}d\xi.$$

(26)

Hence,

$$
\begin{aligned}
&|h_n(x) - \exp(-nx^2)L_n(x)| \\
&\leq \quad c\|w(\sqrt{n}(\circ))T_n\|_{\infty,\mathbb{R}}\oint_{\Gamma_{n,\epsilon}} \frac{|g(\xi)|}{\exp(-n|\xi|^2)|T_n(\xi)|}\frac{|Z_n^2 - \xi^2|}{|x - \xi|}|d\xi|.
\end{aligned}
$$

(27)

Taking into account the fact that $\|w(\sqrt{n}(\circ))T_n\|_{\infty,\mathbb{R}} = \mathcal{E}_n$, Lemma 3.4 implies that

$$
\begin{aligned}
|h_n(x) - \exp(-nx^2)L_n(x)| &\leq c_1\exp(-cn) \max_{\xi \in \Gamma_{n,2,\epsilon} \cup \overline{\Gamma_{n,2,\epsilon}}} |g(\xi)| \\
&+ c_2 n^{-1/2} \max_{\xi \in \Gamma_{n,1,\epsilon} \cup (-\Gamma_{n,1,\epsilon}) \cup \overline{\Gamma_{n,1,\epsilon}} \cup (-\overline{\Gamma_{n,1,\epsilon}})} |g(\xi)|.
\end{aligned}
$$

(28)

In view of the Bernstein inequality [2, Theorem 2.2, Chapter 4],

$$\max_{\xi \in \Gamma_{n,2,\epsilon} \cup \overline{\Gamma_{n,2,\epsilon}}} |g(\xi)| \leq c\exp(c_2\sqrt{n})\|f\|_{\infty,\mathbb{R}}.$$

Along with (24) and (28), this implies for $x \in \mathbb{R}$, $x \neq \pm Z_n$,

$$|h_n(x) - \exp(-nx^2)L_n(x)| \leq c_1\exp(-c\sqrt{n})\|f\|_{\infty,\mathbb{R}}.$$

(29)

Since the functions involved are continuous, (29) holds for all $x \in \mathbb{R}$.

In particular, in view of (23), we have

$$|\exp(-nx^2)L_n(x)| \leq c_1\exp(-c\sqrt{n})\|f\|_{\infty,\mathbb{R}}, \quad x \in [-3, -b] \cup [b, 3].$$

(30)

For $|x| \leq b$, we use (22) to conclude that

$$
\begin{aligned}
|h_n(x) - f(x)| &= |g(x)(Z_n^2 - x^2) - f(x)| \leq c|\mathcal{G}_{\lfloor\sqrt{n}\rfloor}(f_{1,n},x) - f_{1,n}(x)| \\
&\leq c_3 E^*_{c\sqrt{n}}(F) + c_1\exp(-c_2\sqrt{n})\|f\|_{\infty,\mathbb{R}}.
\end{aligned}
$$

Along with (29) and (30), this implies that for $|x| \leq 3$,

$$|f(x) - \exp(-nx^2)L_n(x)| \leq c_3 E^*_{c\sqrt{n}}(F) + c_1\exp(-c_2\sqrt{n})\|f\|_{\infty,\mathbb{R}}.$$

Necessarily, $|\exp(-nx^2)L_n(x)| \le c_1 \exp(-cn)\|f\|_{\infty,\mathbb{R}}$ if $|x| \ge 3$. This completes the proof of the theorem. $\qquad\square$

Next, we apply Theorem 3.2 and a simultaneous approximation theorem [6, Theorem 4.1.7] to arrive at an estimate on the degree of approximation of smooth functions and their derivatives.

Theorem 3.5. *Let $m \ge 1$, $\beta, r \ge 0$, φ be an infinitely often differentiable function on $\mathbb{R}$, supported on $[-1, 1]$, and $\phi(x) := \varphi(x/m)$. There exists $P \in \Pi_{2m^2}$ such that*

$$\|\phi - wP\|_{\infty,r,\mathbb{R}} \le c(\beta, r, \varphi)m^{-\beta}. \tag{31}$$

Proof. In this proof only, let $g(x) = \exp(x^2)\phi(x)$, and $\|\varphi\|_{\infty,r,\mathbb{R}} = 1$. We prove first that for every $\gamma > 0$ there exists a polynomial $P_1 \in \Pi_{2m^2-1}$ such that

$$\|(g' - P_1)w\|_{\infty,\mathbb{R}} \le cm^{-\gamma} \tag{32}$$

We apply Theorem 3.2 with $\varphi((\sqrt{2m^2 - 2}/m)\circ)$ (respectively $\varphi'((\sqrt{2m^2 - 1}/m)\circ)$) in place of f, and make an obvious change of variables to obtain $P_2 \in \Pi_{2m^2-2}$, $R_2 \in \Pi_{2m^2-1}$ such that

$$\|\phi - P_2 w\|_{\infty,\mathbb{R}} \le cm^{-\gamma-1}, \quad \|\phi' - R_2 w\|_{\infty,\mathbb{R}} \le cm^{-\gamma}. \tag{33}$$

Since $wg' = \phi' + 2x\phi$, the polynomial $P_1 := R_2 + 2xP_2 \in \Pi_{2m^2-1}$ satisfies

$$\|(g' - P_1)w\|_{\infty,\mathbb{R}} \le \|\phi' - R_2 w\|_{\infty,\mathbb{R}} + 2\|x\phi - xP_2 w\|_{\infty,\mathbb{R}}. \tag{34}$$

Since $\|P_2 w\|_{\infty,\mathbb{R}} \le c$, $\|xP_2 w\|_{\infty,[-4m,4m]} \le cm$. The infinite-finite range inequality implies that $\|xP_2 w\|_{\infty,\mathbb{R}\setminus[-4m,4m]} \le c_1 \exp(-cm^2)$. Since ϕ is supported on $[-m, m]$, this implies that

$$\|x\phi - xP_2 w\|_{\infty,\mathbb{R}} \le cm\|\phi - P_2 w\|_{\infty,\mathbb{R}} + c_1 \exp(-cm^2) \le cm^{-\gamma}.$$

Together with (33), this leads to (32).

Again, let $P \in \Pi_{cm^2}$, $c \ge 2$, be any polynomial such that

$$\|\phi - wP\|_{\infty,\mathbb{R}} = \|(g - P)w\|_{\infty,\mathbb{R}} \le cm^{-\beta-r}. \tag{35}$$

In view of [6, Theorem 4.1.7] and the estimate (32) applied with $\gamma = \beta + r - 1$, we obtain

$$\|(g' - P')w\|_{\infty,\mathbb{R}} \le cm^{-\beta-r+1}. \tag{36}$$

Arguing as before, we see that $\|x\phi - xPw\|_{\infty,\mathbb{R}} \le cm^{-\beta-r+1}$. Therefore, (36) leads to

$$\|\phi' - (wP)'\|_{\infty,\mathbb{R}} \le cm^{-\beta-r+1}.$$

By induction, we conclude that for any $P \in \Pi_{cm^2}$ for which (35) holds,

$$\|\phi^{(k)} - (wP)^{(k)}\|_{\infty,\mathbb{R}} \le cm^{-\beta-r+k}.$$

The existence of such a polynomial being guaranteed by Theorem 3.2 (applied with $\varphi(\sqrt{c}\circ)$ in place of f), the proof is now complete. $\qquad\square$

4. Gaussian networks

In the sequel, $1 \le p \le \infty$, $A, C > 0$ will be fixed numbers, $r, s \ge 1$ will be fixed integers, and it will be assumed that the reciprocal of the minimal separation satisfies $m \ge 1$. The various constants, denoted by $c, c_1, \cdots$ or by capital letters depend upon the values of those parameters among A, C, r, p, s that are present in the context, but not on m. We will explicitly indicate their dependence only on any other quantities when required. The basis of our proof of Theorem 2.1 is the following proposition.

Proposition 4.1. *Let $\varphi : \mathbb{R} \to [0, \infty)$ be an infinitely many times differentiable function such that $\varphi(x) = 0$ if $x \notin [-1, 1]$. Let $\phi(\mathbf{x}) := \prod_{k=1}^{s} \varphi(x_k/m)$, $\mathbf{x} \in \mathbb{R}^s$. Let $g \in \mathbb{B}(A, C; m, s)$. Then for any $\beta, \gamma > 0$,*

$$\|\phi g\|_{p,r,\mathbb{R}^s} \le c_1 m^r \|\phi g\|_{p,\mathbb{R}^s} + c_2 m^{-\beta} \|g\|_{p,r,[-cm,cm]^s} + c_3 \exp(-\gamma^2 m^2) \|g\|_{p,\mathbb{R}^s}, \quad (37)$$

where c_1, c_2, c_3 are positive constants depending on φ, β, γ.

We begin by reviewing a number of results from [7], and/or proving them in a somewhat modified manner. Thus, results which are not proved here can be found in [7]. In particular, the next proposition is a reformulation of [7, Proposition 3.3].

Proposition 4.2. *There exist positive constants c, B_1 with the following property. If $g \in \mathbb{B}(A, C; m, s)$, then*

$$c\|g\|_{p,\mathbb{R}^s} \le \sum_{k=1}^{N} |a_k| \le c_1 \exp(B_1^2 m^2) \|g\|_{p,\mathbb{R}^s}. \quad (38)$$

Corollary 4.3. *Let $1 \le N \le C \exp(Am^2)$ be an integer, $\mathbf{x}_1, \cdots, \mathbf{x}_N$ be points in $\mathbb{R}^s$, $g = \sum_{k=1}^{N} a_k \exp(-\|\circ -\mathbf{x}_k\|^2) \in \mathbb{G}_{N,\infty,m,s}$, and $S \subseteq \{1, \cdots, N\}$. Then*

$$\left\| \sum_{k \in S} a_k \exp(-\|\circ -\mathbf{x}_k\|^2) \right\|_{p,r,\mathbb{R}^s} \le c \exp(B_1^2 m^2) \|g\|_{p,\mathbb{R}^s}. \quad (39)$$

In particular,

$$\|g\|_{p,r,\mathbb{R}^s} \le c \exp(B_1^2 m^2) \|g\|_{p,\mathbb{R}^s}. \quad (40)$$

The following proposition (cf. [7, Proposition 3.4]) estimates the norm on a cube of a part of a Gaussian network whose centers are away from the cube.

Proposition 4.4. *Let $g := \sum_{k=1}^{N} a_k \exp(-\|\circ -\mathbf{x}_k\|^2) \in \mathbb{B}(A, C; m, s)$, $\mathcal{A}$ be a measurable subset of $\mathbb{R}^s$, and $b > 0$. Let $L_\mathcal{A} := \{k : \text{dist}(\mathbf{x}_k, \mathcal{A}) \ge m\sqrt{2B_1^2 + 2b^2}\}$, and $h_\mathcal{A} := \sum_{k \in L_\mathcal{A}} a_k \exp(-\|\circ -\mathbf{x}_k\|^2)$. Then*

$$\|h_\mathcal{A}\|_{p,r,\mathcal{A}} \le c_3 \exp(-b^2 m^2) \|g\|_{p,\mathbb{R}^s}. \quad (41)$$

In particular, if $a > 0$, there exists a constant B_2 depending only on a, b, (in addition to A, C, p, r, s, but not on g) with the following property: Let $L := \{k : \|\mathbf{x}_k\| \geq B_2 m\}$, and $h := \sum_{k \in L} a_k \exp(-\| \circ -\mathbf{x}_k\|^2)$. Then

$$\|h\|_{p,r,[-am,am]^s} \leq c_3 \exp(-b^2 m^2)\|g\|_{p,\mathbb{R}^s}. \tag{42}$$

Before proving Proposition 4.4, we need to introduce some further notation. For $x > 0$, let $\Pi_{x,s}$ be the class of all polynomials in s real variables with co-ordinatewise degree not exceeding x. The symbol $W\Pi_{x,s}$ denotes the class of all functions of the form $\mathbf{x} \rightarrow \exp(-\|\mathbf{x}\|^2/2)P(\mathbf{x})$, $P \in \Pi_{x,s}$, $\mathbf{x} \in \mathbb{R}^s$. We will make extensive use of the classical Hermite polynomials $\{h_k\}$, defined formally by the generating function (cf. [10, formula (5.5.7)])

$$\exp(2xt - t^2) =: \pi^{1/4} \sum_{k=0}^{\infty} \frac{h_k(x)}{\sqrt{k!}} (\sqrt{2}t)^k, \tag{43}$$

or by means of the Rodrigues' formula (cf. [10, formula (5.5.3)]):

$$\exp(-x^2)h_k(x) = \frac{(-1)^k}{\pi^{1/4}2^{k/2}\sqrt{k!}} \left(\frac{d}{dx}\right)^k \exp(-x^2). \tag{44}$$

The polynomial h_k is of precise degree k, and satisfies (cf. [10, formula (5.5.1)])

$$\int_{\mathbb{R}} h_k(x)h_j(x)\exp(-x^2)dx = \begin{cases} 1, & \text{if } k = j, \ k,j = 0,1,\cdots, \\ 0, & \text{otherwise.} \end{cases} \tag{45}$$

For a multi-integer $\mathbf{k}$, we define

$$h_{\mathbf{k}}(\mathbf{x}) = \prod_{j=1}^{s} h_{k_j}(x_j). \tag{46}$$

Writing $\mathbf{k}! := \prod_{j=1}^{s}(k_j!)$, and using standard multivariate notation, we have

$$\exp(2\mathbf{x} \cdot \mathbf{w} - \|\mathbf{w}\|^2) = \pi^{s/4} \sum_{\mathbf{k} \geq 0} \frac{h_{\mathbf{k}}(\mathbf{x})}{\sqrt{\mathbf{k}!}} (\sqrt{2}\mathbf{w})^{\mathbf{k}}, \qquad \mathbf{x}, \mathbf{w} \in \mathbb{R}^s, \tag{47}$$

and for $\mathbf{k}, \mathbf{j} \geq 0$,

$$\int_{\mathbb{R}^s} h_{\mathbf{k}}(\mathbf{x})h_{\mathbf{j}}(\mathbf{x})\exp(-\|\mathbf{x}\|^2)d\mathbf{x} = \begin{cases} 1, & \text{if } \mathbf{k} = \mathbf{j}, \\ 0, & \text{otherwise.} \end{cases} \tag{48}$$

Proof of Proposition 4.4. We may assume that $\|g\|_{p,\mathbb{R}^s} = 1$. Since $g \in \mathbb{B}(A, C; m, s)$, we conclude from (38) that

$$\sum_{k=1}^{N} |a_k| \leq c_1 \exp(B_1^2 m^2). \tag{49}$$

For $\mathbf{x} \in \mathcal{A}$ and $k \in L_{\mathcal{A}}$, we have from (44) that for $|\mathbf{j}| \le r$,

$$
\begin{aligned}
|\mathbf{D^j} h_{\mathcal{A}}(\mathbf{x})| &\le \sum_{k \in L_{\mathcal{A}}} |a_k| \{\pi^{s/4} 2^{|\mathbf{j}|/2} \sqrt{\mathbf{j}!}\} |h_{\mathbf{j}}(\mathbf{x} - \mathbf{x}_k)| \exp(-\|\mathbf{x} - \mathbf{x}_k\|^2) \\
&\le c \exp\left(-(B_1^2 + b^2)m^2\right) \sum_{k \in L_{\mathcal{A}}} |a_k| |h_{\mathbf{j}}(\mathbf{x} - \mathbf{w}_k)| \exp(-\|\mathbf{x} - \mathbf{w}_k\|^2/2) \\
&\le c \exp\left(-(B_1^2 + b^2)m^2\right) \sum_{k=1}^{N} |a_k| |h_{\mathbf{j}}(\mathbf{x} - \mathbf{w}_k)| \exp(-\|\mathbf{x} - \mathbf{w}_k\|^2/2).
\end{aligned}
$$

Since $\|h_{\mathbf{j}} \exp(-\| \circ \|^2/2)\|_{\infty, \mathbb{R}^s} \le c$, we obtain from (49) that

$$
|\mathbf{D^j} h_{\mathcal{A}}(\mathbf{x})| \le c \exp(-b^2 m^2),
$$

which is (41) with $\|g\|_{p, \mathbb{R}^s} = 1$. The remaining part of the proposition follows by setting $B_2 = \sqrt{s}a + \sqrt{2B_1^2 + 2b^2}$ and $\mathcal{A} := [-am, am]^s$.   □

The following analogue of [7, Proposition 3.5] estimates the rate of approximation of Gaussian networks with weighted polynomials.

Proposition 4.5. *Let $C_1, C_2 > 0$. There exists a constant C_3 depending only on C_1, C_2, with the following property. For any $g \in \mathbb{G}_{\infty, C_1 m, m, s}$, there exists a polynomial $P_g \in \Pi_{C_3 m^2, s}$ such that*

$$
\|g - P_g \exp(-\| \circ \|^2)\|_{p, r, \mathbb{R}^s} \le c_1 \exp(-C_2^2 m^2) \|g\|_{p, \mathbb{R}^s}. \tag{50}
$$

As in [7], the proof of this proposition is immediate from Proposition 4.2, and the following lemma regarding the approximation of basic Gaussians by weighted polynomials.

Lemma 4.6. *For integer $n \ge 1$, $\mathbf{w} \in \mathbb{R}^s$, let*

$$
P_n(\mathbf{x}, \mathbf{w}) := \pi^{s/4} \sum_{0 \le |\mathbf{k}| \le n} \frac{h_{\mathbf{k}}(\mathbf{x}) \exp(-\|\mathbf{x}\|^2)}{\sqrt{\mathbf{k}!}} (\sqrt{2}\mathbf{w})^{\mathbf{k}}. \tag{51}
$$

Then

$$
\| \exp(-\| \circ -\mathbf{w}\|^2) - P_n(\circ, \mathbf{w})\|_{p, r, \mathbb{R}^s} \le c_1 n^c \frac{(\sqrt{2s}\|\mathbf{w}\|)^{n+1} \exp(s\|\mathbf{w}\|^2)}{\sqrt{n!}}. \tag{52}
$$

Proof. Let $|\mathbf{j}| \le r$. In view of the Rodrigues' formula (44),

$$
\mathbf{D^j}\left[h_{\mathbf{k}}(\mathbf{x}) \exp(-\|\mathbf{x}\|^2)\right] = (-\sqrt{2})^{|\mathbf{j}|} \frac{\sqrt{(\mathbf{k}+\mathbf{j})!}}{\sqrt{\mathbf{k}!}} h_{\mathbf{k}+\mathbf{j}}(\mathbf{x}) \exp(-\|\mathbf{x}\|^2).
$$

Therefore, using (47), we obtain

$$
\begin{aligned}
\mathbf{D^j}&\left(\exp(-\|\mathbf{x} - \mathbf{w}\|^2) - P_n(\mathbf{x}, \mathbf{w})\right) \\
&= \pi^{s/4} \sum_{|\mathbf{k}| \ge n+1} (-\sqrt{2})^{|\mathbf{j}|} \frac{\sqrt{(\mathbf{k}+\mathbf{j})!}}{\mathbf{k}!} h_{\mathbf{k}+\mathbf{j}}(\mathbf{x}) \exp(-\|\mathbf{x}\|^2)(\sqrt{2}\mathbf{w})^{\mathbf{k}}. \tag{53}
\end{aligned}
$$

Using the arithmetic-geometric inequality, we see that $|\mathbf{w}^{\mathbf{k}}| \leq \|\mathbf{w}\|^{|\mathbf{k}|}$. Since (cf. [6, Theorem 6.2.10])

$$\|h_{\mathbf{k}+\mathbf{j}} \exp(-\| \circ \|^2/2)\|_{p,\mathbb{R}^s} \leq c_1 |\mathbf{k}|^c,$$

we see that $\|h_{\mathbf{k}+\mathbf{j}} \exp(-\| \circ \|^2)\|_{p,\mathbb{R}^s} \leq c_1 |\mathbf{k}|^c$ as well. Therefore, (53) implies that

$$\left\|\mathbf{D}^{\mathbf{j}}\left(\exp(-\| \circ -\mathbf{w}\|^2) - P_n(\circ,\mathbf{w})\right)\right\|_{p,\mathbb{R}^s} \leq c_1 \sum_{|\mathbf{k}| \geq n+1} \frac{|\mathbf{k}|^c (\sqrt{2}\|\mathbf{w}\|)^{|\mathbf{k}|}}{\sqrt{\mathbf{k}!}}$$

$$= c_1 \sum_{j=n+1}^{\infty} \frac{j^c (\sqrt{2}\|\mathbf{w}\|)^j}{\sqrt{j!}} \sum_{|\mathbf{k}|=j} \left(\frac{j!}{\mathbf{k}!}\right)^{1/2} \leq c_1 \sum_{j=n+1}^{\infty} \frac{j^c (\sqrt{2s}\|\mathbf{w}\|)^j}{\sqrt{j!}}.$$

The estimate (52) now follows from [7, Lemma 3.5]. □

Proof of Proposition 4.5. This proof is verbatim the same as that of [7, Proposition 3.5], except that we use Lemma 4.6 in place of [7, Lemma 3.4]. We omit the details. □

Lemma 4.7. *Let $n \geq 1$ be an integer, and $C_4 > 0$. There exists a positive constant C_5 depending only on C_4 such that for every $P \in \Pi_{n,s}$,*

$$\|P \exp(-\| \circ \|^2)\|_{p,r,\mathbb{R}^s \setminus [-C_5\sqrt{n}, C_5\sqrt{n}]^s} \leq \exp(-C_4 n)\|P \exp(-\| \circ \|^2)\|_{p,\mathbb{R}^s}. \quad (54)$$

Proof of Proposition 4.1. In this proof, we will freely use the following fact, without referring to it explicitly. If F and G are sufficiently smooth functions, then the Leibniz formula implies that for any measurable subset $S \subset \mathbb{R}^s$,

$$\|FG\|_{p,r,S} \leq c\|F\|_{\infty,r,S}\|G\|_{p,r,S}. \quad (55)$$

In this proof, all the constants will depend upon $A, C, \beta, \gamma, r, p, s$, and φ. Without loss of generality, we may assume that $\gamma > B_1$, where B_1 is defined in Proposition 4.2. By a repeated application of Theorem 3.5, we obtain a polynomial $P \in \Pi_{2m^2,s}$ such that

$$\|\phi - P \exp(-\| \circ \|^2)\|_{\infty,r,\mathbb{R}^s} \leq cm^{-\beta-r}. \quad (56)$$

We will write $\mathcal{P} := P \exp(-\| \circ \|^2)$. Using Lemma 4.7 with $C_4 = 3\gamma^2$, we find $a > 1$ such that

$$\|\mathcal{P}\|_{\infty,r,\mathbb{R}^s \setminus [-am,am]^s} \leq c_1 \exp(-3\gamma^2 m^2). \quad (57)$$

Next, we use Proposition 4.4 with $\sqrt{3}\gamma$ in place of b to obtain a number B_2, the set L and the subnetwork h of g such that

$$\|h\|_{p,r,[-am,am]^s} \leq c_1 \exp(-3\gamma^2 m^2)\|g\|_{p,\mathbb{R}^s}. \quad (58)$$

Clearly, the network $g - h$ contains at most cm^{2s} neurons. In view of (39), we observe that

$$\|g - h\|_{p,r,\mathbb{R}^s} \leq c \exp(\gamma^2 m^2)\|g\|_{p,\mathbb{R}^s}. \quad (59)$$

Since $a > 1$ and $\phi(\mathbf{x}) = 0$ outside of $[-m,m]^s$, we see from (58) that

$$\|\phi h\|_{p,r,\mathbb{R}^s} \leq c \exp(-3\gamma^2 m^2)\|g\|_{p,\mathbb{R}^s}. \quad (60)$$

From (58) and the fact that $\|\mathcal{P}\|_{\infty,r,\mathbb{R}^s} \le c$, we obtain that

$$\|\mathcal{P}h\|_{p,r,[-am,am]^s} \le c\exp(-3\gamma^2 m^2)\|g\|_{p,\mathbb{R}^s}. \tag{61}$$

In view of (57) and (40), we see that

$$\|\mathcal{P}g\|_{p,r,\mathbb{R}^s\setminus[-am,am]^s} \le c\exp(-3\gamma^2 m^2)\|g\|_{p,r,\mathbb{R}^s} \le c_1\exp(-2\gamma^2 m^2)\|g\|_{p,\mathbb{R}^s}. \tag{62}$$

Similarly, (57) and (59) imply that

$$\|(g-h)\mathcal{P}\|_{p,r,\mathbb{R}^s\setminus[-am,am]^s} \le c\exp(-2\gamma^2 m^2)\|g\|_{p,\mathbb{R}^s}.$$

Consequently,

$$\|\mathcal{P}h\|_{p,r,\mathbb{R}^s\setminus[-am,am]^s} \le c\exp(-2\gamma^2 m^2)\|g\|_{p,\mathbb{R}^s}.$$

Together with (61), this implies that

$$\|\mathcal{P}h\|_{p,r,\mathbb{R}^s} \le c\exp(-2\gamma^2 m^2)\|g\|_{p,\mathbb{R}^s}.$$

Therefore, (60) implies that

$$\|(\phi-\mathcal{P})h\|_{p,r,\mathbb{R}^s} \le c\exp(-2\gamma^2 m^2)\|g\|_{p,\mathbb{R}^s}.$$

Since (cf. (56), (62))

$$\|(\phi-\mathcal{P})g\|_{p,r,\mathbb{R}^s} \le c_1 m^{-\beta-r}\|g\|_{p,r,[-am,am]^s} + c_2\exp(-2\gamma^2 m^2)\|g\|_{p,\mathbb{R}^s},$$

we conclude that

$$\|(\phi-\mathcal{P})(g-h)\|_{p,r,\mathbb{R}^s} \le c_1 m^{-\beta-r}\|g\|_{p,r,[-am,am]^s} + c_2\exp(-2\gamma^2 m^2)\|g\|_{p,\mathbb{R}^s}. \tag{63}$$

Since $g-h \in \mathbb{G}_{cm^{2s},c_1 m,m,s}$, we may use Proposition 4.5 to obtain a polynomial $Q \in \Pi_{cm^2,s}$, such that with $\mathcal{Q} = Q\exp(-\|\circ\|^2)$,

$$\|g-h-\mathcal{Q}\|_{p,r,\mathbb{R}^s} \le c_1\exp(-3\gamma^2 m^2)\|g-h\|_{p,\mathbb{R}^s} \le c_2\exp(-2\gamma^2 m^2)\|g\|_{p,\mathbb{R}^s}. \tag{64}$$

Since

$$\|\phi g - \mathcal{P}\mathcal{Q}\|_{p,r,\mathbb{R}^s} \le \|\phi h\|_{p,r,\mathbb{R}^s} + \|(\phi-\mathcal{P})(g-h)\|_{p,r,\mathbb{R}^s} + \|(g-h-\mathcal{Q})\mathcal{P}\|_{p,r,\mathbb{R}^s},$$

the estimates (60), (63), (64) lead to

$$\begin{aligned}\|\phi g - \mathcal{P}\mathcal{Q}\|_{p,\mathbb{R}^s} &\le \|\phi g - \mathcal{P}\mathcal{Q}\|_{p,r,\mathbb{R}^s}\\ &\le c_1 m^{-\beta-r}\|g\|_{p,r,[-am,am]^s} + c_2\exp(-2\gamma^2 m^2)\|g\|_{p,\mathbb{R}^s}.\end{aligned} \tag{65}$$

A repeated application of (8), keeping in mind the fact that the derivative of a weighted polynomial in $W\Pi_{n,s}$ is in $W\Pi_{n+1,s}$, implies that $\|\mathcal{P}\mathcal{Q}\|_{p,r,\mathbb{R}^s} \le cm^r\|\mathcal{P}\mathcal{Q}\|_{p,\mathbb{R}^s}$. The estimate (37) now follows easily from (65). □

Proof of Theorem 2.1. We prove the theorem for $1 \le p < \infty$. The same proof works also in the case $p = \infty$, but is simpler. It is enough to prove the theorem for $m \ge c$, where c is some constant depending only on A, C, r, p, s. Let $\varphi : \mathbb{R} \to [0,\infty)$ be an infinitely differentiable function on $\mathbb{R}$, such that $\varphi(x) = 1$ if $|x| \le 1/2$ and $\varphi(x) = 0$ if $|x| \ge 1$. Let

$$\psi(x) := \frac{\varphi(x)}{\sum_{k\in\mathbb{Z}}\varphi(x-k)}$$

and $\phi(\mathbf{x}) := \prod_{j=1}^{s} \psi(x_j/m)$. Then ϕ is supported on $[-m, m]^s$, $0 < c_1 \leq \phi(\mathbf{x}) \leq 1$ if $\mathbf{x} \in [-m/2, m/2]^s$, and $\sum_{\mathbf{k} \in \mathbb{Z}^s} \phi(\mathbf{x} - \mathbf{k}m) = 1$. Let $1 \leq p < \infty$. Since $D^{\mathbf{j}}g = \sum_{\mathbf{k} \in \mathbb{Z}^s} D^{\mathbf{j}}(\phi(\circ - \mathbf{k}m)g)$ for each $\mathbf{j}$, and only finitely many of the supports of $D^{\mathbf{j}}(\phi(\circ - \mathbf{k}m)g)$ may intersect each other (the number being dependent only on s), we conclude that

$$\|g\|_{p,r,\mathbb{R}^s}^p \sim c \sum_{\mathbf{k} \in \mathbb{Z}^s} \|\phi(\circ - \mathbf{k}m)g\|_{p,r,\mathbb{R}^s}^p = c \sum_{\mathbf{k} \in \mathbb{Z}^s} \|\phi g_{\mathbf{k}}\|_{p,r,[-m,m]^s}^p, \qquad (66)$$

where each

$$g_{\mathbf{k}} := g(\circ + \mathbf{k}m) \in \mathbb{B}(A, C; m, s). \qquad (67)$$

Now, let $\gamma^2 > 2A$, and in this proof only, B_1 be the constant defined in Proposition 4.2, $b = \gamma\sqrt{3/p}$, and $B_3 := \sqrt{s} + \sqrt{2B_1^2 + 2b^2}$ (cf. Proposition 4.4). In this proof only, let $\mathcal{C}$ be the set of all the centers of g,

$$\mathbb{Z}^s \setminus S := \{\mathbf{k} \in \mathbb{Z}^s \; : \; \|\mathbf{w} - \mathbf{k}m\| \geq B_3 m, \; \mathbf{w} \in \mathcal{C}\}, \quad \mathcal{A} := \bigcup_{\mathbf{k} \in \mathbb{Z}^s \setminus S} (\mathbf{k}m + [-m, m]^s).$$

With our choice of $\mathcal{A}$ and B_3, it is clear that the subnetwork $h_{\mathcal{A}}$ defined in Proposition 4.4 is the whole network g. We observe also that only finitely many cubes among all the cubes comprising $\mathcal{A}$ intersect each other, the number depending only on s. Therefore, (41) implies that

$$\sum_{\mathbf{k} \in \mathbb{Z}^s \setminus S} \|\phi g_{\mathbf{k}}\|_{p,r,\mathbb{R}^s}^p \sim \|g\|_{p,r,\mathcal{A}}^p \leq c\exp(-3\gamma^2 m^2)\|g\|_{p,\mathbb{R}^s}^p. \qquad (68)$$

In view of (66) and (68), we obtain

$$\|g\|_{p,r,\mathbb{R}^s}^p \leq c \sum_{\mathbf{k} \in S} \|\phi g_{\mathbf{k}}\|_{p,r,[-m,m]^s}^p + c_1 \exp(-3\gamma^2 m^2)\|g\|_{p,\mathbb{R}^s}^p. \qquad (69)$$

In view of the fact that the total number of centers in g is at most $C \exp(Am^2)$, the cardinality of S satisfies

$$|S| \leq cm^{2s} \exp(Am^2) \leq c_1 \exp(2Am^2) \leq c_2 \exp(\gamma^2 m^2). \qquad (70)$$

We now use Proposition 4.1 with each $g_{\mathbf{k}}$, $\mathbf{k} \in S$, and with some $\beta > 0$ to obtain (using (69) and (70)) that

$$\|g\|_{p,r,\mathbb{R}^s}^p \leq c\left\{ m^{rp} \sum_{\mathbf{k} \in S} \|\phi g_{\mathbf{k}}\|_{p,\mathbb{R}^s}^p + m^{-\beta p} \sum_{\mathbf{k} \in S} \|g_{\mathbf{k}}\|_{p,r,[-cm,cm]^s}^p + \|g\|_{p,\mathbb{R}^s}^p \right\}. \qquad (71)$$

Clearly, $\sum_{\mathbf{k} \in S} \|\phi g_{\mathbf{k}}\|_{p,\mathbb{R}^s}^p \leq c\|g\|_{p,\mathbb{R}^s}^p$. Since only finitely many cubes of the form $\mathbf{k}m + [-cm, cm]^s$ intersect each other, we also conclude that $\sum_{\mathbf{k} \in S} \|g_{\mathbf{k}}\|_{p,r,[-cm,cm]^s}^p \leq c_1\|g\|_{p,r,\mathbb{R}^s}^p$. So, (71) implies that

$$\|g\|_{p,r,\mathbb{R}^s}^p \leq c\left\{ m^{rp}\|g\|_{p,\mathbb{R}^s}^p + m^{-\beta p}\|g\|_{p,r,\mathbb{R}^s}^p \right\},$$

and hence, for sufficiently large m,

$$\|g\|_{p,r,\mathbb{R}^s}^p \leq \frac{cm^{rp}}{1 - cm^{-\beta p}}\|g\|_{p,\mathbb{R}^s}^p \leq c_1 m^{rp}\|g\|_{p,\mathbb{R}^s}^p. \qquad \square$$

Acknowledgement

The research of this author was supported, in part, by grant DMS-0204704 from the National Science Foundation and grant W911NF-04-1-0339 from the U.S. Army Research Office. The author would also like to thank the referee for his/her patient and careful reading of the manuscript, and suggestions for the improvement of the presentation in the paper.

References

[1] D. Benko, Approximation by weighted polynomials, J. Approx. Theory, **120** (2003), 153–182.

[2] R.A. DeVore and G.G. Lorentz, *Constructive approximation*, Springer Verlag, Berlin, 1993.

[3] T. Erdélyi, Sharp Bernstein-type inequalities for linear combinations of shifted Gaussians with applications,
www.math.tamu.edu/ tamas.erdelyi/papers-online/list.html.

[4] D. Gaier, Polynomial approximation of piecewise analytic functions, J. Anal., **4** (1996), 67–79.

[5] R. Labahn, On an interesting property of binomial coefficients, Rostock Math. Kolloq., **29** (1986), 21–24.

[6] H.N. Mhaskar, *Introduction to the theory of weighted polynomial approximation*, World Scientific, Singapore, 1996.

[7] H.N. Mhaskar, When is approximation by Gaussian networks necessarily a linear process?, Neural Networks, **17** (2004), 989–1001.

[8] H.N. Mhaskar and J. Prestin, On a sequence of fast decreasing polynomial operators, Applications and computation of orthogonal polynomials (Oberwolfach, 1998), 165–178, Internat. Ser. Numer. Math., 131, Birkhäuser, Basel, 1999.

[9] F. Narcowich and J.D. Ward, Norm estimates for the inverses of a genral class of scattered data radial function interpolation matrices, Journal of Approximation Theory, **69** (1992), 84–109.

[10] G. Szegö, *Orthogonal polynomials*, Amer. Math. Soc. Colloq. Publ. **23**, Amer. Math. Soc., Providence, 1975.

H.N. Mhaskar
Department of Mathematics
California State University
Los Angeles, California, 90032, USA
e-mail: hmhaska@calstatela.edu

Trends and Applications in Constructive Approximation
(Eds.) M.G. de Bruin, D.H. Mache & J. Szabados
International Series of Numerical Mathematics, Vol. 151, 181–194
© 2005 Birkhäuser Verlag Basel/Switzerland

TS Control – The Link between Fuzzy Control and Classical Control Theory

Kai Michels

Abstract. Fuzzy controller can be approximated or generalized respectively by replacing the fuzzy sets in the rule conclusions by real numbers or functions. Such a controller is called a TS controller and can be seen as a classical gain scheduling controller. Therefore, TS control can be interpreted as fuzzy and classical control as well. Besides this, for this type of control during the last years there were methods developed, that make it interesting for practical applications.

The objective of this paper is the introduction of TS control and the discussion of its position at the border between fuzzy and classical control, but also the presentation of suitable methods, approaches and fields of application for this controller type.

1. Introduction

Takagi-Sugeno control was introduced by T. Takagi and M. Sugeno in the middle of the 80s as an additional version of fuzzy control. In principle, it is nothing else but gain scheduling control, which is well known in control theory since decades. But in the following years, in the field of fuzzy control this controller type received intensive research activities, and since the end of the 90s a controller design method was developed, that makes TS control very interesting for practical applications. On the one hand, this state space based method can thoroughly be seen as a classical approach, but on the other hand, a TS controller can also be interpreted as a generalized fuzzy controller, and therefore it can be said, that TS control is a link between fuzzy control and classical control theory.

This paper starts with a short description of the conventional fuzzy controller. This gives the basis to discuss conventional fuzzy control and to point out the special position and advantages of TS control later. In the rest of the paper, TS control with its analysis and design methods is introduced in detail (further details see [5]).

2. Fuzzy control

Let the starting point be a conventional fuzzy controller for temperature control
with the temperature x as input and the actuating variable y as controller output,
given by the fuzzy rules

$$R_1 : \qquad \text{IF } x \text{ IS } cold \text{ THEN } y \text{ IS } big \tag{1}$$

$$R_2 : \qquad \text{IF } x \text{ IS } hot \text{ THEN } y \text{ IS } small \tag{2}$$

The vague expressions *cold, hot, big,* and *small* are defined by the fuzzy sets of
Figure 1.

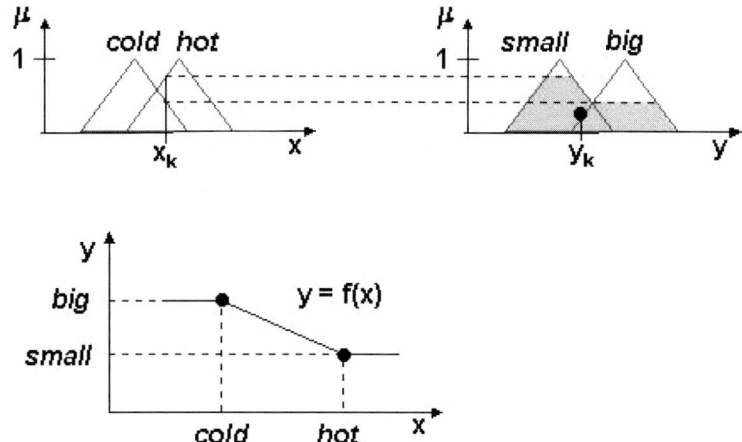

FIGURE 1. Black Forest test.

With a measured temperature x_k at time $t = t_k$ the two fuzzy rules are
activated according to the membership degree of x_k to the fuzzy sets *cold* and *hot*.
The membership degree is equivalent to the value of the respective membership
function at $x = x_k$. In the example of Figure 1 x_k belongs more to the set *hot*
of the hot temperatures than to the set *cold* of the cold temperatures. Following
from that, rule No. 2 is stronger activated than rule No. 1, and the output fuzzy
set *small* of the second rule is higher weighted in the overall fuzzy set. This output
fuzzy set of the entire controller is the set of all possible output values for the input
value x_k, and it is computed by association of all differently activated output fuzzy
sets of all rules.

Then, one value y_k of this controller output fuzzy set has to be determined
by the so-called defuzzification. This value will be the controller response to the
input value x_k. Normally, for defuzzification the center-of-gravitiy method is used,
that means, the y coordinate of the center of gravity of the fuzzy set is used as
defuzzification result.

Considering the transfer function of the entire controller, it can easily be seen, that for any low temperature, that only belongs to the set *cold*, the controller output is always the same, because for each low temperature only the output set *big* is activated more or less while no other set is activated, and therefore the defuzzification result is always the y coordinate of the center of the triangle *big*. Analogously, the controller output for high temperatures is always the y coordinate of the center of the triangle *small*. For all other temperatures in between, that belong to the sets *cold* and *hot* more or less as well, the controller output is always between *big* and *small*. For other input values, that do not belong to any input fuzzy set, the controller output is undefined. But this case does not have to be taken into consideration here, as in practical applications it is easy to define a sufficient number of input fuzzy sets and rules, so that there is always a well-defined controller output for any possible input.

Summarized, the controller transfer function can be defined by a characteristic curve as shown in Figure 1, although the curve between the supporting points is not necessarily linear as in the figure. It depends on the shape of the fuzzy sets and the defuzzification method.

It follows, that the fuzzy controller is a characteristic curve controller, whose supporting points are defined by the center points of the input fuzzy sets, and the type of interpolation between the supporting points is given by the shape of the fuzzy sets and the defuzzification method. This understanding corresponds to the interpretation of fuzzy control with similarity relations as shown in [5].

It is obvious, that the SISO (single input single output) case discussed above with only two fuzzy rules can easily be extended to the MIMO (multi input multi output) case with any number of fuzzy rules. Therefore, regarding to its transfer behaviour, a general fuzzy controller is a characteristic field controller, whose output values depend clearly on its input values. Operations like differentiation or integration of input or output variables can be added to the controller outside the fuzzy system to enable also dynamical transfer behaviour like PI control.

But such a combination of characteristic field and added dynamical operations can be used to represent any classical controller transfer behaviour, too. Therefore, regarding to the transfer behaviour, there is absolutely no difference between a fuzzy and a classical controller. Differences exist only in the design methods and in the representation with fuzzy rules on the fuzzy side and analytical functions on the classical side. Following from that, any discussion about advantages and disadvantages of fuzzy control must concentrate on only these two aspects.

First, it should be noticed that classical controller design has got some principle advantages. The first step, developing an analytical model (differential equations) of the plant to be controlled, as well as the second step, deriving an analytical controller function based on the differential equation system of the plant, can be performed systematically with established methods. Using these methods, stability and selected control performance indices can be guaranteed for the closed loop system implicitly. Model uncertainties can be taken into consideration with norm optimal control methods, so that even for these cases stability can be guaranteed.

In contrast to this, the design of a fuzzy controller is normally heuristic without strict procedure, what may be time consuming and can even make the controller design impossible, if the structure of the plant to be controlled has a certain level of complexity. Besides that, for a heuristically designed controller stability cannot be guaranteed. Since the end of the 80s the systematic fuzzy controller design as well as stability analysis have been object of intensive research, but useful approaches for practical applications have only been developed for fuzzy controllers, that are completely represented as characteristic fields and that do not contain fuzzy sets any more. TS control as discussed in this paper belongs to these approaches.

Another point to be discussed is the robustness. During the 90s it was stated, that a fuzzy controller is more robust than a classical controller because of its principal vagueness. But as explained before, a fuzzy controller has got a well-defined transfer behaviour like any other controller, and therefore its robustness can be discussed like the robustness of any other controller. Before this discussion is started, one point must be mentioned first: The use of the concept *robustness* does only make sense, if it is possible to quantify, how large the differences between the nominal plant model and the real plant may be, so that the controller, whose design is based on the plant model, is also able to stabilize the real plant. The unquantified attribute *robust* holds for any practical controller more or less, and therefore it gives no useful information.

For conventional fuzzy controllers with fuzzy rules, that are designed heuristically for a plant, of which no model exists, it is in principle impossible to quantify their robustness, because for the quantification a model is always needed. But even if a model exists, fuzzy controller robustness can only be investigated by simulation runs with different model parameters, because the controller is defined by fuzzy rules, that do not enable an analytical computation of stability or robustness measures. Obviously, a handful of simulation runs is no proof of robustness. However in classical linear control theory ([6]) there exist controller design methods, in which the uncertainty of any single plant model parameter can be selected and the design method will lead to a controller that can stabilize the plant for any possible parameter constellation within the selected ranges of uncertainty.

Finally, there is the clearness of the fuzzy controller. Undoubtedly, a fuzzy rule is easier to understand especially for non control specialists than a transfer function of a PID controller or a coefficient matrix of a state space controller. But for more complex plants the number of fuzzy rules can increase up to several hundreds. In this case, the clearness of each single fuzzy rule still remains, but the complete controller cannot be overlooked any more. The effect of a change of a certain input variable can only be predicted by computing the complete controller algorithm with computing each fuzzy rule, superimposition of all output fuzzy sets and defuzzification. On the other hand, for a control specialist it is always possible to assess the connection between input and output variables of a PID controller cascade or a state space controller matrix.

Summarized, all points mentioned before give a clear vote for classical control, and therefore the following has to be stated: If an analytical model of the plant is available and if the structure of the plant allows the design of a classical controller, classical control should be preferred to fuzzy control. On the other hand, fuzzy control is useful if

- there is no analytical plant model available.
- the structure of the plant model makes any classical controller design impossible.
- the control goals are defined imprecisely, e.g., the demand for *smooth* switching of an automatic gear in a car ([8]).
- the plant and the necessary control strategy are so simple, that the design of a fuzzy controller with a few fuzzy rules is easier or faster than any classical approach with plant modelling and analytical controller design.

Beside these points fuzzy systems can have a wide range of applications on a higher control level outside any internal closed control loop, e.g., trajectory planning for robots or autonomous vehicles, adaptation and parameterization of classical controllers, control loop supervision or fault diagnosis. In these fields there exist no limits for the application and variety of fuzzy systems. As the fuzzy controller is not working in the closed loop, there are no control-specific problems like the stability problem. For these applications other problems are more relevant, e.g., the protection against a system crash or the question, if all possible situations are covered by the fuzzy rules.

3. TS control

The disadvantages of conventional fuzzy control discussed in the preceding chapter result in essence from the missing systematic design method. But as mentioned before, for a fuzzy controller, that is represented by characteristic fields, since the late 90s there exist systematic design approaches with guaranteed stability. With this feature, theses approaches are equal to any classical approach, but in addition to that, they cover fields of application, that were reserved for fuzzy control until now.

The most interesting of these approaches from a practical point of view is the so-called TS control. This approach was introduced by T. Takagi and M. Sugeno [9, 10] in the middle of the 80s and received further development throughout the 90s by different authors. Although TS control was originally defined just as an additional version of fuzzy control, TS control can also be understood as generalization of fuzzy control, as will be shown in this chapter. On the other hand, TS control corresponds to the gain scheduling method, that is well known in classical control theory since many years. Therefore, TS control is not only interesting because it is easy to use and can be applied to a great variety of systems, but it can also be seen as a link between fuzzy control and classical control theory.

A TS controller is given by fuzzy rules of the form

$$R_i : \quad \text{IF } z_1 \text{ IS } \mu_{R_i}^{(1)} \text{ AND } \ldots \text{ AND } z_l \text{ IS } \mu_{R_i}^{(l)} \text{ THEN } y = f_i(x_1, \ldots, x_n). \quad (3)$$

The $\mu_{R_i}^{(i)}$ in the premises of the rules are fuzzy sets like in any other fuzzy controller. But the conclusion of each rule consists of an analytical function of system variables x_i, that do not necessarily have to be equal to the input variables z_i of the fuzzy rules. The basic idea of TS control is, that in a vague area, that is defined by the premise of a rule, the function in the conclusion is a good description of the output variable. If linear functions are used, the transfer behaviour of the controller is locally (in vague areas) described by linear models.

It remains to be cleared, how the conclusions have to be supcrimposed, depending on the degrce of activation of their fuzzy rules. In a conventional fuzzy controller, the output fuzzy set of each rule is weighted according to the degree of activation of that rule and associated with all output fuzzy sets of all rules, to create one common controller output fuzzy set, of which one single value has to be selected by defuzzification as controller output value.

For TS control, there is the same principle. The difference is, that here is no superimposition of differently weighted fuzzy sets. As here each rule already gives a real value, these values only have to be added to a weighted sum, with the weights corresponding to the degrees of activation of each rule:

$$y = \frac{\sum_i c_i(z_1, \ldots, z_l) \cdot f_i(x_1, \ldots, x_n)}{\sum_i c_i(z_1, \ldots, z_l)}$$

$$y = \sum_i k_i(z_1, \ldots, z_l) \cdot f_i(x_1, \ldots, x_n)$$

$$\text{with} \qquad k_i(z_1, \ldots, z_l) = \frac{c_i(z_1, \ldots, z_l)}{\sum_i c_i(z_1, \ldots, z_l)} \qquad (4)$$

$c_i(z_1, \ldots, z_l)$ is the degree of activation of rule R_i, and $k_i(z_1, \ldots, z_l)$ is the weighting factor of the function f_i. Obviously it holds

$$\sum_i k_i = 1 \qquad \text{and} \qquad 0 \leq k_i \leq 1 \qquad (5)$$

With these formulas, it is easy to see that a TS controller is a gain scheduling controller. The f_i are the controller transfer functions for single operating points, that are activated by the weighting factors depending on the operating point.

Besides this, TS control can also be seen as a generalization of conventional fuzzy control, because for constant functions f_i the TS controller is a characteristic field, whose values f_i and weighting factors $k_i(z_1, \ldots, z_l)$ can be chosen in a way, so that any conventional fuzzy controller transfer behaviour can be represented in the form (4).

Furthermore, in practical applications the characteristic field representation of a fuzzy controller with linear interpolation between the supporting points is the usual way to reduce computation time in every time step. While for the computation of a conventional fuzzy controller in every single time step all rules have to be worked through, the output fuzzy sets have to be superimposed, and the output value has to be computed by defuzzification, the computational effort for a characteristic field is reduced to a simple computation according to eq. (4). And the difference between the original fuzzy controller curve and the linearly interpolated characteristic curve can be kept small for a sufficient number of supporting points.

Because of these points TS control has got a particular position in control theory, as it belongs to the fields of fuzzy and classical control as well, and besides that it can be seen as a generalization of conventional fuzzy control.

In the following only one specific type of TS control will be discussed. At first, the considerations shall be extended to multi input multi output (MIMO) systems. This is easy, because the scalar functions f_i just have to be replaced by the vector functions $\mathbf{f}_i$. Furthermore, the variables of the functions $\mathbf{f}_i$ shall be restricted to state variables of the plant to be controlled, and the functions themselves shall be only linear functions of these state variables. With these restrictions the TS controller is given by

$$\mathbf{u} = \sum_i k_i(\mathbf{z}) \, \mathbf{F}_i \mathbf{x} \tag{6}$$

$\mathbf{z} = (z_1, \ldots, z_l)^T$ is the vector of input variables, $\mathbf{u} = (u_1, \ldots, u_m)^T$ is the controller output vector (actuating variable), $\mathbf{x} = (x_1, \ldots, x_n)^T$ is the vector of state variables, and $\mathbf{F}_i$ are linear coefficient matrices of dimension $m \times n$. Obviously, the TS controller of eq. (6) is a superimposition of linear state space controllers, whose weighting factors depend on the operating points.

The same type of representation can be used for plants:

$$\dot{\mathbf{x}}(t) = \sum_i k_i(\mathbf{z}(t)) \left[\mathbf{A}_i \mathbf{x}(t) + \mathbf{B}_i \mathbf{u}(t) \right] \tag{7}$$

$\mathbf{A}_i$ of dimension $n \times n$ and rank n is the system matrix for operating point i, and $\mathbf{B}_i$ of dimension $n \times m$ is the input matrix of the plant. Such a TS model can be used to describe any linear or non-linear transfer behaviour with any degree of exactness depending on the number of supporting points, except systems containing hysteresis or time delay effects. The TS model consists of linear models, whose output values are superimposed with different weighting factors depending on the input values.

A TS model can be obtained for instance with an identification algorithm. In the first step, suitable operating points have to be chosen, that cover the complete range of operation with sufficient resolution, that means, it must be possible to represent the nonlinear transfer characteristic of the plant by the superimposition of linear models at the operating points. After that, for every operating point a classical system identification algorithm has to be performed, e.g., least error squares methods, to compute the linear model matrices $\mathbf{A}_i, \mathbf{B}_i$ for that operating point.

If a nonlinear analytical model of the plant already exists, but a classical controller design is impossible because of the unsuitable model structure, it can be useful to transfer this model into a TS model and to design a TS controller for that plant. One approach for that strategy is discussed in [13] and [5].

Inserting the TS controller equation (6) into (7) gives the state space equation of the closed loop system in TS representation:

$$\dot{\mathbf{x}}(t) \;=\; \sum_i \sum_j k_i(\mathbf{z}(t))k_j(\mathbf{z}(t))\left[\mathbf{A}_i + \mathbf{B}_i\mathbf{F}_j\right]\mathbf{x}(t) \tag{8}$$

$$\dot{\mathbf{x}}(t) \;=\; \sum_i \sum_j k_i(\mathbf{z}(t))k_j(\mathbf{z}(t))\mathbf{G}_{ij}\mathbf{x}(t) \tag{9}$$

with $\mathbf{G}_{ij} = \mathbf{A}_i + \mathbf{B}_i\mathbf{F}_j$. Re-indexing with $\mathbf{A}_{g,l} = \mathbf{G}_{ij}$ and $k_l(\mathbf{z}(t)) = k_i(\mathbf{z}(t))k_j(\mathbf{z}(t))$ leads to

$$\dot{\mathbf{x}}(t) = \sum_l k_l(\mathbf{z}(t))\mathbf{A}_{g,l}\mathbf{x}(t) \tag{10}$$

The index g shall make clear, that here the system matrices stand for the closed loop system and not just for the model of the plant like the $\mathbf{A}_i$ before.

4. Stability analysis of TS systems

With the model (10) of the closed loop system, the stability of a system in TS representation can be investigated. The following theorem holds (see [12]):

Theorem 4.1. *A time-continuous system of the form*

$$\dot{\mathbf{x}} = \sum_i k_i(\mathbf{z}(t))\mathbf{A}_i\mathbf{x}(t) \tag{11}$$

possesses a globally asymptotically stable rest position $\mathbf{x} = \mathbf{0}$, *if there exists a common positive definite matrix* $\mathbf{P}$ *for all partial systems* $\mathbf{A}_i$, *so that*

$$\mathbf{M}_i = \mathbf{A}_i^T\mathbf{P} + \mathbf{P}\mathbf{A}_i \tag{12}$$

is negative definite for all i *(*$\mathbf{M}_i < 0$*)*.

For the proof, the Direct Method of Lyapunov shall be used. In the first step, a Lyapunov function $V = \mathbf{x}^T\mathbf{P}\mathbf{x}$ with positive definite matrix $\mathbf{P}$ shall be defined. This scalar function is zero for $\mathbf{x} = \mathbf{0}$, and its values get bigger with increasing distance to the origin. If it can be shown, that the time derivative $\dot{V}(\mathbf{x}(t))$ of that function is negative for all $\mathbf{x}(t) \neq \mathbf{0}$, it follows, that for the actual system the distance between the state vector and the origin is decreasing until the rest position $\mathbf{0}$ is reached, independently from the starting state $\mathbf{x}(t = 0)$. But this means global asymptotic stability for the rest position $\mathbf{x} = \mathbf{0}$.

For the given system, the time derivative of the Lyapunov function is

$$\begin{aligned}
\dot{V} &= \dot{\mathbf{x}}^T\mathbf{P}\mathbf{x} + \mathbf{x}^T\mathbf{P}\dot{\mathbf{x}} \\
&= \sum_i k_i\mathbf{x}^T\mathbf{A}_i^T\mathbf{P}\mathbf{x} + \sum_i k_i\mathbf{x}^T\mathbf{P}\mathbf{A}_i\mathbf{x} \\
&= \sum_i k_i\mathbf{x}^T(\mathbf{A}_i^T\mathbf{P} + \mathbf{P}\mathbf{A}_i)\mathbf{x} < 0
\end{aligned} \tag{13}$$

From the condition of the theorem it follows, that all terms in brackets are negative definite, and therefore every term of the sum is negative. Therefore, $\dot{V}(\mathbf{x}(t)) < 0$ and the system is stable.

Until the end of the 90s the problem to apply theorem 4.1 was the necessity of a non-systematic search for a suitable, common positive definite matrix $\mathbf{P}$, with which all terms (12) would be negative definite. This is not a specific problem of this theorem, it is a principle problem of the Direct Method of Lyapunov: If a suitable Lyapunov function can be found, the proof of stability with this method is absolutely easy. But if no Lyapunov function is found, no statement regarding the stability is possible, because the fact, that one could not find that function, does not necessarily mean, that is does not exist.

Finally, in the 90s a systematic (numerical) approach to solve this problem was presented, based on linear matrix inequalities (LMI's). With this approach it is possible, to check the existence of the matrix $\mathbf{P}$ at least with numerical methods clearly.

5. LMI algorithms

The basic problem and starting point of all solution algorithms in the theory of linear matrix inequalities (LMI's) can be formulated in the following way (see [7]): A symmetrical matrix is given, whose coefficients depend linearly on free parameters. The question is, if these free parameters can be chosen in a way, so that the symmetrical matrix will be negative definite. With $\mathbf{p}$ is the vector of free parameters, and $\mathbf{H}(\mathbf{p})$ the symmetrical matrix, whose coefficients depend linearly on the free parameters, the formal definition of the basic problem is:

Does a vector $\mathbf{p}$ exist, that makes

$$\mathbf{H}(\mathbf{p}) < 0 \qquad (14)$$

negative definite?

Because of the linearity of the matrix function $\mathbf{H}(\mathbf{p})$ the solution set for $\mathbf{p}$ is always convex, and therefore, the question of the existence of a solution can be answered with numerical algorithms clearly. One example for such algorithms is the Matlab LMI Toolbox, that can answer the question of the existence, but besides that, it can also compute a solution $\mathbf{p}$, if one exists.

With such a tool, a given problem just has to be transferred to the form (14). For this step, the following remarks give some support:

- An LMI of the form $\mathbf{G}(\mathbf{p}) > 0$ is equivalent to (14) with $\mathbf{G} = -\mathbf{H}$.
- A block diagonal matrix

$$\begin{pmatrix} \mathbf{H}_1 & & & \\ & \mathbf{H}_2 & & \\ & & \ddots & \\ & & & \mathbf{H}_n \end{pmatrix} = \mathrm{diag}(\mathbf{H}_1, \ldots, \mathbf{H}_n) \qquad (15)$$

is negative definite, iff every single block is negative definite.

Based on these remarks, for the stability proof following theorem 4.1 the block diagonal matrix

$$\mathbf{H}(\mathbf{P}) = \begin{pmatrix} \mathbf{A}_1^T\mathbf{P} + \mathbf{P}\mathbf{A}_1 & & & \\ & \ddots & & \\ & & \mathbf{A}_n^T\mathbf{P} + \mathbf{P}\mathbf{A}_n & \\ & & & -\mathbf{P} \end{pmatrix} = \mathbf{H}(\mathbf{p}) < \mathbf{0} \qquad (16)$$

is defined as input matrix of an LMI algorithm. The matrix $\mathbf{P}$ is unknown, an its coefficients form the free parameter vector $\mathbf{p}$ of the basic problem. The coefficients of $\mathbf{H}$ obviously depend linearly on the unknown coefficients of $\mathbf{P}$. To save the symmetry of $\mathbf{H}$, the coefficients of $\mathbf{P}$ have to be defined in a way, that solutions for $\mathbf{P}$ are restricted to symmetrical matrices, that means $p_{ij} = p_{ji}$.

The LMI algorithm determines, if a parameter vector $\mathbf{p}$ or a coefficient matrix $\mathbf{P}$ respectively exists, so that $\mathbf{H}(\mathbf{P})$ is negative definite. In that case, also every block matrix $\mathbf{A}_i^T\mathbf{P} + \mathbf{P}\mathbf{A}_i$ is negative definite as remarked before. And because $-\mathbf{P}$ is negative definite, too, $\mathbf{P}$ is positive definite. It follows, that all conditions of theorem 4.1 are fulfilled, and the stability of the closed loop system is proven. If the algorithm shows the non-existence of a suitable matrix $\mathbf{P}$, the system is unstable.

Besides that, the LMI algorithm does not only show the existence of a solution, it can also compute one possible solution. For the proof of stability this is not relevant, but it can be used for the controller design, as shown in the next chapter.

6. Design of a TS controller

As mentioned in the last chapter, with LMI algorithms not only the stability of a closed loop system with a given TS controller can be proven, they can also be used to design a TS controller ([11]). The only requirement is, that a TS model of the plant is given.

The starting point of the reflection is the stability condition of the closed loop system

$$\mathbf{A}_{g,l}^T\mathbf{P} + \mathbf{P}\mathbf{A}_{g,l} < \mathbf{0} \qquad (17)$$

with the system matrix $\mathbf{A}_{g,l}$ of the closed loop system of eq. (10). If the proof of theorem 4.1 would have been based on the double sum of eq. (8) instead of the re-indexed representation of eq. (10), this would have been lead to the following stability condition, as can be checked easily:

$$\mathbf{A}_i^T\mathbf{P} + (\mathbf{B}_i\mathbf{F}_j)^T\mathbf{P} + \mathbf{A}_j^T\mathbf{P} + (\mathbf{B}_j\mathbf{F}_i)^T\mathbf{P}$$
$$+\mathbf{P}\mathbf{A}_i + \mathbf{P}\mathbf{B}_i\mathbf{F}_j + \mathbf{P}\mathbf{A}_j + \mathbf{P}\mathbf{B}_j\mathbf{F}_i \quad < \quad \mathbf{0} \qquad \text{for all} \qquad i,j \qquad (18)$$

For a given plant model $(\mathbf{A}_i, \mathbf{B}_i)$ the unknown variables of this inequality system are the controller matrices $\mathbf{F}_i$ and the symmetrical matrix $\mathbf{P}$. Obviously, this inequality system is not linear for the unknown variables, so that the left sides of these inequalities cannot be used as input block matrices for an LMI algorithm.

But, with a simple conversion this problem can be transferred into a linear inequality system. At first, every ineqality is multiplied by the left and the right side with $\mathbf{P}^{-1}$ (as $\mathbf{P}$ is symmetrical, its inversion is not critical):

$$\mathbf{P}^{-1}\mathbf{A}_i^T + \mathbf{P}^{-1}\mathbf{F}_j^T\mathbf{B}_i^T + \mathbf{P}^{-1}\mathbf{A}_j^T + \mathbf{P}^{-1}\mathbf{F}_i^T\mathbf{B}_j^T$$
$$+\mathbf{A}_i\mathbf{P}^{-1} + \mathbf{B}_i\mathbf{F}_j\mathbf{P}^{-1} + \mathbf{A}_j\mathbf{P}^{-1} + \mathbf{B}_j\mathbf{F}_i\mathbf{P}^{-1} \quad < \quad \mathbf{0} \qquad \text{for all} \quad i,j \quad (19)$$

After this, with the definitions $\mathbf{X} = \mathbf{P}^{-1}$ and $\mathbf{H}_i = \mathbf{F}_i\mathbf{P}^{-1}$, and under consideration of the symmetry of $\mathbf{P}$ or $\mathbf{X}$ respectively, the inequality system can be written as

$$\mathbf{X}\mathbf{A}_i^T + \mathbf{H}_j^T\mathbf{B}_i^T + \mathbf{X}\mathbf{A}_j^T + \mathbf{H}_i^T\mathbf{B}_j^T$$
$$+\mathbf{A}_i\mathbf{X} + \mathbf{B}_i\mathbf{H}_j + \mathbf{A}_j\mathbf{X} + \mathbf{B}_j\mathbf{H}_i \quad < \quad \mathbf{0} \qquad \text{for all} \quad i,j \qquad (20)$$

Now, this inequality system is linear for the unknown matrices $\mathbf{X}$ and $\mathbf{H}_i$, and therefore, the left sides of the inequalities can be used as input block matrices for an LMI algorithm. The coefficients of the unknown matrices $\mathbf{X}$ and $\mathbf{H}_i$ are the unknown free parameters. With symmetrical $\mathbf{X}$, the overall block diagonal input matrix for the algorithm will also be symmetrical.

After the construction of the block diagonal input matrix, the LMI algorithm does not only clear the existence of a solution, it also computes a solution for $\mathbf{X}$ and the $\mathbf{H}_i$, if one exists. With this solution, the controller matrices of the TS controller can be computed by

$$\mathbf{F}_i = \mathbf{H}_i\mathbf{X}^{-1} \qquad (21)$$

As these controller matrices follow directly from the stability conditions, it is guaranteed, that the closed loop system with this controller is stable. To sum it up, the following steps are necessary for the design of a TS controller:

- Get a TS model of the plant, for example with system identification based on least error squares methods in the different operating points.
- Construct a block diagonal matrix as input matrix for an LMI algorithm, with the blocks consisting of the left sides of the inequality system (20). The coefficients of the unknown matrices $\mathbf{X}$ and $\mathbf{H}_i$ form the vector of the unknown parameters of the LMI algorithm.
- If a solution for the unknown matrices $\mathbf{X}$ and $\mathbf{H}_i$ exists, the LMI algorithm will compute it.
- The TS controller matrices can be obtained by $\mathbf{F}_i = \mathbf{H}_i\mathbf{X}^{-1}$. The closed loop system will be stabilized by this controller.

7. Extensions

In addition to stability there exist also other criteria, that can be worked into the stability analysis or controller design with LMI algorithms. For example, the demand for a fast system state approximation to the rest point after a disturbance (control velocity) is equivalent to the demand for a high change rate of the Lyapunov function. Instead of just being negative, this change rate can be demanded to depend on the distance of the system state to the rest point ([11]):

$$\dot{V}(\mathbf{x}(t)) \leq -\alpha V(\mathbf{x}(t)) \tag{22}$$

$\alpha > 0$ can be chosen independently, just limited by technical restrictions of the plant regarding the maximum possible control values. The greater its value is, the faster is the approximation of the system state to the rest point. Besides this, this demand will lead to higher approximation rates, if the system is far away from the rest point. This makes sense from a practical point of view, because you need strong control actions to drive a system close to the rest point, and soft control actions to drive the system into the rest point.

Equation (13) changes with (22) to

$$\dot{V} = \sum_i k_i \mathbf{x}^T (\mathbf{A}_i^T \mathbf{P} + \mathbf{P}\mathbf{A}_i)\mathbf{x} \quad < \quad -\alpha V(\mathbf{x}) = -\alpha \mathbf{x}^T \mathbf{P}\mathbf{x}$$

$$\sum_i k_i \mathbf{x}^T (\mathbf{A}_i^T \mathbf{P} + \mathbf{P}\mathbf{A}_i)\mathbf{x} + \alpha \mathbf{x}^T \mathbf{P}\mathbf{x} \quad < \quad 0$$

$$\sum_i k_i \mathbf{x}^T (\mathbf{A}_i^T \mathbf{P} + \mathbf{P}\mathbf{A}_i + \alpha \mathbf{P})\mathbf{x} \quad < \quad 0 \quad \text{with} \quad \sum_i k_i = 1 \tag{23}$$

and therefore, the stability condition (12) changes to

$$\mathbf{A}_i^T \mathbf{P} + \mathbf{P}\mathbf{A}_i + \alpha \mathbf{P} < \mathbf{0} \tag{24}$$

Obviously, just $\alpha \mathbf{P}$ has to be added to each inequality. The inequality system remains linear with $\mathbf{P}$, so that also the LMI algorithms remain applicable.

The discussion of the preceding chapters can even be extended to systems with model parameter uncertainties. One example is described in [2]. The system discussed there is a time-discrete representation of a system with time-varying model parameters:

$$\mathbf{x}(k+1) = \sum_i k_i (\mathbf{A}_i + \Delta \mathbf{A}_i(k))\mathbf{x}(k) \tag{25}$$

with the time-varying part of the system matrix

$$\Delta \mathbf{A}_i(k) = \mathbf{E}_i \mathbf{F}_i(k)\mathbf{H}_i \tag{26}$$

$\mathbf{E}_i$ and $\mathbf{H}_i$ are constant matrices of suitable dimension, while the matrix $\mathbf{F}_i(k)$ contains the time variability. The matrices may be chosen independently, just $\mathbf{F}_i(k)$ has to fulfil the condition

$$\mathbf{F}_i^T(k)\mathbf{F}_i(k) \leq \mathbf{I} \tag{27}$$

In order not to complicate the equations unnecessarily, $\mathbf{F}_i(k)$ can be chosen as a diagonal matrix, whose diagonal elements vary between -1 and 1:

$$\mathbf{F}_i(k) = \begin{pmatrix} f_1 & 0 \\ 0 & f_2 \end{pmatrix} \quad \text{with} \quad -1 < f_1(k), f_2(k) < 1 \tag{28}$$

In [2] the conditions are developed, under which the closed-loop system stability can be guaranteed for a system of the type (25). The conditions are summarized in matrix inequalities, so that an LMI algorithm can be used again to prove the stability. In addition, as performed in the last chapter for the time-constant case, the equations for the controller design are also derived. Obviously, these equations are rather extensive, and therefore, they are not presented in this paper.

Finally, it should be remarked, that any method discussed in this paper can be extended to closed-loop systems with observers. The state vector of such a system contains not only the state variables of the plant, but also the state variables of the observer. By suitable combination of the state space equations for the plant and the observer, the equations can be transferred to a form similar to (9), and based on this, the stability conditions can be defined (see [11, 4, 1]). However, the resulting equations get really extensive.

8. Conclusion

In this paper it was shown, that for TS systems there exist approaches for stability analysis and controller design, that are suitable for practical applications and that can be extended for control velocity and robustness. Even the inclusion of observers is possible.

A special aspect of TS control is the fact, that on the one hand it can be seen as classical gain scheduling control, but on the other hand it belongs to fuzzy control as well and can even be interpreted as generalization of conventional fuzzy control. However, the presented methods for stability analysis and controller design belong clearly to the field of classical control, because in the first step an analytical plant model has to be generated, followed by a systematic analysis or design procedure including stability, robustness and control velocity. But just with this systematic procedure the decisive disadvantage of conventional fuzzy control, the unsystematic control design procedure, can be avoided.

At the sight of these rather formal aspects the high use for practical applications should not be forgotten. A TS model as a basis for the presented procedures can be got easily with classical least error squares methods for each operating point, and the rest of the analysis or design can be performed by LMI algorithms. In addition, TS models are suitable to approximate any nonlinear multi input multi output (MIMO) system behaviour without time delay or hysteresis quite well, so that there is a wide range of application for the presented methods.

Besides that, from the point of fuzzy control the presented stability analysis method is generally very interesting, because an existing fuzzy controller can be represented as TS controller and checked for stability with the presented method, if a TS model of the plant is available.

The critical point of all presented methods lies in the high dimension of the resulting LMI problem. This could cause numerical problems, especially for high order systems with a large number of operating or supporting points. After all, this point will always define the limits of any TS approach in practical applications.

References

[1] Y.-Y. Cao, P.M. Frank: *Analysis and Synthesis of Nonlinear Time-Delay Systems via Fuzzy Control Approach*, IEEE Transactions on Fuzzy Systems 8 (2000), 200–211

[2] Y.-Y. Cao, P.M. Frank: *Robust H_∞ Disturbance Attenuation for a Class of Uncertain Discrete-Time Fuzzy Systems*, IEEE Transactions on Fuzzy Systems 8 (2000), 406–415

[3] E. Kim, H. Lee: *New Approaches to Relaxed Quadratic Stability Condition of Fuzzy Control Systems*, IEEE Transactions on Fuzzy Systems 8 (2000), 523–534

[4] K. Kiriakidis: *Fuzzy Model-Based Control of Complex Plants*, IEEE Transactions on Fuzzy Systems 6 (1998), 517–530

[5] K. Michels, F. Klawonn, R. Kruse, A. Nuernberger: *Fuzzy Control*, textbook in German, Springer-Verlag, Berlin, 2002 (to be published in English at Wiley and Sons 2004)

[6] K. Mueller: *Robust Control Design*, textbook in German, Teubner-Verlag, Stuttgart, 1996

[7] C.W. Scherer: *Linear Matrix Inequalities in Robust Control Theory*, in German, Automatisierungstechnik 45 (1997), 306–318

[8] M. Schroeder, R. Petersen, F. Klawonn, R. Kruse: *Two Paradigms of Automotive Fuzzy Logic Applications*. In M. Jamshidi, A. Titli, L.A. Zadeh, S. Boveri (Editors): Applications of Fuzzy Logic – Towards High Machine Intelligence Quotient Systems, Prentice Hall, Upper Saddle River, 1997

[9] M. Sugeno: *An Introductory Survey of Fuzzy Control*, Information Sciences 36 (1985), 59–83

[10] T. Takagi, M. Sugeno: *Fuzzy Identification of Systems and its Applications to Modeling and Control*, IEEE Transactions on Systems, Man, and Cybernetics 15 (1985), 116–132

[11] K. Tanaka, T. Ikeda, H.O. Wang: *Fuzzy Regulators and Fuzzy Observers: Relaxed Stability Conditions and LMI-Based Designs*, IEEE Transactions on Fuzzy Systems 6 (1998), 250–265

[12] K. Tanaka, M. Sugeno: *Stability analysis and design of fuzzy control systems*, Fuzzy Sets and Systems 45 (1992), 135–156

[13] M.C.M. Teixeira, S.H. Zak: *Stabilizing Controller Design for Uncertain Nonlinear Systems Using Fuzzy Models*, IEEE Transactions on Fuzzy Systems 7 (1999), 133–142

Kai Michels
Fichtner Engineering GmbH
Power Supply Department
Sarweystrasse 3
D-70191 Stuttgart, Germany
e-mail: michelsk@fichtner.de

Trends and Applications in Constructive Approximation
(Eds.) M.G. de Bruin, D.H. Mache & J. Szabados
International Series of Numerical Mathematics, Vol. 151, 195–205
© 2005 Birkhäuser Verlag Basel/Switzerland

Polynomial Bases for Continuous Function Spaces

Josef Obermaier and Ryszard Szwarc⋆

Abstract. Let $S \subset \mathbb{R}$ denote a compact set with infinite cardinality and $C(S)$ the set of real continuous functions on S. We investigate the problem of polynomial and orthogonal polynomial bases of $C(S)$.

In case of $S = \{s_0, s_1, s_2, \ldots\} \cup \{\sigma\}$, where $(s_k)_{k=0}^{\infty}$ is a monotone sequence with $\sigma = \lim_{k \to \infty} s_k$, we give a sufficient and necessary condition for the existence of a so-called Lagrange basis. Furthermore, we show that little q-Jacobi polynomials which fulfill a certain boundedness property constitute a basis in case of $S_q = \{1, q, q^2, \ldots\} \cup \{0\}$, $0 < q < 1$.

1. Introduction

One important goal in approximation theory is the representation of functions with respect to a set of simple functions. Here, we focus on the Banach space $C(S)$ of real continuous functions on a compact set $S \subset \mathbb{R}$ with infinite cardinality. Among the continuous functions polynomials are the most simple to deal with. Hence, further on we discuss the representation of $f \in C(S)$ with respect to a sequence of polynomials $(P_k)_{k=0}^{\infty}$. Moreover, it is profitable to look for a sequence with

$$\deg P_k = k \quad \text{for all } k \in \mathbb{N}_0, \tag{1}$$

which guarantees that every polynomial has a finite representation.

Of special interest are orthogonal polynomial sequences with respect to a probability measure π on S, where a representation is based on the Fourier coefficients

$$\hat{f}(k) = \int_S f(x) P_k(x) \, d\pi(x), \quad k \in \mathbb{N}_0, \tag{2}$$

of $f \in C(S)$.

⋆ Partially supported by KBN (Poland) under grant 2 P03A 028 25.

Let us recall some important facts about orthogonal polynomials, see [3]. An orthogonal polynomial sequence $(P_k)_{k=0}^{\infty}$ with compact support S and property (1) satisfies a three term recurrence relation

$$P_1(x)P_k(x) = a_k P_{k+1}(x) + b_k P_k(x) + c_k P_{k-1}(x), \quad k \in \mathbb{N}, \tag{3}$$

starting with

$$P_0(x) = a_0 \text{ and } P_1(x) = (x - b)/a, \tag{4}$$

where the coefficients are real numbers with $c_k a_{k-1} > 0$, $k \in \mathbb{N}$, and $(c_k a_{k-1})_{k=1}^{\infty}$, $(b_k)_{k=1}^{\infty}$ are bounded sequences. The other way around, if we construct $(P_k)_{k=0}^{\infty}$ by (3) with coefficients satisfying the conditions above, then we get an orthogonal polynomial sequence with compact support S.

The sequence of kernels $(K_n)_{n=0}^{\infty}$ is defined by

$$K_n(x,y) = \sum_{k=0}^{n} P_k(x)P_k(y)h(k) = \sum_{k=0}^{n} p_k(x)p_k(y), \tag{5}$$

where

$$h(k) = \left(\int_S P_k^2(x)\,d\pi(x)\right)^{-1} = \frac{1}{a_0^2}\frac{\prod_{i=0}^{k-1} a_i}{\prod_{i=1}^{k} c_i}, \quad k \in \mathbb{N}_0, \tag{6}$$

and $(p_k)_{k=0}^{\infty}$ is the orthonormal polynomial sequence with respect to π defined by

$$p_k = \sqrt{h(k)}P_k. \tag{7}$$

For $z \in S$ it holds

$$(K_n(z,z))^{-1} = \min_{Q \in \mathcal{P}_{(n)}, Q(z)=1} \int_S (Q(x))^2\,d\pi(x), \tag{8}$$

where $\mathcal{P}_{(n)}$ denotes the set of polynomials with degree less or equal n. One of the most important tools is the Christoffel-Darboux formula

$$\begin{aligned} K_n(x,y) &= a_n h(n)\frac{P_{n+1}(x)P_n(y) - P_n(x)P_{n+1}(y)}{P_1(x) - P_1(y)} \\ &= a\sqrt{c_{n+1}a_n}\frac{p_{n+1}(x)p_n(y) - p_n(x)p_{n+1}(y)}{x - y}. \end{aligned} \tag{9}$$

The linearization coefficients $g(i,j,k)$ are defined in terms of

$$P_i P_j = \sum_{k=0}^{\infty} g(i,j,k)P_k = \sum_{k=|i-j|}^{i+j} g(i,j,k)P_k, \quad i,j \in \mathbb{N}_0, \tag{10}$$

where $g(i,j,|i-j|), g(i,j,i+j) \neq 0$. The nonnegativity of the linearization coefficients is sufficient for a special boundedness property, which we will introduce in Section 4.

2. Polynomial bases for $C(S)$

Let us first refer to the concept of a basis.

Definition 2.1. *A sequence $(\Phi_k)_{k=0}^{\infty}$ in $C(S)$ is called basis if for every $f \in C(S)$ there exists a unique sequence of $(\varphi_k)_{k=0}^{\infty}$ of real numbers such that*

$$f = \sum_{k=0}^{\infty} \varphi_k \Phi_k, \qquad (11)$$

where $\lim_{n \to \infty} \sum_{k=0}^{n} \varphi_k \Phi_k$ is with respect to the sup-norm. A basis $(\Phi_k)_{k=0}^{\infty}$ of polynomials is called polynomial basis. A polynomial basis with (1) is called Faber basis.

There is a famous result of Faber [4] in 1914 that in case of S being an interval $[c, d]$ there does not exist a polynomial basis $(P_k)_{k=0}^{\infty}$ of $C([c, d])$ with property (1). Concerning $C([c, d])$ great efforts have been made in constructing polynomial bases and to minimize the degrees as far as possible. In 1977 Temlyakov [20] has investigated a method of construction, where the growth of the degrees fulfills $\deg P_k \leq C\, k \log \log(k)$. Later on, in 1985 Bochkarev [2] has used the Fejér kernel to construct a basis with linear bounds, that is $\deg P_k \leq 4k$. In 1987 Privalov [14] published a somehow negative result, which implies the result of Faber. Namely, if there is a polynomial basis $(P_k)_{k=0}^{\infty}$ of $C([c, d])$, then there exists a $\delta > 0$ such that $\deg P_k \geq (1 + \delta)\, k$ for all $k \geq k_0$, where k_0 is a proper integer. Also, in 1991 Privalov gave a positive result, see [15]. He proved that for any $\epsilon > 0$ there exists a polynomial basis of $C([c, d])$ with $\deg P_k \leq (1 + \epsilon)\, k$. Such a basis is called *polynomial basis of optimal degree* (with respect to ϵ).

If we are searching for Banach spaces $C(S)$ equipped with a Faber basis, then we have to choose S different from an interval. In this setting spaces $C(S)$ with a so-called Lagrange basis are discussed in [16]. In Section 3 we investigate a basic class of compact sets S and give a sufficient and necessary condition for the existence of a Lagrange basis.

Note, that the results mentioned above are not based upon the fact of orthogonality. In case of orthogonality the question "Does $(P_k)_{k=0}^{\infty}$ constitute an orthogonal polynomial basis of $C(S)$" is equivalent to the question if any function $f \in C(S)$ is represented by its Fourier series

$$\sum_{k=0}^{\infty} \hat{f}(k) P_k h(k). \qquad (12)$$

In this particular branch of study there also are some positive results. In 1996 Kilgore, Prestin and Selig [7] constructed an orthogonal polynomial basis of optimal degree with respect to the Chebyshev weight of first kind ($\alpha = \beta = -\frac{1}{2}$) using wavelet methods. Later on, in 1998 Girgensohn [6] gave optimal polynomial bases for all of the four Chebyshev weights ($\alpha = \pm\frac{1}{2}$, $\beta = \pm\frac{1}{2}$) and in 2001 Skopina [17] succeeded for Legendre weights ($\alpha = \beta = 0$). The general problem for Jacobi weights $(1 - x)^{\alpha}(1 + x)^{\beta} dx$, $\alpha, \beta > -1$, seems still to be open.

In order to check if an orthogonal polynomial sequence $(P_k)_{n=0}^{\infty}$ constitutes a basis of $C(S)$ we have to show

$$\sup_{x \in S} \int_S |K_n(x, y)| \, d\pi(y) \le C \quad \text{for all } n \in \mathbb{N}_0. \tag{13}$$

We should mention that the sequence $(P_k)_{k=0}^{\infty}$ is a basis of $C(S)$ if and only if it is a basis of $L^1(S, \pi)$, see [11]. For the discussion of an example based on little q-Jacobi polynomials see Section 5.

3. Lagrange bases

In [16] we have introduced the concept of a Lagrange basis. Let $S \subset \mathbb{R}$ be a compact set and $(s_k)_{k=0}^{\infty}$ a sequence of distinct points in S. Define as usual the Lagrange basic functions L_n^k as

$$L_n^k(x) = \frac{\prod_{i=0, i \neq k}^n (x - s_i)}{\prod_{i=0, i \neq k}^n (s_k - s_i)} \quad \text{for all} \quad n \in \mathbb{N}_0, \ k = 0, 1, \dots, n. \tag{14}$$

and

$$l_k(x) = L_k^k(x) \quad \text{for all} \quad k \in \mathbb{N}_0. \tag{15}$$

Definition 3.1. *The sequence $(l_k)_{k=0}^{\infty}$ is called sequence of Lagrange polynomials with respect to $(s_k)_{k=0}^{\infty}$. If $(l_k)_{k=0}^{\infty}$ is a basis of $C(S)$, then we call $(l_k)_{k=0}^{\infty}$ a Lagrange basis of $C(S)$ with respect to $(s_k)_{k=0}^{\infty}$.*

In case of a Lagrange basis it holds $f = \sum_{k=0}^{\infty} \varphi_k(f) l_k$ with

$$\varphi_0(f) = f(s_0); \quad \varphi_k(f) = f(s_k) - \sum_{j=0}^{k-1} \varphi_j(f) l_j(s_k) \quad \text{for all } k \in \mathbb{N}. \tag{16}$$

A sequence $(v_n)_{n=0}^{\infty}$ of linear operators from $C(S)$ into $C(S)$ is defined by

$$v_n(f) = \sum_{k=0}^{n} \varphi_k(f) l_k. \tag{17}$$

By simple means we have

$$\sum_{k=0}^{n} \varphi_k(f) l_k(s_i) = f(s_i) \quad \text{for all } i = 0, 1, \dots, n, \tag{18}$$

which implies

$$\sum_{k=0}^{n} \varphi_k(f) l_k = \sum_{k=0}^{n} f(s_k) L_n^k. \tag{19}$$

The sequence of Lagrange polynomials $(l_k)_{k=0}^{\infty}$ constitutes a basis of $C(S)$ if and only if

$$\|v_n\| = \max_{x \in S} \sum_{k=0}^{n} |L_n^k(x)| \le C \quad \text{for all } n \in \mathbb{N}_0, \tag{20}$$

see [16].

Further let us assume that $(s_k)_{k=0}^{\infty}$ is strictly increasing or strictly decreasing and

$$S = \{s_0, s_1, s_2, \ldots\} \cup \{\sigma\}, \tag{21}$$

where

$$\sigma = \lim_{k \to \infty} s_k. \tag{22}$$

Using this assumption we derive

$$\|v_n\| = \sum_{k=0}^{n} |L_n^k(\sigma)|, \tag{23}$$

see [16].

Let us now give the main result of this section.

Theorem 3.2. *Assume $(s_k)_{k=0}^{\infty}$ is a strictly increasing or strictly decreasing sequence and $S = \{s_0, s_1, s_2, \ldots\} \cup \{\sigma\}$.*

Then $(l_k)_{k=0}^{\infty}$ is a basis of $C(S)$ if and only if there exists $0 < q < 1$ with

$$|\sigma - s_{k+1}| \le q\,|\sigma - s_k| \quad \text{for all } k \in \mathbb{N}_0. \tag{24}$$

Proof. Let $(l_k)_{k=0}^{\infty}$ be a basis of $C(S)$. Then there exists $C > 1$ such that $|l_k(\sigma)| < C$ for all $k \in \mathbb{N}_0$, which implies

$$|\sigma - s_{k-1}| < C|s_k - s_{k-1}| \quad \text{for all } k \in \mathbb{N}. \tag{25}$$

If $(s_k)_{k=0}^{\infty}$ is strictly increasing, then $(\sigma - s_{k-1}) < C(s_k - \sigma + \sigma - s_{k-1})$, which is equivalent to

$$(\sigma - s_k) < \frac{C-1}{C}(\sigma - s_{k-1}) \quad \text{for all } k \in \mathbb{N}. \tag{26}$$

In case of a strictly decreasing sequence we get the inequality the other way around. Choose $q = (C-1)/C$.

Let us now assume that (24) holds and $(s_k)_{k=0}^{\infty}$ is strictly decreasing. If $k > i$ we get

$$\frac{|\sigma - s_i|}{|s_k - s_i|} = \frac{s_i - \sigma}{s_i - \sigma - (s_k - \sigma)} < \frac{s_i - \sigma}{s_i - \sigma - q^{k-i}(s_i - \sigma)} = \frac{q^i}{q^i - q^k}, \tag{27}$$

and if $k < i$ we get

$$\frac{|\sigma - s_i|}{|s_k - s_i|} < \frac{q^i}{q^k - q^i}. \tag{28}$$

Furthermore, it holds

$$\sum_{k=0}^{n} \prod_{i=0,i\neq k}^{n} \frac{q^i}{|q^k - q^i|} = \sum_{k=0}^{n} \prod_{j=1}^{n-k} \frac{1}{1-q^j} \prod_{i=1}^{k} \frac{1}{1-q^i} q^{k(k+1)/2} \tag{29}$$

$$\leq \left(\prod_{j=1}^{\infty} \frac{1}{1-q^j}\right)^2 \sum_{k=0}^{\infty} q^k < \infty. \tag{30}$$

Hence, $\sum_{k=0}^{n} |L_n^k(\sigma)| \leq C$ for all $n \in \mathbb{N}_0$, which implies that $(l_k)_{k=0}^{\infty}$ constitutes a basis of $C(S)$. The case $(s_k)_{k=0}^{\infty}$ is strictly increasing is quite similar. $\qquad\square$

The standard example due to the geometric sequence is

$$S_q = \{1, q, q^2, \ldots\} \cup \{0\}, \tag{31}$$

where $0 < q < 1$. Now, by Theorem 3.2 it follows that the Lagrange polynomials $(l_k)_{k=0}^{\infty}$ with respect to $(s_k)_{k=0}^{\infty}$, where $s_k = q^k$, constitute a Lagrange basis of $C(S_q)$.

For instance, if one is rearranging the sequence $(\frac{1}{2^k})_{k=0}^{\infty}$ in the way that

$$s_k = \begin{cases} 1 & \text{if } k = 0, \\ 2^{-(k+1)/2} & \text{if } k \neq 0 \text{ and } \log_2(k+1) \in \mathbb{N}, \\ 2^{-(k+1)} & \text{else}, \end{cases} \tag{32}$$

then the Lagrange polynomials $(l_k)_{k=0}^{\infty}$ with respect to $(s_k)_{k=0}^{\infty}$ don't constitute a Lagrange basis of $C(S_{\frac{1}{2}})$. The proof is left to the reader.

If we set $S_q^{(i)} = S_q \setminus \{q^i\}$ and $(l_k)_{k=0}^{\infty}$ denotes the sequence of Lagrange polynomials with respect to the sequence $(q^k)_{k=0}^{\infty}$ then $C(S_q^{(i)})$ in companion with $(l_k)_{k=0}^{\infty}$ states an example where a representation (11) exists but is not unique. To show this first notice that any $f^{(i)} \in C(S_q^{(i)})$ could be easily extended to a function $f \in C(S_q)$, where $f|_{S_q^{(i)}} = f^{(i)}$ and $f(q^i)$ is arbitrary. A representation of f in $C(S_q)$ also represents $f^{(i)}$ in $C(S_q^{(i)})$. Choose $f_1, f_2 \in C(S_q)$ with $f_1(q^i) \neq f_2(q^i)$ and $f_1|_{S_q^{(i)}} = f_2|_{S_q^{(i)}}$ to show that the representation is not unique. Of course, by Theorem 3.2 there is a Lagrange basis of $C(S_q^{(i)})$ with respect to the sequence $(q^k)_{k=0,k\neq i}^{\infty}$.

Let

$$S^r = \{1, \frac{1}{2^r}, \frac{1}{3^r}, \ldots\} \cup \{0\}, \tag{33}$$

where $0 < r < \infty$. With $s_k = \frac{1}{(k+1)^r}$ we get $\lim_{k\to\infty} s_{k+1}/s_k = 1$. By Theorem 3.2 the Lagrange polynomials with respect to $(s_k)_{k=0}^{\infty}$ do not constitute a basis of $C(S^r)$.

4. A boundedness property for orthogonal polynomial sequences

The following boundary property for orthogonal polynomial sequences is important for many reasons, see for instance [12], and is used in Section 5.

Definition 4.1. *We say that a polynomial sequence* $(R_k)_{k=0}^{\infty}$ *fulfills property (B), if there exists* $\xi \in S$ *such that*

$$|R_k(x)| \leq R_k(\xi) = 1 \quad \text{for all } x \in S, \ k \in \mathbb{N}_0. \tag{34}$$

There is a condition on the linearization coefficients which yields that property (B) holds with respect to a proper normalization of the system.

Lemma 4.2. *Assume that the linearization coefficients* $g(i, j, k)$ *belonging to the sequence* $(P_k)_{k=0}^{\infty}$ *are nonnegative for all* $i, j, k \in \mathbb{N}_0$, *then there exists a normalization* $R_k = \gamma_k P_k$ *such that property (B) holds.*

Proof. The assumption yields $g(i, i, 2i) > 0$ for all $i \in \mathbb{N}_0$. Hence, by (10) it follows $P_0 > 0$ and $\lim_{x \to -\infty} P_{2i}(x) = \lim_{x \to \infty} P_{2i}(x) = \infty$ for all $i \in \mathbb{N}$. Regarding $P_1 P_{2i}$ we get $\lim_{x \to \infty} P_1(x) = \lim_{x \to \infty} P_{2i+1}(x)$ for all $i \in \mathbb{N}$.

All zeros of the polynomials P_k are in the open interval $(\min S, \max S)$, see [3]. Hence there are two cases to handle. Namely, $P_k(\min S) > 0$ for all $k \in \mathbb{N}_0$, or $P_k(\max S) > 0$ for all $k \in \mathbb{N}_0$. Depending on this put $\xi = \min S$ or $\xi = \max S$ and define

$$R_k(x) = \frac{P_k(x)}{P_k(\xi)} \quad \text{for all } k \in \mathbb{N}_0. \tag{35}$$

Then the linearization coefficients g_R of $(R_k)_{k=0}^{\infty}$ are also nonnegative because

$$g_R(i, j, k) = \frac{P_k(\xi)}{P_i(\xi) P_j(\xi)} g(i, j, k) \quad \text{for all } i, j, k \in \mathbb{N}_0, \tag{36}$$

and it holds

$$\sum_{k=|i-j|}^{i+j} g_R(i, j, k) = 1 \quad \text{for all } i, j \in \mathbb{N}_0. \tag{37}$$

Hence, a hypergroup structure is associated with the orthogonal polynomial sequence $(R_k)_{k=0}^{\infty}$ which yields property (B), see [10]. $\square$

There are well-known criteria by Askey [1] or [18, 19] implying the nonnegativity of the linearization coefficients. In case of a discrete measure π and nonnegative linearization coefficients we refer to Koornwinder [9] and [13].

In the next section we use the fact that in case of property (B) it holds

$$|K_n(x, y)| \leq K_n(\xi, \xi) \quad \text{for all } n \in \mathbb{N}_0, \ x, y \in S. \tag{38}$$

5. Little q-Jacobi polynomials

In all that follows we keep $0 < q < 1$ fixed and S_q is defined by (31). For $\alpha > -1$ we define a probability measure $\pi^{(\alpha)}$ on S_q by

$$\pi^{(\alpha)}(\{q^j\}) = (q^{\alpha+1})^j(1 - q^{\alpha+1}), \quad \pi^{(\alpha)}(\{0\}) = 0. \tag{39}$$

The orthogonal polynomial sequence $(R_k^{(\alpha)})_{k=0}^\infty$ with respect to $\pi^{(\alpha)}$ are special little q-Jacobi polynomials, see [8]. They fulfill the following orthogonality relation

$$\int_{S_q} R_k^{(\alpha)} R_l^{(\alpha)} \, d\pi^{(\alpha)} = \sum_{j=0}^\infty R_k^{(\alpha)}(q^j) R_l^{(\alpha)}(q^j)(q^{\alpha+1})^j(1 - q^{\alpha+1})$$

$$= \frac{(q^{\alpha+1})^k(1 - q^{\alpha+1})}{1 - q^{2k+\alpha+1}} \left(\prod_{i=1}^k \frac{1 - q^i}{1 - q^{\alpha+i}} \right)^2 \delta_{k,l}. \tag{40}$$

Starting with

$$R_0^{(\alpha)} = 1 \text{ and } R_1^{(\alpha)}(x) = 1 - \frac{1 - q^{\alpha+2}}{1 - q^{\alpha+1}} x \tag{41}$$

they are defined by the three term recurrence relation (3) with coefficients

$$a_k = \frac{1 - q^{\alpha+2}}{1 - q^{\alpha+1}} A_k, \tag{42}$$

$$b_k = 1 - \frac{1 - q^{\alpha+2}}{1 - q^{\alpha+1}} (A_k + C_k), \tag{43}$$

$$c_k = \frac{1 - q^{\alpha+2}}{1 - q^{\alpha+1}} C_k, \tag{44}$$

where

$$A_k = q^k \frac{(1 - q^{k+\alpha+1})(1 - q^{k+\alpha+1})}{(1 - q^{2k+\alpha+1})(1 - q^{2k+\alpha+2})} \tag{45}$$

$$C_k = q^{k+\alpha} \frac{(1 - q^k)(1 - q^k)}{(1 - q^{2k+\alpha})(1 - q^{2k+\alpha+1})}. \tag{46}$$

In case of $\alpha \geq 0$ the orthogonal polynomial sequence $(R_k^{(\alpha)})_{k=0}^\infty$ has nonnegative linearization coefficients, see [9] ($\alpha = 0$) and [5] ($\alpha > 0$).

Theorem 5.1. *If $0 \leq \alpha$, then the sequence $(R_k^{(\alpha)})_{k=0}^\infty$ of little q-Jacobi polynomials constitutes a basis of $C(S_q)$.*

Proof. The nonnegativity of the linearization coefficients implies property (B) with $\xi = 0$. Let $(p_k^{(\alpha)})_{k=0}^\infty$ denote the corresponding orthonormal polynomial sequence. Using (40) we get

$$\sqrt{\frac{1 - q^{2k+\alpha+1}}{(q^{\alpha+1})^k(1 - q^{\alpha+1})}} \prod_{i=1}^k \frac{1 - q^{\alpha+i}}{1 - q^i} = p_k^{(\alpha)}(0) \geq \max_{x \in S_q} |p_k^{(\alpha)}(x)|. \tag{47}$$

Note that $S_q \subset [0,1]$. In order to prove

$$\sup_{x \in S_q} \int_{[0,1]} |K_n(x,y)| \, d\pi^{(\alpha)}(y) \leq C \quad \text{for all } n \in \mathbb{N}_0, \tag{48}$$

we split the integration domain into two parts $[0,\epsilon]$ and $[\epsilon,1]$. For the first it holds

$$\int_{[0,\epsilon]} |K_n(x,y)| \, d\pi^{(\alpha)}(y) \leq K_n(0,0)\pi^{(\alpha)}([0,\epsilon]). \tag{49}$$

By the Christoffel-Darboux formula (9) and property (B) we get

$$|K_n(x,y)| \leq \sqrt{c_{n+1}a_n} \, \frac{p_{n+1}^{(\alpha)}(0)|p_n^{(\alpha)}(y)| + p_n^{(\alpha)}(0)|p_{n+1}^{(\alpha)}(y)|}{|x-y|}, \quad x \neq y. \tag{50}$$

Hence, setting $\lambda_n = \sqrt{c_{n+1}a_n}$ and applying $|x-y| \geq (1-q)\,y$, $x \neq y$, it follows

$$\int_{[\epsilon,1]} |K_n(x,y)| \, d\pi^{(\alpha)}(y) \;\leq\; \frac{\lambda_n p_{n+1}^{(\alpha)}(0)}{1-q} \int_{[\epsilon,1]} \frac{|p_n^{(\alpha)}(y)|}{y} \, d\pi^{(\alpha)}(y) \tag{51}$$

$$+ \; \frac{\lambda_n p_n^{(\alpha)}(0)}{1-q} \int_{[\epsilon,1]} \frac{|p_{n+1}^{(\alpha)}(y)|}{y} \, d\pi^{(\alpha)}(y) \tag{52}$$

$$+ \; \sum_{k=0}^{n} (p_k^{(\alpha)}(x))^2 \pi^{(\alpha)}(\{x\}). \tag{53}$$

By (8) we obtain

$$\sum_{k=0}^{n} (p_k^{(\alpha)}(x))^2 \pi^{(\alpha)}(\{x\}) \leq 1. \tag{54}$$

It is simple to derive that

$$K_n(0,0) = \mathcal{O}(q^{-(\alpha+1)n}). \tag{55}$$

Now, we fix $\epsilon = q^n$ to get

$$\pi^{(\alpha)}([0,\epsilon]) = \mathcal{O}(q^{(\alpha+1)n}), \tag{56}$$

which yields a uniform bound for the integral on the left-hand side of (49) not depending on $x \in S$. Next, note that

$$\lambda_n = \mathcal{O}(q^n), \tag{57}$$

and by (47) we get

$$p_n^{(\alpha)}(0) = \mathcal{O}(q^{-(\alpha+1)\frac{n}{2}}). \tag{58}$$

In order to obtain a uniform bound for the integral on the left-hand side of (51) it remains to prove

$$\int_{[\epsilon,1]} \frac{|p_n^{(\alpha)}(y)|}{y} \, d\pi^{(\alpha)}(y) = \mathcal{O}(q^{(\alpha-1)\frac{n}{2}}). \tag{59}$$

For that purpose let $k \in \mathbb{N}_0$ with $\alpha < 2k+1$.

By the Cauchy-Schwarz inequality we get

$$\int_{[\epsilon,1]} \frac{|p_n^{(\alpha)}(y)|}{y} \, d\pi^{(\alpha)}(y) \leq \left(\int_{[\epsilon,1]} \frac{d\pi^{(\alpha)}(y)}{y^{2(k+1)}}\right)^{\frac{1}{2}} \left(\int_{[0,1]} (p_n^{(\alpha)}(y)y^k)^2 \, d\pi^{(\alpha)}(y)\right)^{\frac{1}{2}}.$$
(60)

By simple means it follows

$$\int_{[\epsilon,1]} \frac{d\pi^{(\alpha)}(y)}{y^{2(k+1)}} = \mathcal{O}(q^{(\alpha-2k-1)n}).$$
(61)

The three term recurrence formula for $p_n^{(\alpha)}$ is

$$yp_n^{(\alpha)} = -\Lambda_n p_{n+1}^{(\alpha)} + (A_n + C_n)p_n^{(\alpha)} - \Lambda_{n-1}p_{n-1}^{(\alpha)},$$
(62)

where $\Lambda_n = \sqrt{C_{n+1}A_n}$, see (45) and (46). So the coefficients behave like q^n. The minus sign comes from the fact that $p_n(0) > 0$. By applying the recurrence relation k times we get

$$y^k p_n^{(\alpha)} = \sum_{n-k}^{n+k} d(k,n,i)p_i^{(\alpha)},$$
(63)

where each coefficient $d(k,n,i)$ behaves like q^{kn}. Therefore, by orthogonality

$$\int_{[0,1]} (p_n^{(\alpha)}(y)y^k)^2 \, d\pi^{(\alpha)}(y) = \sum_{n-k}^{n+k} (d(k,n,i))^2 = \mathcal{O}(q^{2kn}).$$
(64)

So we have shown (59) and the proof is complete.   □

The little q-Legendre case ($\alpha = 0$) is also investigated in [16].

So in case of S_q we are able to give orthogonal Faber basis for $C(S_q)$. For the set S^r, see (33), the existence of an orthogonal Faber basis or even a Faber basis for $C(S^r)$ seems still to be open.

References

[1] R. Askey, *Linearization of the product of orthogonal polynomials*, in: Problems in Analysis, Princeton University Press, Princeton, 1970, 223–228.

[2] S.V. Bochkarev, Construction of a dyadic basis in the space of continuous functions on the basis of Fejér kernels, Tr. Mat. Inst. Akad. Nauk SSSR 172, (1985) 29–59.

[3] T.S. Chihara, *An introduction to orthogonal polynomials*, Gordon and Breach, New York, 1978.

[4] G. Faber, *Über die interpolatorische Darstellung stetiger Funktionen*, Jahresber. Deutsch. Math. Verein. **23** (1914), 192–210.

[5] P.G.A. Floris, *A noncommutative discrete hypergroup associated with q-disk polynomials*, J. Comp. Appl. Math. **68** (1996), 69–78.

[6] R. Girgensohn, *Polynomial Schauder bases for C[−1,1] with Chebiseff orthogonality*, preprint (1998).

[7] T. Kilgore, J. Prestin and K. Selig, *Orthogonal algebraic polynomial Schauder bases of optimal degree*, J. Fourier Anal. Appl. **2** (1996), 597–610.

[8] R. Koekoek and R.F. Swartouw, *The Askey-scheme of hypergeometric orthogonal polynomials and its q-analogue*, Technical Report 98-17, Delft University of Technology, 1998.

[9] T.H. Koornwinder, *Discrete hypergroups associated with compact quantum Gelfand pairs*, in: Applications of hypergroups and related measure algebras, Contemp. Math. **183**, Amer. Math. Soc., 1995, 213–235.

[10] R. Lasser, *Orthogonal polynomials and hypergroups*, Rend. Mat. **3** (1983), 185–209.

[11] R. Lasser and J. Obermaier, *On the convergence of weighted Fourier expansions*, Acta. Sci. Math. **61** (1995), 345–355.

[12] R. Lasser, D.H. Mache and J. Obermaier, *On approximation methods by using orthogonal polynomial expansions*, in: Advanced Problems in Constructive Approximation, Birkhäuser, Basel, 2003, 95–107.

[13] W. Młotkowski and R. Szwarc, *Nonnegative linearization for polynomials orthogonal with respect to discrete measures*, Constr. Approx. 17 (2001), 413–429.

[14] Al.A. Privalov, *Growth of the degrees of polynomial basis and approximation of trigonometric projectors*, Mat. Zametki 42, (1987) 207–214.

[15] Al.A. Privalov, *Growth of degrees of polynomial basis*, Mat. Zametki 48, (1990) 69–78.

[16] J. Obermaier, *A continuous function space with a Faber basis*, J. Approx. Theory **125** (2003), 303–312.

[17] M. Skopina, *Orthogonal polynomial Schauder bases in $C[-1,1]$ with optimal growth of degrees*, Mat. Sbornik, **192**:3 (2001), 115–136.

[18] R. Szwarc, *Orthogonal polynomials and discrete boundary value problem I*, SIAM J. Math. Anal. **23** (1992), 959–964.

[19] R. Szwarc, *Orthogonal polynomials and discrete boundary value problem II*, SIAM J. Math. Anal. **23** (1992), 965–969.

[20] V.N. Temlyakov, On the order of growth of the degrees of a polynomial basis in the space of continuous functions, Mat. Zametki 22, (1977) 711–727.

Josef Obermaier
Institute of Biomathematics and Biometry
GSF-National Research Center for Environment and Health
Ingolstädter Landstrasse 1
D-85764 Neuherberg, Germany
e-mail: josef.obermaier@gsf.de

Ryszard Szwarc
Institute of Mathematics
Wrocław University
pl. Grunwaldzki 2/4
50-384 Wrocław, Poland
e-mail: szwarc@math.uni.wroc.pl

Tea & Coffee-Break in Haus Bommerholz, Witten

Trends and Applications in Constructive Approximation
(Eds.) M.G. de Bruin, D.H. Mache & J. Szabados
International Series of Numerical Mathematics, Vol. 151, 207–223
© 2005 Birkhäuser Verlag Basel/Switzerland

Novel Simulation Approaches for Cyclic Steady-state Fixed-bed Processes Exhibiting Sharp Fronts and Shocks

Frank Platte, Dmitri Kuzmin, Christoph Fredebeul and Stefan Turek

Abstract. Over the past decades, the field of chemical engineering has witnessed an increased interest in unsteady-state processes. Multifunctional, as well as intensified chemical processes, may exhibit instationary behaviour especially when based on periodical operating conditions. Ideally, instationary processes lead to a higher yield and increased selectivities compared to conventional steady-state fixed-bed processes. Typical candidates among these are the reverse-flow-reactor, the chromatographic reactor and the adsorptive reactor. Since the underlying regeneration strategy is nearly always based on cycles – e.g., a reaction cycle is followed by a regeneration cycle and so on – the overall temporal behaviour of such processes eventually develops into cyclic steady-states (after a transient phase). Experiments reveal a slow transient behaviour into the cyclic steady-state. This can also be observed in simulation based on conventional numerical treatment such as the method of lines. In addition to this problem many instationary processes exhibit sharp fronts or even shocks which require stabilisation of the convective terms. In this work we present a method of combining the idea of global discretisation with modern stabilisation techniques of type FEM-FCT and FEM-TVD in order to obtain an efficient, well approximating and robust tool for the general simulation of instationary and in particularly cyclic steady-state processes.

1. Instationary fixed-bed processes

Fixed-bed reactors are suggested for many chemical engineering applications. In contrast to batch processes, fixed-bed processes offer the possibility (over the length of the apparatus) of taking influence on the physical processes and chemical reactions inside – in order to obtain higher overall performance. Very often, it is the goal to achieve higher yield and selectivities based on the value product, larger space-time-yield and – in the case of exothermic reactions – to guarantee better heat-integration. Depending on how these modifications are made, one has to distinguish between intensified and multifunctional processes. In multifunctional reactors (beside the reaction itself) additional process functions, e.g., mixing/separation and/or heat-accumulation are integrated in the apparatus, whereas intensified processes are justified on the fact that most chemical engineering processes are rather limited by heat and mass transport or thermodynamics and not so much by the reaction itself. Hence, intensified as well as multifunctional processes, lead to higher product quality and purity making them economically preferable.

More recently, instationary processes moved into the focus of chemical engineering. This instationary behaviour results from internal or external recursion of heat and/or mass within the process. From the operator's point of view, continuous instationary variants are the most interesting.

In general, one can distinguish between forced-cyclic and autonomous-instationary processes. In both of these, one can monitor moving temperature and concentration fronts which follow reoccurring patterns of cycles after the start-up phase. Multifunctional and intensified processes are either inherently instationary or forced instationary. Their behaviour based on properly chosen operating conditions allow for additional enhancement of the performance. Moreover, the time dependent behaviour leads to more data which can be exploited for model evaluation. Unfortunately, due to the non-linearity and stiffness of many of these instationary fixed-bed processes, experiments and simulations are rather time-consuming projects.

1.1. Application examples

In the following section, three examples of instationary fixed-bed processes are briefly described:

- Example I: Catalytic combustion in reverse-flow operation
- Example II: Adsorptive reactor
- Example III: Coupled endothermic/exothermic reaction in reverse-flow operation

For each example, we present a schematic view of the process function and consider a typical balance (transport) equation either for energy or mass of type

'ACCUMULATION' + 'CONVECTION' = 'DIFFUSION' + 'REACTION' + 'SOURCES/SINKS'.

Then, in each equation we highlight the term that causes numerical instabilities and discuss appropriate numerical treatments.

Example I: Catalytic combustion in reverse-flow operation

One important instationary process is the reverse-flow-reactor (RFR) which is operated in a forced periodical way by switching the side/direction of the inflow (cf. Fig. 1) [14]. One of the most notable advantages of the reverse-flow concept is certainly that due to the regenerative heat recovery, a hot reaction zone surrounded by two cold zones is trapped in the centre of the fixed-bed. The classical RFR operates with two identical half-cycles, i.e., the two functions "reaction" and "regeneration" are fulfilled simultaneously. Due to an inherently low heat-loss, even weak exothermic processes (or alternatively processes with trace gases) can remain ignited without additional external heat or fuel gas. Suggested examples for industrial application are the catalytic treatment of waste gases in air, oxidation of SO_2 [15] and many more [16].

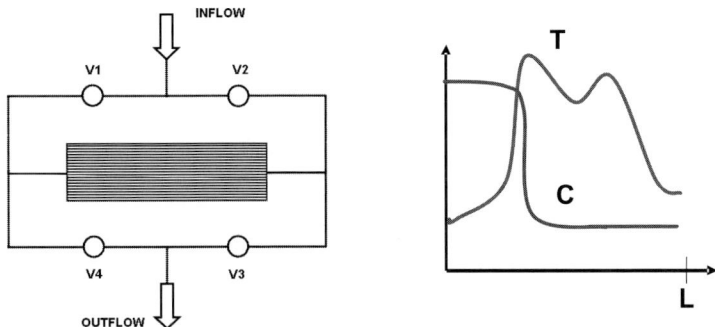

FIGURE 1. Scheme of a reverse-flow reactor: Cyclic opening and closing of the valves-pairs V1/V3 and V2/V4 (left). Typical temperature-fronts and concentration-distribution within the fixed-bed for a fixed time (right).

Numerical demands for catalytic combustion in reverse-flow operation: It is well known that the RFR reaches the cyclic steady-state after a long operation time and a large number of flow-reversals. Moreover, high reaction rates at elevated temperature levels lead to sharp fronts in the distribution of temperature and concentration. Theretore, the numerical algorithms should incorporate a direct calculation of cyclic steady-states and an appropriate stabilisation of the convective terms. High reaction rates in the energy equation (1) may cause sharp profiles.

$$(\varrho c_p) \cdot \frac{\partial T}{\partial t} + \varepsilon_F \varrho c_p \frac{\partial (w_F T)}{\partial z} = \lambda_{ax} \frac{\partial^2 T}{\partial z^2} + \varepsilon_F \underbrace{\sum_j (-\Delta H_{R,i}) r_j}_{\text{high reaction rates!}} + \frac{\alpha}{r} (T^* - T). \quad (1)$$

Example II: Adsorptive reactor
The adsorptive reactor is currently considered for enhancing the yield of equilibrium limited reactions. A survey of chemical reactions investigated in gas-phase adsorptive reactors is given in [6]. Fig. 2 illustrates the two functions "reaction + adsorption" and "regeneration" in two separate sequential half-cycles. Most of the authors use adsorptive reaction processes as means to enhance equilibrium conversions by the uptake of one of the products according to LE CHATELIER's principle: The equilibrium of the two reactants A and B

$$A + B \Leftrightarrow C + D$$

is shifted to the right-hand side. By adsorbing the by-product

$$C \Leftrightarrow C*$$

the conversion of the value product D is increased. Two suggested adsorptive fixed-bed processes can be found in [6].

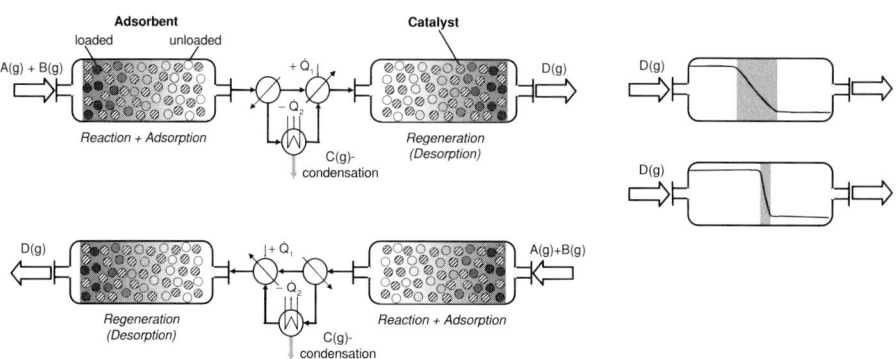

FIGURE 2. Adsorptive reactor with production- and regeneration-cycle (left) and typical shifted fronts (right) due to the self-sharpening effect within the cycle.

Numerical demands for adsorptive reactor: Nonlinear adsorption isotherms could cause problems during the calculation due to steep gradients. For example convex isotherms $q(C)$ lead to a so-called self-sharpening effect noticeable in the concentration profiles. In the mass balance (2) one can locate the non-linear accumulation term. Therefore, the numerical algorithms should consider an appropriate stabilisation of the accumulation term.

$$\left(1 + \frac{1 - \varepsilon_F}{\varepsilon_F} \cdot \underbrace{\frac{dq(c)^*}{dc}}_{\text{nonlinear accumulation!}}\right) \cdot \frac{\partial c}{\partial t} + w_F \frac{\partial c}{\partial z} = D_{ax} \frac{\partial^2 c}{\partial z^2} + \sum_j \nu_{ij} r_j. \quad (2)$$

Example III: Coupled endothermic/exothermic reaction in reverse-flow operation
This example is closely related to example I. Here the reverse-flow operation can also be established by switching the inflow direction. But in contrast to example I, two consecutive half-cycles now fulfil different functions, namely "endothermic reaction" and "exothermic regeneration".

Numerical demands for Coupled endothermic/exothermic reaction in reverse-flow operation: Nonlinear equilibrium may cause numerical problems during the calculation due to shock fronts. In the heat balance (3) one can locate the non-linear

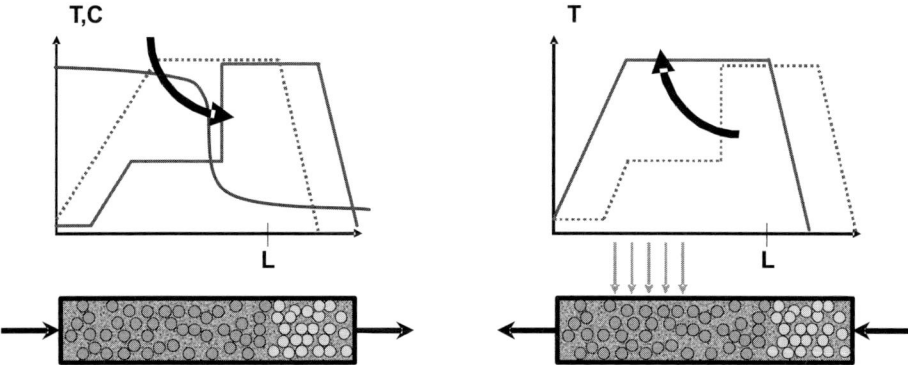

FIGURE 3. Schematic view of Coupled reaction in RFR: Function of endothermic (left) and exothermic half-cycle (right). Dotted lines denote the distribution at the beginning of each half-cycle. Full lines show where the distributions end up.

convective term. Therefore, the numerical algorithms should consider an appropriate stabilisation of the convective terms.

$$\left(\varrho^*c_p^* + \varepsilon_F \varrho c_p\right) \cdot \frac{\partial T}{\partial t} + w_F \varrho c_p \left(1 + \Delta T_{ad} \cdot \underbrace{\frac{dX(T)}{dT}}_{\text{nonlinear convection!}}\right) \frac{\partial T}{\partial z} = \lambda_{ax} \frac{\partial^2 T}{\partial z^2}.$$

(3)

In conclusion, for all three presented processes the numerical algorithms must account for

- direct calculation of cyclic steady-states and
- stabilisation of convective terms.

2. Numerical treatment

Compared to batch-processes, stationary fixed-bed processes are in general more complex to design and to build. Nevertheless, the mathematical description of these two process classes is of similar complexity. In the case of batch processes, one has to appropriately handle (large) systems of time-dependent ODEs, whereas stationary fixed-bed processes cause difficulties due to the spatial distribution of the variables. Comparing stationary and instationary processes leads to a similar conclusion: The additional temporal behaviour implies stronger demands on the experimental equipment and the operation of such apparatus. From the mathematical point of view, chemical engineers are still commonly tackling these problems by discretising the analytically insolvable PDEs in spaces and then integrating the resulting ODE-system in time. This method is referred as method-of-lines (mol).

Unfortunately, this approach suffers very often from slow transient behaviour due to the nonlinear nature and stiffness of the problem. It should be stressed that modern time-integrators – based on multi-step or extrapolation schemes – which allow for large time steps, lose most of their efficiency after every switch. Strictly speaking, a new initial value problem arises and for reasons of accuracy and stability most algorithms start over with a one-step scheme, e.g., an implicit one-step Euler scheme in combination with a small time step. Additionally, the exact transient behaviour into the cyclic steady-state is not of major interest for a systematic process-design (cf. Fig. 4). For the analysis, design and optimisation of instationary processes one merely requires knowledge of cyclic steady-states. When using conventional dynamical simulations, sometime up to a few hundred cycles need to be simulated [21]. On the other hand, due to the inherently instationary behaviour one cannot find steady-state solutions by just setting the time-derivatives to zero and solving the resulting system. This becomes clear when looking at a typical (cf. Fig. 4, middle) behaviour in time.

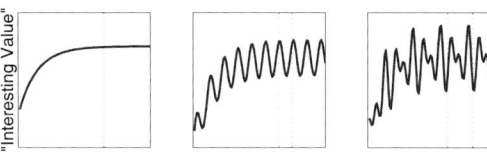

FIGURE 4. Principle transient behaviour of a stationary, (cyclic) instationary and chaotic processes (left to right)

For the direct calculation of cyclic steady-states modern algorithms make use of the periodicity condition, such as symmetries of the solution at the beginning and the end of full cycle. As a result, the problem is reformulated from an initial value problem (IVP) to a (stationary) boundary value problem (BVP) in space and time which can consequentially be solved, e.g., by the shooting method or by a global discretisation over a period. We have restricted our research to latter.

2.1. Space-time finite-difference discretisation

An efficient approach to solve the governing equations is the direct calculation method which can be based on global discretisation. Alternatively, a direct calculation of cyclic steady-states can be solved with a dynamical simulation wrapped by a shooting method algorithm. Both methods exhibit characteristic advantages. We choose the global discretisation approach since we believe that it allows better implementation of modern mathematical algorithms (mesh-generation, discretisation and solvers) developed for 2D/3D problems, whereby the shooting method is essentially restricted to method of lines.

Earlier research clearly shows that either approach is far more efficient than a simple dynamical simulation – but only when cyclic steady-states are of major interest [21, 22]. In particular, this applies to the case of parameter studies in which hundreds of cyclic steady-states must be calculated. To obtain a well-posed problem, in addition to the usual Danckwerts boundary conditions in space, we require either an initial value condition or boundary value conditions in time. A initial value is usually a prescribed distribution, e.g., of the temperature

$$T(z, t = 0) = T_0(x).$$

A typical boundary condition in the case of the RFR can be a mirror symmetric profile for the temperature in time:

$$T(z, t) = T(L - z, t + \Delta t_{cyc}).$$

It should be noted that this (mirror) symmetry condition formulated in the direct calculation is precisely the commonly used stopping criterion in the dynamical simulation. Tab. 1 shows a comparison of the two approaches.

TABLE 1. Comparison of state of the art methods for direct calculation

	Shooting method	Global discretisation
memory consumption	moderate	high
flexibility	comparably high	low
stability of iteration	possibly problematic	high
discretisation	adaptive	adaptive
2D/3D	unclear	simple
Stabilising conv. terms	possibly FCT	TVD

2.2. Treatment of convection-dominated cases

There are several general demands on a "good" numerical algorithm, e.g., high accuracy, robustness and moderate consumption of computer resources. These prerequisites are rather difficult to satisfy for the above-mentioned class of problems due to following effects:

- Strongly exothermic reactions in RFR → **steep gradients**.
- Adsorptive/desorptive reactions → **self-sharpening phenomena**.
- Endothermic/exothermic coupling in RFR → **shock-like fronts**.

In light of the above effects, the direct calculation should be based on a numerical method that provides a proper stabilization of the "bad-behaved" convective terms. Let us consider a generic transport equation typical of non-stationary processes in a fixed bed reactor

$$a_1 \frac{\partial u}{\partial t} + a_2 \frac{\partial u}{\partial z} = a_3 \frac{\partial^2 u}{\partial z^2} + a_4(u - u^*) + f(u, v, \dots). \tag{4}$$

It is well known that discretisation of the first derivatives in the left-hand side is a potential source of numerical troubles. Standard high-order methods give rise to non-physical oscillations, while the results produced by low-order ones are corrupted by excessive numerical diffusion. Unfortunately, there is no way out of this dilemma as long as the discretisation technique is **linear** [10]. Therefore,

modern high-resolution schemes are typically based on a **nonlinear** approximation of the convective fluxes. Roughly speaking, a high-order method is employed
in regions where the solution is sufficiently smooth but in the vicinity of steep
gradients it is replaced by a non-oscillatory first-order scheme like "upwind". The
far-reaching idea of adaptive switching between high- and low-order discretisations
can be traced back to the concepts of *flux-corrected transport* (FCT) which were
introduced in the early 1970s by Boris and Book [4].

In the limit of pure convection, any physically admissible solution to a scalar
transport problem proves *total variation diminishing* (TVD). In one dimension,
the total variation is defined as

$$TV(u) = \int \left| \frac{\partial u}{\partial x} \right| dx. \tag{5}$$

As long as this quantity does not increase with time, it can be shown that

- there is no formation and/or enhancement of local extrema,
- positivity and/or monotonicity of initial data is preserved.

Hence, it is natural to require that numerical solutions also possess these properties, which lead to the following constraint to be imposed at the fully discrete
level

$$TV(u^{n+1}) \leq TV(u^n), \quad \text{where} \quad TV(u^n) = \sum_i |u_i^n - u_{i-1}^n|. \tag{6}$$

Here and below u_i stand for the values of the approximate solution at the mesh
nodes z_i and the superscripts refer to the time level at which it is evaluated.

To illustrate the derivation of classical TVD schemes, consider the linear
convection equation

$$\frac{\partial u}{\partial t} + v \frac{\partial u}{\partial z} = 0, \qquad v > 0 \tag{7}$$

discretised in space by a conservative finite difference/volume method which yields

$$\frac{du_i}{dt} + \frac{f_{i+1/2} - f_{i-1/2}}{\Delta z} = 0. \tag{8}$$

The neighbouring grid points x_i and x_{i+1} exchange the conserved quantities via
numerical fluxes $f_{i\pm1/2}$ which are supposed to be consistent with the underlying
continuous flux $f = vu$. Harten [11] proved that such a semi-discrete scheme is
TVD if it can be rewritten in the form

$$\frac{du_i}{dt} = c_{i-1/2}(u_{i-1} - u_i) + c_{i+1/2}(u_{i+1} - u_i) \tag{9}$$

with (possibly nonlinear) nonnegative coefficients $c_{i-1/2} \geq 0$ and $c_{i+1/2} \geq 0$. To
meet these requirements, the numerical flux for a TVD method can be constructed
by blending a high-order approximation $f_{i+1/2}^H$ and its low-order counterpart $f_{i+1/2}^L$
as follows

$$f_{i+1/2} = f_{i+1/2}^L + \Phi_{i+1/2}[f_{i+1/2}^H - f_{i+1/2}^L], \qquad 0 \leq \Phi_{i+1/2} \leq 2, \tag{10}$$

where the value of $\Phi_{i+1/2}$ depends on the local smoothness of the solution and on the choice of the **limiter** function. If the linear flux approximations are given by

$$f^L_{i+1/2} = vu_i \quad \text{and} \quad f^H_{i+1/2} = v\frac{u_{i+1} + u_i}{2}, \tag{11}$$

then it is easy to verify that the standard upwind, central, and downwind discretisation of the convective term are recovered in case of $\Phi_{i+1/2} = 0, 1$, and 2, respectively. The most popular flux limiters are known as MINMOD, VAN LEER, MC, and SUPERBEE. All of them yield non-oscillatory results but their numerical behaviour may be quite different. For a detailed presentation and a comparative study of classical TVD methods, the interested reader is referred to [24].

A fully multidimensional flux limiter of TVD type was proposed by Kuzmin and Turek [12]. Their novel approach to the design of high-resolution schemes is based on the principle of *algebraic flux correction*. In essence, a centred space discretisation of the convective terms is rendered *local extremum diminishing* (LED → TVD in the 1D case) by a conservative elimination of negative off-diagonal coefficients from the discrete operator. This straightforward "postprocessing" technique is very flexible and can be readily integrated into existing CFD codes.

The flow chart of required algebraic manipulations is sketched in Fig. 5. First, the governing equation is discretised in space by an arbitrary linear high-order

1. Linear high-order scheme   (e.g., Galerkin FEM)

$$M_C\frac{du}{dt} = Ku \quad \text{such that} \quad \exists\, j \neq i:\ k_{ij} < 0$$

2. Linear low-order scheme $L = K + D$

$$M_L\frac{du}{dt} = Lu \quad \text{such that} \quad l_{ij} \geq 0,\ \forall j \neq i$$

3. Nonlinear high-resolution scheme $K^* = L + F$

$$M_L\frac{du}{dt} = K^*u \quad \text{such that} \quad \exists\, j \neq i:\ k^*_{ij} < 0$$

Equivalent representation $L^*u = K^*u$ is LED

$$M_L\frac{du}{dt} = L^*u \quad \text{such that} \quad l^*_{ij} \geq 0,\ \forall j \neq i$$

FIGURE 5. Roadmap of matrix manipulations.

method (e.g., central differences or the Galerkin FEM) which yields a system of ordinary differential equations for the vector of time-dependent nodal values. In the finite element context, the consistent mass matrix M_C is replaced by its "lumped" counterpart M_L. Furthermore, the high-order transport operator K is transformed into a non-oscillatory low-order one by adding a *discrete diffusion operator* D (i.e., a symmetric matrix with zero row and column sums) designed so as to get rid of all negative off-diagonal coefficients. In order to prevent excessive smearing, it is necessary to remove as much artificial diffusion as possible without generating wiggles. To this end, a limited amount of compensating *antidiffusion* F is added in the next step. In practice, both diffusive and antidiffusive terms are represented as a sum of internodal fluxes which are constructed edge-by-edge and inserted into the global vectors. Even though the final transport operator K^* does have negative off-diagonal coefficients, they are harmless as long as there exists an equivalent LED representation of the modified scheme. That is: for a given solution vector u, there should exist a matrix L^* such that all off-diagonal entries l_{ij}^* are nonnegative and $L^*u = K^*u$.

Remarkably, this methodology is directly applicable to steady-state problems as well as to time-dependent PDEs reformulated as stationary ones in space-time domain. To put it another way, it is possible to solve the discretised equations for all time levels simultaneously instead of doing it step-by-step as usual. Consider a scalar conservation law discretised in space and time by a linear second-order scheme (e.g., central differences / leapfrog time-stepping)

$$M_L \frac{u^{n+1} - u^{n-1}}{2\Delta t} = Ku^n, \qquad n = 2, \ldots, M \qquad (12)$$

where $u^n = [u_1^n, \ldots, u_N^n]^T$ denotes the vector of nodal values at the time level $t^n = (n-1)\Delta t$.

Combining the $N \times M$ equations, we obtain an algebraic system of the form $Au = f$, where

$$A = \begin{bmatrix} -M_L - \Delta t K & M_L & & & & \\ -M_L & -2\Delta t K & M_L & & & \\ & -M_L & -2\Delta t K & M_L & & \\ & & \cdots & \cdots & & \\ & & & -M_L & -2\Delta t K & M_L \\ & & & & -M_L & M_L - \Delta t K \end{bmatrix}, \qquad u = \begin{bmatrix} u^1 \\ u^2 \\ u^3 \\ \cdots \\ u^{M-1} \\ u^M \end{bmatrix}.$$
$$(13)$$

Since the coefficients of A and K have opposite signs, all **positive** off-diagonal ones need to be eliminated. If the order of nodes is such that $a_{ji} < a_{ij}$, then the "optimal" artificial diffusion coefficient is given by $d_{ij} = \max\{0, a_{ij}\}$ [12]. The required matrix modification is as follows:

$$a_{ii} := a_{ii} + d_{ij}, \qquad a_{ij} := a_{ij} - d_{ij},$$
$$a_{ji} := a_{ji} - d_{ij}, \qquad a_{jj} := a_{jj} + d_{ij}.$$

As a result, we obtain a non-oscillatory low-order operator which has no positive off-diagonal entries and no negative diagonal ones (a so-called *M-matrix*)

$$
A^* = \begin{bmatrix}
-\Delta t L & 0 & & & & \\
-2M_L & 2M_L - 2\Delta t L & 0 & & & \\
& -2M_L & 2M_L - 2\Delta t L & 0 & & \\
& & \ddots & & & \\
& & & -2M_L & 2M_L - 2\Delta t L & 0 \\
& & & & -2M_L & 2M_L - \Delta t L
\end{bmatrix}. \quad (14)
$$

Here L represents the least diffusive linear LED counterpart of K [12]. Interestingly enough, the fully implicit upwind difference scheme is recovered in the case of pure convection in 1D.

The converged low-order solution serves as a reasonable initial guess for the space-time TVD solution to be computed in an iterative way within an outer defect correction loop. Each defect/flux correction cycle consists of the following algorithmic steps [12]

1. Compute the residual of the low-order scheme $r = f - Au$.
2. Evaluate the limited *antidiffusive fluxes* $f_{ij}^a = \Phi_{ji} d_{ij}(u_i - u_j)$ and insert them into the global defect vector r, see [12] for details.
3. Solve the linear sub-problem $A\Delta u = f$ and compute $u := u + \Delta u$.

All the necessary information is extracted from the original matrix A, while its low-order counterpart A^* constitutes an excellent preconditioner. In each outer iteration, the quality of the solution improves but intermediate results may exhibit spurious undershoots/overshoots. In order to secure the convergence, it is worthwhile to perform implicit underrelaxation (divide the diagonal entries of the preconditioner by a suitably chosen parameter $0 < \omega \leq 1$ so as to enhance the diagonal dominance) which can be interpreted as a local time-stepping method [7].

3. Numerical results

3.1. Prestudy: Cauchy problem

As a prestudy a pure transport problem was considered (15). A step (initial value) moves with a constant positive velocity of 0.5 in space:

$$
u_t + 0.5 \cdot u_x = 0 \qquad (x, t) \in \Omega = (0, 1)^2, \qquad (15)
$$

$$
u(z, t = 0) = \begin{cases}
0, & \text{for} \quad 0.0 \leq z \leq 0.2 \\
1, & \text{for} \quad 0.2 < z < 0.4 \\
0, & \text{for} \quad 0.4 \leq z \leq 1.0 \, .
\end{cases}
$$

Although there is an analytical solution for this problem, the numerical treatment is very hard to solve due to the two discontinuities. Therefore, this Cauchy-problem is an adequate test for the quality of the numerical method. The computational domain was chosen to be a unit-square. An equidistant mesh comprising 100 points in time and 50 in space was applied which corresponds to 5000 as overall number of unknowns. The global discretisation was based on the leapfrog-scheme (LF) which tends to exhibit unacceptable oscillations for non-smooth solutions

$$\frac{u_i^{j+1} - u_i^{j-1}}{2\Delta t} + v\frac{u_{i+1}^{j} - u_{i-1}^{j}}{2\Delta z} = 0$$

and hence, results in a system of linear algebraic equations

$$M^{LF}\mathbf{U} = b. \tag{16}$$

The nodal unknowns u_i^j for $i = 1, \ldots, 1/\Delta z - 1$ and $j = 1, \ldots, 1/\Delta t - 1$ approximate the solution of (15) in the points (z_i, t^j) for $z_i = i \cdot \Delta z$ and $t^j = j \cdot \Delta t$. The matrix M^{LF} depicts the so-called discrete transport operator and the right-hand side b contains the initial condition (step) and spatial boundary conditions. Solving the linear system (16) directly by any linear solver, e.g., by direct solvers, leads to a solution exhibiting the mentioned oscillations throughout the domain (cf. Fig.6, left). To suppress these numerical or unphysical oscillations and to present the power of non-linear stabilisation techniques we applied the FEM-TVD method suggested by Kuzmin and Turek [12]. Starting from the linear system (16), this method first substitutes the high order transport operator by the help of *discrete upwinding*. The new transport matrix M^{DU} already fulfils TVD-properties but it is also very diffusive at the same time (cf. Fig.6, middle). Secondly, in a defect-correction-loop the amount of additional admissible antidiffusive flux for each node is detected by the help of limiter functions and then added nodewise. By this measure the solution can essentially be extended to second order accuracy and still fulfilling TVD properties at the same time. This correction is carried out in the right-hand side vector to prevent a costly matrix update in each step (cf. previous section). Fig.6 shows a considerable low amount of numerical diffusion in the plotted solution vector. (In this case the SUPERBEE-limiter function was chosen which leads to extreme low numerical diffusion.)

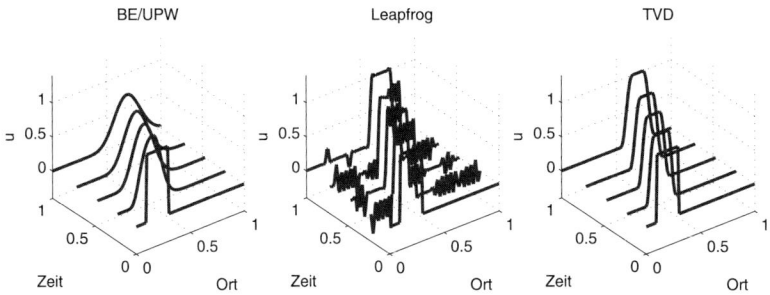

FIGURE 6. Pure transport of a step. Solved with upwinding (left), non-stabilised *Leap-frog* (middle) and space-time-TVD (right)

3.2. Catalytic combustion of N_2O in reverse-flow operation

In the case of the catalytic combustion of N_2O in reverse-flow operation we are also able to compare our simulations with experimental results retrieved from our laboratory of Bio- and chemical engineering [17]. Looking at Fig.7 one can find a qualitative good agreement of calculated and measured temperature profiles. It can clearly be seen that the non-adiabatic condition leads to heat losses forming "M"-shaped temperature profiles. Although the reaction rates are quite large there was no need for the application of TVD-stabilisation. In the present case, the convective terms were discretised and sufficiently solved by a linear LUDS approach [24].

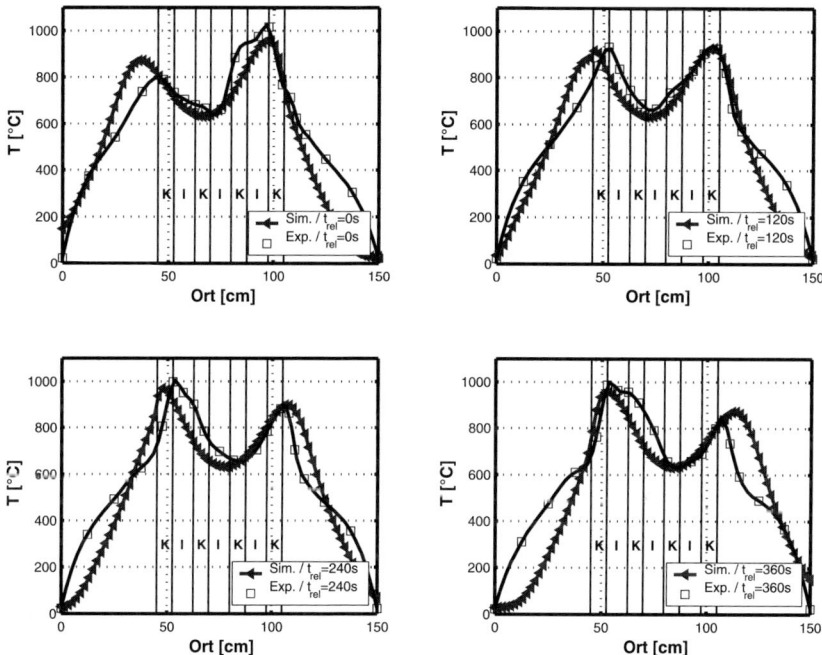

FIGURE 7. Comparison between simulation and experiment. Four spatial temperature distributions at four time-points within the cycle are depicted. The measured "M"-shaped profiles and the local extrema are qualitatively well predicted by the direct calculation. Cycle-time is $\Delta t_{cyc} = 180s$ and inlet is concentration $C = 2,0$ mol/m^3.

3.3. Endothermic steam reforming in reverse-flow operation

To present the power of the implemented TVD-algorithms not only for the test case, but also for a technically relevant problem, we now consider a novel fixed-bed process for the methane-steam-reforming under instationary conditions, most recently suggested in [9]. Here only the reaction cycle was simulated from which it is known that it can exhibit shock fronts. In [9] a simplified model is derived under the assumption that the reaction dominates, compared to any physical diffusion of mass and heat. The resulting model consists only of one heat balance

$$((1 - \varepsilon)\varrho^* c_p^*) \cdot \frac{\partial T}{\partial t} + \dot{m} c_p \left(1 + \Delta T_{ad} \frac{dX(T)}{dT}\right) \frac{\partial T}{\partial z} = 0. \tag{17}$$

The crucial part for the simulation is the knowledge of the temperature dependent equilibrium $X(T)$. In [9] this function was found to exhibit a saturating behaviour. For low temperatures X adopts values just above zero increasing fast around 800 K and then approaching the unity for higher temperatures which thermodynamically corresponds to nearly full conversion of the (value) product. In this case the equilibrium function shows an inflexion point at approx. 820 K. As a result the first derivative $dX(T)/dT$ possesses an extreme value (maximum) at the same temperature. With respect to the simplified model (17) one can find the derivative $dX(T)/dT$ incorporated in the convective term which leads to the mentioned shock fronts. We also applied a global discretisation based on the leap-frog stencil for this problem. All physical properties were taken from [9]. The computational rectangular domain has the size $0 \leq z \leq 0.7$ for the space coordinate in meters and the size $0 \leq t \leq 60$ for the time in seconds. We choose 200 grid points in each dimension. For the initial temperature distribution, T(t=0,z), we selected a ramp which graduately increases from 400 to 1500 K in the first 10 cm and then remains constant at 1500 K (cf. Fig. 8, right, first line).

$$T(z, t = 0) = \begin{cases} 400 + 11000 \cdot z, & \text{for} \quad 0.0 \leq z \leq 0.1 \\ 1500, & \text{for} \quad 0.1 \leq z \leq 0.7 \end{cases}.$$

This initial condition enables shocks to develop because there are temperatures lower and higher than 820 K. Since only the first half-cycle (reaction cycle) was considered there was no need to formulate periodicity conditions in time. Hence the calculation is more of a dynamical simulation – but in which all time steps are simultaneously solved – than a true direct calculation. The presented test case can be regarded as worst-case approximation as the model equation includes no physical diffusion at all. With some justification it can be claimed that any amount of additional diffusion in the physical model will lead to better convergence behaviour in the nonlinear solution loop.

With respect to Fig. 8 one can clearly see the formation of the shock front in the temperature distribution during the course of the cycle. The solution profiles exhibit a considerable low amount of diffusion so that the gradients are resolved very accurately.

Numerical behaviour

From the three presented test cases it can be concluded that the global discretisation in combination with stabilisation techniques for convective dominated transport problems delivers a flexible and accurate numerical treatment of the underlying mathematical model. But is the method also efficient and fast? To tell the

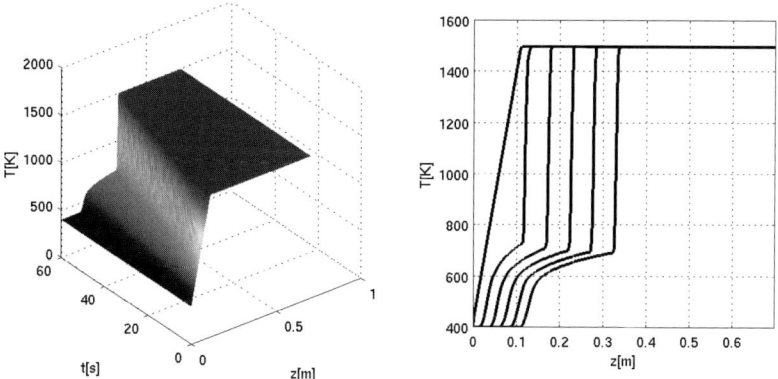

FIGURE 8. Thermodynamically determined endothermic reaction forms shocks

truth there are still unresolved numerical problems. The sharper the fronts become the more non-linear iterations must be carried out in order the bring the residual close to zero. In our future work we shall focus on an appropriate treatment of the non-linear flux correction in the TVD algorithm. These could for example be based on pseudo time-stepping or methods of quasi-Newton type.

4. Summary and discussion

Many intensified and multifunctional fixed bed processes in the field of chemical engineering exhibit cyclic steady-states due to the underlying operation scheme – reaction cycle followed by regeneration cycle and so on. Since standard numerical simulations based on method of lines reach the cyclic steady-state only after a long simulation time and many simulated cycles, modern numerical methods for the direct calculation of cyclic steady-states have attracted an increased interest. In particularly in the case of parameter studies, where many cyclic steady-states are to be calculated, this approach is a must. We have presented a modified direct calculation which is based on space-time global discretisation of the governing model equations. This technique enables the user to calculate initial value problems as well as boundary value problems. In addition to this, modern non-linear stabilisation techniques of type FEM-TVD have been incorporated. This way not only sharp fronts are resolved with a high accuracy, but also shocks do not lead to a break-down of calculation.

Acknowledgement

This work became possible due to funding of Deutsche Forschungsgemeinschaft (DFG). Special thanks belong to R. Dyll.

References

[1] Agar, D.W., Galle, M., Watzenberger, O., 2001, Thermal N_2O Decomposition in Regenerative Heat Exchanger Reactors. *Chem. Ing. Sci.*

[2] Baerns, M., Hofmann, H., Renken, A., 1987, Chemische Reaktionstechnik – Lehrbuch der Techischen Chemie, Band **1 15**.

[3] Björck, A., Dahlquist, G., 1972, Numerische Methoden. *R. Oldenbourg Verlag, München* **Auflage 1**.

[4] Boris, J.P. and Book, D.L., 1973, Flux-corrected transport. I. SHASTA, A fluid transport algorithm that works. *J. Comput. Phys.* **11** 38–69.

[5] Davis, T.A., Duff, I.S., 1997, *A Combined Unifrontal/Multifrontal Method for unsymmetric Sparse Matrices*, Technical Report TR-97-016, Computer and Information Science and Engineering Department, University of Florida.

[6] Elsner, M.P., Dittrich, C., Agar, D.W., 2002, Adsorptive reactors for enhancing equilibrium gas-phase reactions – Two case studies, *Chemical Engineering Science* **57** (9), 1607.

[7] Ferziger J.H., and Peric, M.,1996, *Computational Methods for Fluid Dynamics.* Springer.

[8] Galle, M., 2001, Hybride homogene und heterogene Reaktionsführung in Hochtemperatursystemen am Beispiel der Lachgaszersetzung, *Dissertation, University of Dortmund.*

[9] Glöckler, B., Kolios, G., Eigenberger, G., 2003, Analysis of a novel reverse-flow reactor concept for autothermal methane steam reforming, *Chem. Eng. Sci.* **58** 593–601.

[10] Godunov, S.K., 1959, Finite difference method for numerical computation of discontinuous solutions of the equations of fluid dynamics. *Mat. Sbornik* **47** 271–306.

[11] Harten, A., 1983, High resolution schemes for hyperbolic conservation laws. *J. Comput. Phys.* **49** 357–393.

[12] Kuzmin, D. and Turek,S., 2004, High-resolution FEM-TVD schemes based on a fully multidimensional flux limiter. *J. Comput. Phys.* **198** 131–158.

[13] Kuzmin, D., Möller, M. and Turek, S., 2004, High-resolution FEM-FCT schemes for multidimensional conservation laws. Submitted to *Comput. Methods Appl. Mech. Engrg.*

[14] Matros, Yu.Sh., 1989, Catalytic processes under unsteady-state conditions. *Studies of surface Science and catalysis* **43**, Amsterdam, Elsevier.

[15] Matros Yu.Sh., Boreskov, G.K., Lahmostov, V.S., Volkov, Yu.V., & Ivanov, A.A., 1984, *Method of producing sulfur trioxide.* US Patent 4,478,808.

[16] Matros, Yu.Sh., & Bunimovich, G.A., 1996, Reverse-flow operation in fixed-bed catalytic reactors. *Catalysis Review-Science and Engineering*, 38, 1–68.

[17] Nalpantidis, K., 2003, Untersuchung der heterogen/homogen Zersetzung von N_2O in einem periodisch betriebenen Festbettreaktor, *Diploma thesis, University of Dortmund.*

[18] Platte, F., 2000, Untersuchungen zum nichtlinearen dynamischen Verhalten des Reverse Flow Reactors im Hinblick auf die Diskriminierung gekoppelter homogener und heterogener Reaktionsbeiträge bei der N_2O-Zersetzung. *Diploma thesis, University od Dortmund.*

[19] Platte, F., Fredebeul, C., 2000, *Dynamische und stationäre Simulation des Re-aktionsverhaltens eines Strömungsumkehrreaktors*, Ergebnisberichte Angewandte Mathematik Nr. 190-T, University of Dortmund.

[20] Platte, F., Fredebeul, C., 2001, *Zur Anwendung direkter Löser bei der direkten Berchnung periodisch stationärer Zustände eines Strömungsumkehrreaktors*, Ergebnisberichte Angewandte Mathematik Nr. 197-T, Uni Dortmund.

[21] Salinger, A.G., 1996, The Direct Calculation of Periodic States of the Reverse Flow Reactor – I. Methodology and Propane Combustion Results. *Chem. Eng. Sci.* **51**, 4903–4913.

[22] Salinger, A.G., 1996, The Direct Calculation of Periodic States of the Reverse Flow Reactor – II. Multiplicity and Instabiliy. *Chem. Eng. Sci.* **51**, 4915–4922.

[23] Seydel, R., 1990, Practical bifurcation and stability analysis - from equilibrium to chaos. *Springer Verlag, New York* **2nd Edition**.

[24] Sokolichin, A. 2003, *Mathematische Modellbildung und numerische Simulation von Gas-Flüssigkeits-Blasenströmungen.* Habilitation thesis, University of Stuttgart.

[25] Unger, J., Kolios, G., Eigenberger, G., 1997, On the Efficient Simulation and Analysis of Regenerative Processes in cyclic operation, *Comp. Chem. Eng.* **21** 167–172.

F. Platte, D. Kuzmin, C. Fredebeul and S. Turek
University of Dortmund
Department of Applied Mathematics
Vogelpothsweg 87
D-44221 Dortmund, Germany
e-mail: `Frank.Platte@math.uni-dortmund.de`

Excursion Group to the Research Institute of the
Scientific Society Gottfried Wilhelm Leibniz e.V.

Trends and Applications in Constructive Approximation
(Eds.) M.G. de Bruin, D.H. Mache & J. Szabados
International Series of Numerical Mathematics, Vol. 151, 225–228
© 2005 Birkhäuser Verlag Basel/Switzerland

Semigroups Associated to Mache Operators (II)

Ioan Rasa

Abstract. This short note contains some supplementary results concerning the operators introduced by D.H. Mache and the semigroup associated with them. Special attention is paid to the action of the operators and the semigroup on monomials.

1. Introduction

This short note is a continuation of [3], where the semigroups associated with Mache operators were investigated. Other results in this direction can be found in [1] and [4].

In Section 2 we study the images of the monomials under Mache operators, in relation with the images under the classical Bernstein operators. The action of the semigroup on the monomials is explicitly described in Section 3. The result can be used to find the moments of the solution of the stochastic differential equation associated with the generator of the semigroup.

2. The operators

Let $a, b > -1$ and $\alpha \geq 0$ be real numbers. For $n \geq 1$, $k = 0, 1, \ldots, n$ and $x \in [0, 1]$ let $p_{nk}(x) := \binom{n}{k} x^k (1-x)^{n-k}$. Set $c := [n^\alpha]$ and consider the functionals $A_{nk} : C[0, 1] \longrightarrow \mathbb{R}$,

$$A_{nk}(f) = \left(\int_0^1 t^{ck+a}(1-t)^{cn-ck+b} f(t) dt \right) / B(ck + a + 1, cn - ck + b + 1),$$

where B is Euler's Beta function.

Let $P_n : C[0, 1] \longrightarrow C[0, 1]$,

$$P_n f = \sum_{k=0}^{n} A_{nk}(f) p_{nk}, \ f \in C[0, 1].$$

The positive linear operators P_n were introduced by D.H. Mache; see [2], [3], and the references given there.

Let $e_m(x) := x^m$, $m = 0, 1, \ldots$, $x \in [0, 1]$.

The classical Bernstein operators are defined by

$$B_n f(x) = \sum_{k=0}^n p_{nk}(x) f\left(\frac{k}{n}\right), \ f \in C[0,1], \ x \in [0,1].$$

By $||\cdot||$ we denote the uniform norm on $C[0,1]$.

Proposition 2.1. *Let* $a, b > -1$, $n \geq 2$ *and* $m \geq 0$ *be given. Then there exists a constant* K *such that*

$$||P_n e_m - B_n e_m|| \leq K n^{-\alpha}, \ \alpha \geq 0.$$

Proof. Consider the function

$$\varphi(x) = \frac{(cnx + a + 1) \ldots (cnx + a + m)}{(cn + a + b + 2) \ldots (cn + a + b + m + 1)} \ , \ x \in [0, 1].$$

It is not difficult to see that there exists a constant $K = K(a, b, m, n)$ such that $||\varphi - e_m|| \leq K n^{-\alpha}$, $\alpha \geq 0$.

On the other hand, $\varphi(k/n) = A_{nk}(e_m)$, $k = 0, 1, \ldots, n$. We infer that $P_n e_m = B_n \varphi$, and thus

$$||P_n e_m - B_n e_m|| = ||B_n(\varphi - e_m)|| \leq ||\varphi - e_m|| \leq K n^{-\alpha}. \qquad \square$$

Corollary 2.2. *If* $n \geq 2$, *then*

$$\lim_{\alpha \to \infty} P_n f = B_n f \ , \ f \in C[0, 1].$$

Concerning the behavior of the operators P_n as $\alpha \to \infty$, see also [2].

3. The semigroup

For $\alpha = 0$, $a, b \geq 0$, consider the differential operator

$$W u(x) = x(1-x)u''(x) + (a + 1 - (a + b + 2)x)u'(x), \ u \in C^2(0,1), \ x \in (0, 1),$$

with domain

$$D(W) = \{u \in C^1[0, 1] \cap C^2(0, 1) : \lim_{x \to 0, 1} x(1-x)u''(x) = 0\}.$$

For $\alpha > 0$, $a, b > -1$, set

$$W u(x) = \frac{x(1-x)}{2} u''(x), \ u \in C^2(0, 1), \ x \in (0, 1),$$

$$D(W) = \{u \in C[0, 1] \cap C^2(0, 1) : \lim_{x \to 0, 1} W u(x) = 0\}.$$

Then (see [1], [3], and the references therein):

Theorem 3.1. $(W, D(W))$ *is the infinitesimal generator of a positive contractive* C_0 *semigroup* $(T(t))_{t\geq 0}$ *on* $C[0,1]$. *For* $f \in C[0,1]$ *and* $t \geq 0$,

$$T(t)f = \lim_{n\to\infty} P_n^{[nt]}f \, , \text{ uniformly on } [0,1].$$

We shall use the following notation.

$$(y)_k = y(y-1)\ldots(y-k+1) \, , \text{ if } k \geq 1; (y)_0 = 1.$$

$$c_{m-k}(t) = \frac{(m)_k(m+a)_k}{k!} \sum_{i=0}^{k}(-1)^i \binom{k}{i} \frac{2m-2k+a+b+1+2i}{(2m-k+a+b+1+i)_{k+1}} \times$$
$$exp((k-m-i)(m-k+i+a+b+1)t) \, , \; k = 0,1,\ldots,m; t \geq 0.$$

$$d_{m-k}(t) = \frac{(m)_k(m-1)_k}{k!} \sum_{i=0}^{k}(-1)^i \binom{k}{i} \frac{2m-2k-1+2i}{(2m-k-1+i)_{k+1}} \times$$
$$exp((k-m-i)(m-k+i-1)t/2) \, , \; k = 0,1,\ldots,m-1; t \geq 0.$$
$$d_0(t) = 0, \; t \geq 0.$$

Theorem 3.2.

(i) *For* $\alpha = 0$, $a, b \geq 0$,

$$T(t)e_m = \sum_{j=0}^{m} c_j(t)e_j \, , \; m \geq 0, \; t \geq 0.$$

(ii) *For* $\alpha > 0$, $a, b > -1$,

$$T(t)e_m = \sum_{j=0}^{m} d_j(t)e_j \, , \; m \geq 0, \; t \geq 0.$$

Proof. In both cases $e_m \in D(W)$.

In the first case set $u(t,x) = \sum_{j=0}^{m} c_j(t)x^j$, and in the second,

$u(t,x) = \sum_{j=0}^{m} d_j(t)x^j, \; t \geq 0 \; x \in [0,1]$.

Then $u(t, \cdot) \in D(W)$, $t \geq 0$. We shall prove that

$$\begin{cases} u_t(t,x) = Wu(t,x), \\ u(0,\cdot) = e_m. \end{cases} \tag{1}$$

In the case (i), (1) is equivalent to the following four relations:

$$c_m' + m(m+a+b+1)c_m = 0; \tag{2}$$

$$c_{m-k}' + (m-k)(m-k+a+b+1)c_{m-k} =$$
$$(m-k+1)(m-k+1+a)c_{m-k+1}, \quad k = 1,\ldots,m; \tag{3}$$

$$c_m(0) = 1; \tag{4}$$

$$c_{m-k}(0) = 0, \quad k = 1,\ldots,m. \tag{5}$$

(2), (3) and (4) can be verified by a straightforward calculation. To prove (5) we need the following identity which can be established by using elementary algebra:

$$\sum_{i=0}^{k}(-1)^i\binom{k}{i}\frac{x-k+2i}{(x+i)(x+i-1)\ldots(x+i-k)}=0. \tag{6}$$

Set $x := 2m - k + a + b + 1$; then (5) follows from (6).

The proof of (1) in the case (ii) is similar.

Since the function $u(t,x)$ is a solution of (1), it coincides with $T(t)e_m$, and the proof is finished.   □

The expressions of $T(t)e_1$ and $T(t)e_2$ in the case $\alpha = 0$, $a,b \geq 0$ can be found also in [1], where they are interpreted as moments of order one and two for the solution of the stochastic differential equation associated with the generator of the semigroup. The moments of higher orders can be determined using the above theorem.

References

[1] Altomare, F. and Rasa, I., *On some classes of diffusion equations and related approximation problems,* these Proceedings.

[2] Mache, D.H., *A link between Bernstein polynomials and Durrmeyer polynomials with Jacobi weights,* in: Approximation Theory VIII, Vol. 1, Approximation and Interpolation, C.K. Chui and L.L. Schumaker (Eds.), World Scientific Publ.Co., 1995, 403–410.

[3] Rasa, I., *Semigroups associated to Mache operators, in:* Advanced Problems in Constructive Approximation, (Eds.) M.D. Buhmann and D.H. Mache, Int. Series of Numerical Mathematics Vol.142, Birkhäuser Verlag, Basel (2002), 143–152.

[4] Rasa, I., *Positive operators, Feller semigroups and diffusion equations associated with Altomare projections,* Conf. Sem. Mat. Univ. Bari 284 (2002), 1–26.

Ioan Rasa
Technical University of Cluj-Napoca
Department of Mathematics
Str.C.Daicoviciu, 15
3400 Cluj-Napoca(Romania)
e-mail: `Ioan.Rasa@math.utcluj.ro`

Trends and Applications in Constructive Approximation
(Eds.) M.G. de Bruin, D.H. Mache & J. Szabados
International Series of Numerical Mathematics, Vol. 151, 229–245
© 2005 Birkhäuser Verlag Basel/Switzerland

Recent Progress on Univariate and Multivariate Polynomial and Spline Quasi-interpolants

Paul Sablonnière

Abstract. Polynomial and spline quasi-interpolants (QIs) are practical and effective approximation operators. Among their remarkable properties, let us cite for example: good shape properties, easy computation and evaluation (no linear system to solve), uniform boundedness independently of the degree (polynomials) or of the partition (splines), good approximation order. We shall emphasize new results on various types of univariate and multivariate polynomial or spline QIs, depending on the nature of coefficient functionals, which can be differential, discrete or integral. We shall also present some applications of QIs to numerical methods.

1. Introduction

A quasi-interpolant of f has the general form

$$Qf = \sum_{\alpha \in A} \mu_\alpha(f) B_\alpha,$$

where $\{B_\alpha, \alpha \in A\}$ is a family of polynomials or B-splines forming a partition of unity, and $\{\mu_\alpha(f), \alpha \in A\}$ is a family of linear functionals which are local in the sense that they only use values of f in some neighbourhood of $\Sigma_\alpha = supp(B_\alpha)$. The main interest of QIs is that they provide excellent approximants of functions without solving any linear system of equations. In the literature, one can find the three following types of QIs:

(i) Differential QIs (abbr. DQIs): the linear functionals are linear combinations of values of derivatives of f at some point in Σ_α.

(ii) Discrete QIs (abbr. dQIs): the linear functionals are linear combinations of values of f at some points in the neighbourhood of Σ_α.

(iii) Integral QIs (abbr. iQIs): the linear functionals are linear combinations of weighted mean values of f in the neighbourhood of Σ_α.

We shall present various types of univariate and multivariate polynomial and spline QIs, mainly dQIs and iQIs, which were recently introduced in the literature. For polynomial QIs, we only present QIs which are close to the orifginal Bernstein or Durrmeyer operators (for other types of QIs, see for example [35], [36]).

The prototype of *polynomial dQIs* is the classical Bernstein operator

$$B_n f = \sum_{i=0}^{n} f(\frac{i}{n}) b_i^{(n)}$$

where $\{b_i^{(n)}(x) = C_n^i x^i (1-x)^{n-i}, 0 \le i \le n\}$ is the Bernstein basis of the space $\mathbb{P}_n$ of polynomials of degree at most n (the C_n^i are binomial coefficients).

The prototype of *polynomial iQIs* is the Durrmeyer operator [33]

$$M_n f = \sum_{i=0}^{n} \langle f, \tilde{b}_i^{(n)} \rangle b_i^{(n)}$$

where $\tilde{b}_i^{(n)} = b_i^{(n)} / \int_0^1 b_i^{(n)} = (n+1) b_i^{(n)}$ and $\langle f, g \rangle = \int_0^1 fg$. Both can be extended to the multivariate case, either on the hypercube or on the simplex. Another extension consists in adding a Jacobi weight in the scalar product.

The prototypes of *spline DQIs* are de Boor-Fix QIs [11] and their various univariate and multivariate extensions

$$Q f = \sum_{j \in J} \lambda_j(f) B_j.$$

Here $\{B_j \; j \in J\}$ is a family of univariate B-splines of degree m on a nonuniform sequence of knots $\{t_k\}$. Assuming that $\Sigma_j = supp(B_j) = [t_{j-m}, t_{j+1}]$, we set $E_m = \{-m+1, \ldots, 0\}$ and we define $\psi_j(t) = \prod_{r \in E_m}(t_{j+r} - t) \in \mathbb{P}_m$ for all $j \in J$. For any $\tau \in \Sigma_j$, the coefficient functionals are

$$\lambda_j(f) = \frac{1}{(m-1)!} \sum_{l=0}^{m-1} (-1)^{m-l-1} D^{m-l-1} \psi_j(\tau) D^l f(\tau).$$

The prototypes of *spline dQIs* are the various univariate and multivariate extensions of Schoenberg-Marsden operators [52], [53].

$$S f = \sum_{j \in J} f(\tau_j) B_j$$

where τ_j is an interior point of $\Sigma_j = supp(B_\alpha)$.

The prototypes of *spline iQIs* are the various univariate and multivariate extensions of operators [21], [63]

$$T f = \sum_{j \in J} \langle f, M_j \rangle B_j,$$

where M_j is a B-spline (which can be different from B_j) normalized by $\int_{\Sigma_j} M_j = 1$.

As emphasized by de Boor ([10], chapter XII), a spline QI defined on non uniform partitions has to be *uniformly bounded independently of the partition* (abbr. UB) in order to be interesting for applications. Therefore, with some coworkers, we have defined various families of QIs satisfying this property and having an infinite norm as small as possible. In general it is difficult to minimize the true norm of the operator, however, it is often possible to minimize an upper bound of this norm: this gives rise to what we have called near-best (abbr. NB) QIs (see [1], [2]–[4], [40]).

Numerical applications are still not very much developed. However, QIs can be useful in approximation and estimation [21], [22], [85], in numerical quadrature [23], [73], [75], and for the numerical solution of integral or partial differential equations.

2. Univariate polynomial QIs

2.1. Basic operators

1) The *Bernstein-Stancu QI* [83] is defined for $x \in [0,1]$ by

$$S_n^{(\alpha)} f(x) = \sum_{i=0}^{n} f(\frac{i}{n}) b_i^{(n)}(x, \alpha)$$

where the *Bernstein-Stancu basis* is defined by

$$b_k^{(n)}(x, \alpha) = C_n^k \frac{(x)_\alpha^k (1-x)_\alpha^{n-k}}{(1)_\alpha^n}.$$

Here $(x)_\alpha^k = x(x+\alpha)\ldots(x+(k-1)\alpha)$, for $\alpha \in \mathbb{R}$. For $\alpha = 0$, we recover the classical Bernstein basis.

2) The *Bernstein- Phillips* (or *q-Bernstein*) *QI* ([56]–[59]) is defined for $x \in [0,1]$ by

$$B_n^q f(x) = \sum_{i=0}^{n} f\left(\frac{[i]}{[n]}\right) b_k^{(n)}(x, q)$$

where the *Bernstein-Phillips or q-Bernstein basis* is defined for $q \neq 1$ by $b_k^{(n)}(x, q) = \Gamma_n^k x^k (1-x)_q^{n-k}$. Here $[i] = \frac{1-q^i}{1-q}$, $[i]! = \prod_{s=1}^{i}[s]$, $\Gamma_n^k = \frac{[n]!}{[k]![n-k]!}$ and $(x)_q^k = \prod_{s=0}^{k-1}(1-q^s x)$. For $q = 1$, we recover the classical Bernstein basis.

Using the notation $e_s(x) = x^s$ for monomials, it is easy to prove that all the above QIs $\mathcal{B}_n$ are exact on $\mathbb{P}_1$, i.e., $\mathcal{B}_n e_s = e_s$ *for* $s = 0, 1$. Moreover they are degree preserving since $\mathcal{B}_n e_s(x) = e_s(x) + r_{s-1}(x, n)$ where $r_{s-1}(x, n)$ is some polynomial of degree at most $s - 1$ depending on n (and eventually on the parameters α, N or q).

2.2. Left and right BQIs

All operators $\mathcal{B}_n$ defined above are isomorphisms of $\mathbb{P}_n$. Moreover $\mathcal{B}_n$ and $\mathcal{A}_n = \mathcal{B}_n^{-1}$ can be expressed as linear differential operators with polynomial coefficients

$$\mathcal{B}_n = \sum_{k=0}^{n} \beta_k^{(n)} D^k, \quad \mathcal{A}_n = \sum_{k=0}^{n} \alpha_k^{(n)} D^k,$$

where $D = \frac{d}{dx}$ and the polynomials $\beta_k^{(n)} \in \mathbb{P}_k$ and $\alpha_k^{(n)} \in \mathbb{P}_k$ are defined by simple recursions (see, e.g., [64]–[67] for partial results in this sense).

For $0 \leq r \leq n$, we introduce the partial inverses:

$$\mathcal{A}_n^{(r)} = \sum_{k=0}^{r} \alpha_k^{(n)} D^k,$$

and we consider the two families of right and left BQIs:

(RBQI) The right BQIs $\mathcal{B}_n^{[r]} = \mathcal{B}_n \circ \mathcal{A}_n^{(r)}$ are defined for C^r-functions f by

$$\mathcal{B}_n^{[r]} f = \mathcal{B}_n(\mathcal{A}_n^{(r)} f) = \mathcal{B}_n\left(\sum_{k=0}^{r} \alpha_k^{(n)} D^k f\right).$$

(LBQI) The left BQIs $\mathcal{B}_n^{(r)} = \mathcal{A}_n^{(r)} \circ \mathcal{B}_n$ are defined on any (e.g., continuous) function

$$\mathcal{B}_n^{(r)} f = \mathcal{A}_n^{(r)}(\mathcal{B}_n f) = \sum_{k=0}^{r} \alpha_k^{(n)} D^k (\mathcal{B}_n f).$$

By construction, for $0 \leq r \leq n$, the BQIs $\mathcal{B}_n^{[r]}$ and $\mathcal{B}_n^{(r)}$ are exact on the space $\mathbb{P}_r$. Moreover, in many cases, the LBQIs have a uniformly bounded infinite norm, independent on n for each $0 \leq k \leq n$ fixed (see, e.g., [70], [86] for some results of this type). From this property are deduced some convergence results (see [30], [67]).

2.3. Kageyama QIs

Kageyama [44], [45] considers Stancu operators for $\alpha \in [-\frac{1}{n}, 0]$

$$S_n^{(-\frac{1}{n})} = \mathcal{L}_n, \quad S_n^{(0)} = \mathcal{B}_n,$$

where $\mathcal{L}_n$ is the Lagrange interpolation operator on the uniform partition of $[0, 1]$ (this result is due to Mühlbach). Then he truncates at order s the Maclaurin series of $S_n^{(\alpha)} f$ w.r.t. α and he takes the value of this polynomial at $\alpha = -\frac{1}{n}$:

$$\mathcal{K}_n^{(s)} f = \sum_{j=0}^{s} \frac{1}{j!} \frac{(-1)^j}{n^j} \frac{\partial^j}{\partial \alpha^j} \left[S_n^{(\alpha)} f \right]_{\alpha=0}$$

$\mathcal{K}_n^{(0)} = S_n^{(0)} = \mathcal{B}_n$ and $\mathcal{K}_n^{(\infty)} = S_n^{(-\frac{1}{n})} = \mathcal{L}_n$. He also gives expansions of $\mathcal{K}_n f$ in terms of derivatives of $\mathcal{B}_n f$ and in powers of $\frac{1}{n}$. He proves that, for all s fixed,

$\|\mathcal{K}_n^{(s)}\|_\infty$ is uniformly bounded and give Voronovskaja type results, e.g.,

$$\lim_{n\to\infty} n^{s+1}(\mathcal{K}_n^{(s)}f - f) = -\sum_{k=0}^{2s+2} \frac{1}{k!}\Upsilon_{s+1,k}D^k f$$

where the polynomials $\Upsilon_{s+1,k}$ can be computed by recursion. He also compares the expansions of $\mathcal{L}_n$, the BQIs $B_n^{(r)}$ and $\mathcal{K}_n^{(s)}$ interms of derivatives of $B_n f$ with polynomial coefficients. Numerical experiments done by the author suggest that these operators are in general better approximants than BQIs of Section 3.1.

2.4. Univariate Durrmeyer and Goodman-Sharma QIs

A straightforward generalization of the Durrmeyer operator M_n consists in introducing a Jacobi weight on $[0,1]$ in the associated scalar product

$$\langle f, g \rangle = \int_0^1 w_{\alpha,\beta}(t) f(t) g(t) dt, \quad w_{\alpha,\beta}(t) = t^\alpha (1-t)^\beta, \quad for \ \alpha, \beta > -1$$

The extended Durrmeyer-Jacobi operator ([6], [61]) is then defined by

$$M_n^{(\alpha,\beta)} f = \sum_{i=0}^n \frac{\langle f, b_i^{(n)} \rangle}{\langle e_0, b_i^{(n)} \cdot \rangle} b_i^{(n)}$$

The limit case $(\alpha, \beta) = (-1, -1)$, corresponding to the weight $\tilde{w}(x) = \dfrac{1}{x(1-x)}$, gives a QI with very attractive properties. It has been introduced by Goodman and Sharma [38], [39] for polynomial (and a variant for spline) QIs. It can be written as follows, with $Lf(x) = (1-x)f(0) + xf(1)$:

$$G_n f = Lf + (n-1) \sum_{i=1}^{n-1} \langle f - Lf, b_{i-1}^{(n-2)} \rangle b_i^{(n)}.$$

This operator is exact on $\mathbb{P}_1$ and its behaviour is quite similar to that of the classical Bernstein operator. For example, one has for $f \in C^2(I)$

$$\lim n(f(x) - G_n f(x)) = x(1-x)f''(x).$$

It also preserves the positivity, the monotonicity and the convexity of f. As discrete Bernstein operators, the above operators G_n have associated QIs in the sense of Section 3.1 [70] .

2.5. Extrapolation

All operators $\mathcal{B}_n$ described in this section have asymptotic expansions of type

$$\mathcal{B}_n f(x) \sim f(x) + \sum_{k \geq r} \frac{\varphi_k^{(n)}(f, x)}{n^k}$$

for some index r, the $\varphi_k^{(n)}(f, x)$ being linear differential operators depending on n and k. Therefore they are good candidates for extrapolation methods (see, e.g., [14] and [82]). Numerical experiments done by the author show that Richardson

234   P. Sablonnière

extrapolation is efficient while the use of variants of epsilon or Δ^2 algorithms often introduce spurious poles in the interval of definition.

3. Polynomial QIs on a simplex

3.1. Bernstein operator and associated QIs

The simplex S of dimension $d-1$ is defined in barycentric coordinates as
$S = \{\mathbf{x} = (x_1, x_2, \ldots, x_d) : |\mathbf{x}| = 1\}$ with $|\mathbf{x}| - \sum_{i=1}^{d} |x_i|$.
The associated simplex of indices, monomials and partial derivatives are defined by $\Sigma_n = \{\mathbf{i} = (i_1, i_2, \ldots, i_d) : |\mathbf{i}| = n\}$, $X_n = \{\frac{\mathbf{i}}{n} : \mathbf{i} \in \Sigma_n\} \subset S$,
$\mathbf{i}! = i_1! i_2! \ldots, i_d!$, $\quad \mathbf{x^i} = x_1^{i_1} x_2^{i_2} \ldots x_d^{i_d}$, $D^{\mathbf{i}} = D_1^{i_1} D_2^{i_2} \ldots D_d^{i_d}$ with $D_s = \frac{\partial}{\partial x_s}$.
The Bernstein basis of $\mathbb{P}_n$ (space of polynomials of total degree at most n) and the Bernstein operator are defined respectively by:

$$b_{\mathbf{i}}^{(n)}(\mathbf{x}) = \frac{n!}{\mathbf{i}!}\mathbf{x^i} \quad for \ \mathbf{i} \in \Sigma_n, \quad B_n f(\mathbf{x}) = \sum_{\mathbf{i} \in \Sigma_n} f(\frac{\mathbf{i}}{n}) b_{\mathbf{i}}^{(n)}(\mathbf{x})$$

As $\sum_{\mathbf{i} \in \Sigma_n} b_{\mathbf{i}}^{(n)}(\mathbf{x}) = 1$ and $\sum_{\mathbf{i} \in \Sigma_n} \frac{i_s}{n} b_{\mathbf{i}}^{(n)}(\mathbf{x}) = \mathbf{x}^{\varepsilon_s} = x_s$, for $1 \le s \le d$, where $\varepsilon_s = (0, 0, \ldots, 1, \ldots, 0)$, then B_n is exact on $\mathbb{P}_1$.

Let $\{l_{\mathbf{i}}^n, \mathbf{i} \in \Sigma_n\}$ be the Lagrange basis of $\mathbb{P}_n$ associated with the data points X_n. Then $l_{\mathbf{i}}^n(\frac{\mathbf{j}}{n}) = \delta_{\mathbf{ij}}$ implies $B_n l_{\mathbf{i}}^n = b_{\mathbf{i}}^{(n)}$, hence B_n is an isomorphism of $\mathbb{P}_n$. For $f \in C^2(S)$, we have the Voronovskaja type result ([47], [79], [85]),

$$\lim n \left[B_n f - f \right] = \frac{1}{2} \bar{\mathcal{D}} f$$

where $\mathcal{D} f$ is the differential operator

$$\bar{\mathcal{D}} f(x) = \sum_{i<j} x_i x_j (\partial_i - \partial_j)^2.$$

B_n and its inverse $A_n = B_n^{-1}$ in $\mathbb{P}_n$ can be expressed as linear differential operators

$$B_n = \sum_{\mathbf{i} \in \Sigma_n} \beta_{\mathbf{i}}^{(n)} D^{\mathbf{i}}, \quad A_n = \sum_{\mathbf{i} \in \Sigma_n} \alpha_{\mathbf{i}}^{(n)} D^{\mathbf{i}}$$

whose coefficients can be computed by recursion. For $0 \le k \le n$, define partial inverses

$$A_n^{(k)} = \sum_{\mathbf{i} \in \Sigma_k} \alpha_{\mathbf{i}}^{(n)} D^{\mathbf{i}}.$$

As in Section 2.2 for univariate QIs, we can consider the two families of operators:
left Bernstein quasi-interpolants (LBQIs) $B_n^{(k)} = A_n^{(k)} \circ B_n$, and right Bernstein quasi-interpolant (RBQIs) $B_n^{[k]} = B_n \circ A_n^{(k)}$, where $B_n^{(0)} = B_n^{[0]} = B_n =$ and $B_n^{(n)} = B_n^{[n]} = \mathcal{L}_n =$ Lagrange interpolation on X_n.

We have proved [65] that $\|B_n^{(2)}\|_\infty \le 2d+1$ for all $n \ge 2$, and we conjecture that for all $k \ge 0$, there exists a constant $C_k(d)$ such that for all $n \ge k$,

$$\|B_n^{(k)}\|_\infty \le C_k(d).$$

We also conjecture the Voronovskaja-type results

$$\lim n^{r+1}(B_n^{(2r)}f - f) = \mathcal{A}_{2r}f, \quad \lim n^{r+1}(B_n^{(2r+1)}f - f) = \mathcal{A}_{2r+1}^*f,$$

where $\mathcal{A}_{2r}$ and $\mathcal{A}_{2r+1}^*$ are linear differential operators, and the asymptotic expansions

$$B_n^{(2r)}f \text{ and } B_n^{(2r+1)}f \sim f + \frac{c_{r+1}}{n^{r+1}} + \frac{c_{r+2}}{n^{r+2}} \cdots$$

3.2. Durrmeyer-Jacobi QIs on a simplex

One can introduce a Jacobi weight on the simplex in the scalar product of $L_w^2(S)$: $w_\alpha(x) = x^\alpha$, $\langle f, g \rangle = \int_S w_\alpha(x)f(x)g(x)dx$, and define the Durrmeyer-Jacobi quasi-interpolants (DJQIs)

$$M_n f = \sum_{i \in \Sigma_n} \frac{\langle f, b_i^{(n)} \rangle}{\langle e_0, b_i^{(n)} \rangle} b_i^{(n)}.$$

Its eigenvectors are the Jacobi polynomials on the simplex. There holds a Voronovskaja type result [13], [79]

$$\lim n(M_n f(x) - f(x)) = \mathcal{D}_\alpha f(x)$$

where the differential operator $\mathcal{D}_\alpha$ is defined by

$$\mathcal{D}_\alpha = x^{-\alpha} \sum_{i<j} (\partial_i - \partial_j)x_i x_j^\alpha (\partial_i - \partial_j).$$

As M_n is an isomorphism of $\mathbb{P}_n$, one can expand $M_n = \sum_{k=0}^{n} \sum_{i \in \Sigma_k} \beta_i^{(n)} D^i$ and $L_n = M_n^{-1} = \sum_{k=0}^{n} \sum_{i \in \Sigma_n} \alpha_i^{(n)} D^i$. As in the univariate case [71], the polynomials $\beta_i^{(n)}$ and $\alpha_i^{(n)}$ are probably linear combinations of Jacobi polynomials on S ([26]) Setting $L_n^{(r)} = \sum_{k=0}^{r} \sum_{i \in \Sigma_n} \alpha_i^{(n)} D^i$, one can define the left DJQIs $M_n^{(r)} = L_n^{(r)} \circ M_n$, and the right DJQIs $M_n^{[r]} = M_n \circ L_n^{(r)}$, with $M_n^{(0)} = M_n$ and $M_n^{(n)} = P_n = $ orthogonal projector on $\mathbb{P}_n$ in $L^2(S)$. They have the same properties as univariate QIs, and it would be interesting to have detailed proofs, those of [65], [66] being only sketched. However, the author thinks that the following operators are still more attractive.

3.3. Jetter-Stöckler operators on a triangle

For the sake of simplicity, we describe them over a triangle (with barycentric coordinates $\{\lambda_1, \lambda_2, \lambda_3\}$) in the case of the Legendre weight ($w = 1$, see [42] for the general study on a simplex with Jacobi weight). Using the following notations:

$$D_{ij} = \partial_j - \partial_i, \ i < j, \ D = \{D_{12}, D_{13}, D_{23}\}, \ \Lambda = \{\lambda_1\lambda_2, \lambda_1\lambda_3, \lambda_2\lambda_3\},$$

$$\mathbf{k} = (k_{12}, k_{13}, k_{23}) \in \mathbb{N}^3, \ \mathcal{D}^{\mathbf{k}} = D_{12}^{k_{12}} D_{13}^{k_{13}} D_{23}^{k_{23}},$$

$$\Lambda^{\mathbf{k}} = (\lambda_1\lambda_2)^{k_{12}}(\lambda_1\lambda_3)^{k_{13}}(\lambda_2\lambda_3)^{k_{23}},$$

the authors define the following basic differential operators:

$$U_{\mathbf{k}} = \frac{1}{\mathbf{k}!}(-1)^{|\mathbf{k}|}\mathcal{D}^{\mathbf{k}}\Lambda^{\mathbf{k}}\mathcal{D}^{\mathbf{k}}, \quad \mathcal{U}_\ell = \frac{1}{\ell!}\sum_{|\mathbf{k}|=\ell} U_{\mathbf{k}}, \quad \mathcal{Y}_n = \sum_{\ell=0}^{n}(C_n^\ell)^{-1}\mathcal{U}_\ell$$

Let M_n be the Durrmeyer operator, then they prove that $U_{\mathbf{k}}$ commute with M_n for all pairs $(\mathbf{k}, n)$ and that $\mathcal{Y}_n$ is the inverse of M_n in the space of polynomials $\mathbb{P}_n$. Now, for $0 \le r \le n$ fixed, they define partial inverses and left Jetter-Stöckler quasi-interpolants (LJSQIs)

$$\mathcal{Y}_n^{(r)} = \sum_{\ell=0}^{r}(C_n^\ell)^{-1}\mathcal{U}_\ell, \quad M_n^{(r)} = \mathcal{Y}_n^{(r)}M_n.$$

One can also define right JSQIs $M_n^{[r]} = M_n\mathcal{Y}_n^{(r)}$. Both operators $M_n^{(r)}$ and $M_n^{[r]}$ are exact on $\mathbb{P}_r$. Moreover, for r fixed, the left JSQIs have uniformly bounded infinite norms w.r.t. n. Finally, the authors prove Voronovskaja-type results:

$$\lim_{n\to\infty} C_n^r(f - M_n^{(r-1)}f) = \mathcal{U}_r f$$

3.4. Extrapolation

All operators $\mathcal{B}_n$ described in this section have asymptotic expansions of type

$$\mathcal{B}_n f(x) \approx f(x) + \sum_{k\ge r} \frac{\varphi_k^{(n)}(f, x)}{n^k}$$

(see, e.g., [47] and [85]). In particular, the latter reports interesting numerical results on Richardson extrapolation of classical Bernstein operators on the triangle. It would be interesting to compare these results with those which could be obtained by extrapolating the above QIs.

4. Univariate spline QIs on uniform partitions

4.1. Univariate differential and discrete QIs

For the construction of QIs with optimal approximation order, we refer to [15] and [16], where general solutions are given, thus completing the initial work by Schoenberg in [80].

4.2. Near-best spline dQIs

Consider the family of spline dQIs of order $2m$ depending on $n + 1$ arbitrary parameters $a = (a_0, a_1, \ldots, a_n)$, $n \ge m$:

$$Q_a f = \sum_{i\in\mathbb{Z}} \Lambda f(i) M_{2m}(x - i)$$

with coefficient functionals

$$\Lambda f(i) = a_0 f(i) + \sum_{j=1}^{n} a_j \left(f(i+j) + f(i-j)\right).$$

Setting $\nu(a) = |a_0| + \sum_{j=1}^{n} |a_j|$, then we have $\|Q_a\|_\infty \leq \nu(a)$. By imposing that Q_a be exact on $\mathbb{P}_r$, with $0 \leq r \leq 2m - 1$, we obtain a set of linear constraints: $a \in V_r \subset \mathbb{R}^{n+1}$. We say that $Q^* = Q_{a^*}$ is a *near best dQI* if

$$\nu(a^*) = \min\{\nu(a); a \in V_r\}.$$

There is existence, but in general not unicity, of solutions.

Example: cubic splines (see [40]). There is a unique optimal solution for $n \geq 2$:

$$a_0^* = 1 + \frac{1}{3n^2}, \quad a_n^* = -\frac{1}{6n^2}, \quad a_j^* = 0 \; for \; 1 \leq j \leq n - 1$$

Moreover, for all $n \geq 4$, $\|Q^*\|_\infty \leq 1 + \frac{2}{3n^2}$. Here are the first values of $\|Q^*\|_\infty$ and $\nu(a^*)$; $n = 1 : 1.222$ & 1.666; $n = 2 : 1.139$ & 1.166; $n = 3 : 1.074$ & 1.074.

4.3. Near-best spline iQIs

A similar study can be done for integral spline QIs. We refer to [2], [40] and we only give an example given in these papers. Setting $a = (a_0, a_1, \ldots, a_n)$, $n \geq m$ and $M_i(x) = M_{2m}(x-i)$, we consider $Q_a f = \sum_{i \in \mathbb{Z}} \Lambda f(i) M_i$ with coefficient functionals

$$\Lambda f(i) = a_0 \langle f, M_i \rangle + \sum_{j=1}^{n} a_j \left(\langle f, M_{i-j} \rangle + \langle f, M_{i+j} \rangle \right).$$

As in Section 4.2, we have $\|Q_a\|_\infty \leq \nu(a)$ and we say that $Q^* = Q_{a^*}$ is a *near best iQI* if $\nu(a^*) = \min\{\nu(a); a \in V_r\}$. There is existence, but in general not unicity, of solutions.

Example: cubic splines (see [40]). There is a unique optimal solution for $n \geq 2$:

$$a_0^* = 1 + \frac{2}{3n^2}, \quad a_n^* = -\frac{1}{3n^2}, \quad a_j^* = 0 \; for \; 1 \leq j \leq n - 1$$

Moreover, for all $n \geq 4$, $\|Q^*\|_\infty \leq 1 + \frac{4}{3n^2}$. Here are the first values of $\|Q^*\|_\infty$ and $\nu(a^*)$; $n = 1 : 1.5278$ & 2.333; $n = 2 : 1.2778$ & 1.333; $n = 3 : 1.1481$ & 1.1482.

5. Bivariate spline dQIs on uniform partitions

5.1. A general construction of dQIs

Let φ be any kind of bivariate B-spline on one of the two classical three- or four-directional meshes of the plane (e.g., box-splines, see [7], [12], [19]). Let $\Sigma = supp(\varphi)$ and $\Sigma^* = \Sigma \cap \mathbb{Z}^2$. Let a be the hexagonal (or lozenge=rhombus) sequence formed by the values $\{\varphi(i), i \in \Sigma^*\}$. The associated central difference operator $\mathcal{D}$ is an isomorphism of $\mathbb{P}(\varphi)$, the maximal subspace of "complete " polynomials in the space of splines $\mathcal{S}(\varphi)$ generated by the integer translates of the B-spline φ (see [12], [69], [71], [72]). Computing the expansion of a in some basis of the space of hexagonal (or lozenge) sequences amounts to expand $\mathcal{D}$ in some basis of central difference operators. Then, computing the formal inverse $\mathcal{D}^{-1}$ allows to define the dQI

$$Qf = \sum_{k \in \mathbb{Z}^2} \mathcal{D}^{-1} f(k) \varphi(\cdot - k)$$

which is exact on $\mathbb{P}(\varphi)$. Let us now give two examples which are detailed in [40].

5.2. Near-best spline dQIs on a three direction mesh

Example: let φ be the C^2 quartic box-spline. Let H_s be the regular hexagon with edges of length $s \geq 1$, centered at the origin (here $\Sigma = H_2$) and let $H_s^* = H_s \cap \mathbb{Z}^2$. The near-best dQIs have coefficient functionals with supports consisting of the center and the 6 vertices of $H_s^*, s \geq 1$. The coefficients of values of f at those points are respectively $1 + \frac{1}{2s^2}$ and $-\frac{1}{12s^2}$, therefore the infinite norm of the optimal dQIs Q_s^* is bounded above by $\nu_s^* = 1 + \frac{1}{s^2}$. Here are the first values of $\|Q^*\|_\infty$ and ν_s^*; $n = 1 : 1.34028$ & 2; $n = 2 : 1.22917$ & 1.25; $n = 3 : 1.10185$ & 1.111.

5.3. Near-best spline dQIs on a four direction mesh

Example: let φ be the C^1 quadratic box-spline. Let Λ_s be the lozenge (rhombus) with edges of length $s \geq 1$, centered at the origin, and let $\Lambda_s^* = \Lambda_s \cap \mathbb{Z}^2$. The near-best dQIs have coefficient functionals with supports consisting of the center and the 4 vertices of $\Lambda_s^*, s \geq 1$. The coefficients of values of f at those points are respectively $1 + \frac{1}{2s^2}$ and $-\frac{1}{8s^2}$, therefore the infinite norm of the optimal dQIs Q_s^* is bounded above by $\nu_s^* = 1 + \frac{1}{s^2}$. Here are the first values of $\|Q^*\|_\infty$ and ν_s^*; $n = 1 : 1.5$ & 2; $n = 2 : 1.25$ & 1.25; $n = 3 : 1.111$ & 1.111.

6. Univariate spline QIs on non uniform partitions

6.1. Uniformly bounded dQIs

Let us only give an example: we start from a family of DQIs of degree m which are *exact on* $\mathbb{P}_2$.

$$Q_2 f = \sum_{j \in J} \lambda_j^{(2)}(f) B_j, \quad \lambda_j^{(2)}(f) = f(\theta_j) - \frac{1}{2}(\theta_j^2 - \theta_j^{(2)}) D^2 f(\theta_j).$$

We recall the expansion [52], [53]

$$A_j^{(2)} = \theta_j^2 - \theta_j^{(2)} = \frac{1}{(m-1)^2(m-2)} \sum_{(r,s) \in E_m^2, r \neq s} (t_{j+r} - t_{j+s})^2 > 0.$$

On the other hand, $\frac{1}{2} D^2 f(\theta_j)$ can be replaced on the space $\mathbb{P}_2$ by the second order divided difference $[\theta_{j-1}, \theta_j, \theta_{j+1}]f$, therefore the dQI defined by

$$Q_2^* f = \sum_{j \in J} \mu_j^{(2)}(f) B_j, \quad \mu_j^{(2)}(f) = f(\theta_j) - A_j^{(2)}[\theta_{j-1}, \theta_j, \theta_{j+1}]f,$$

is also exact on $\mathbb{P}_2$. Moreover, one can write

$$\mu_i^{(2)}(f) = a_i f_{i-1} + b_i f_i + c_i f_{i+1}$$

with $a_i = -A_i^{(2)}/\Delta\theta_{i-1}(\Delta\theta_{i-1} + \Delta\theta_i)$, $c_i = -A_i^{(2)}/\Delta\theta_i(\Delta\theta_{i-1} + \Delta\theta_i)$, and $b_i = 1 + A_i^{(2)}/\Delta\theta_{i-1}\Delta\theta_i$, So, according to the introduction

$$\|Q_2^*\|_\infty \le \max_{i\in J}(|a_i| + |b_i| + |c_i|) \le 1 + 2\max_{i\in J}\frac{A_i^{(2)}}{\Delta\theta_{i-1}\Delta\theta_i}.$$

The following theorem [4] extends a result given for quadratic splines in [4], [73], [75].

Theorem 6.1. *For any degree m, the dQIs Q_2^* are UB. More specifically, for all partitions of I:*

$$\|Q_2^*\|_\infty \le [\frac{1}{2}(m + 4)]$$

6.2. Uniformly bounded iQIs

General types of integral QIs are studied in [21], [63], [68]. Here, we have chosen to study a family of QIs that we call Goodman-Sharma type iQIs, as they first appear in [38]. They seem simpler and more interesting than those we have studied in [68]. The simpler GS-type IQI can be written as follows

$$G_1 f = f(t_0)B_0 + \sum_{i=1}^{n+m-2} \tilde{\mu}_i(f)B_i + f(t_n)B_{n+m-1},$$

where the integral coefficient functionals are defined by

$$\tilde{\mu}_i(f) = \int_0^1 \tilde{M}_{i-1}(t)f(t)dt,$$

$\tilde{M}_{i-1}(t)$ being the B-spline of degree $m - 2$ with support $\tilde{\Sigma}_{i-1} = [t_{i-m+1}, t_i]$, normalized by $\tilde{\mu}_i^{(0)} = \tilde{\mu}_i(e_0) = \int_0^1 \tilde{M}_{i-1}(t) = 1$. It is easy to verify that G_1 is exact on $\mathbb{P}_1$ and that $\|G_1\|_\infty = 1$. We shall study the family of GS-type iQIs defined by

$$G_2 f = f(t_0)B_0 + \sum_{i=1}^{n+m-2} [a_i\tilde{\mu}_{i-1}(f) + b_i\tilde{\mu}_i(f) + c_i\tilde{\mu}_{i+1}(f)]B_i + f(t_n)B_{n+m-1},$$

which are *exact* on $\mathbb{P}_2$. The three constraints $G_2 e_k = e_k$, $k = 0, 1, 2$, lead to the following system of equations, for $1 \le i \le n + m - 2$:

$$a_i + b_i + c_i = 1, \quad \theta_{i-1}a_i + \theta_i b_i + \theta_{i+1}c_i = \theta_i, \quad \tilde{\mu}_{i-1}^{(2)}a_i + \tilde{\mu}_i^{(2)}b_i + \tilde{\mu}_{i+1}^{(2)}c_i = \theta_i^{(2)}.$$

This is a consequence of the following facts

$$\tilde{\mu}_i(e_1) = \int_0^1 t\tilde{M}_{i-1}(t)dt = \frac{1}{m}\sum_{s=1}^m t_{i-m+s} = \theta_i,$$

$$\tilde{\mu}_i^{(2)} = \mu_i(e_2) = \int_0^1 t^2\tilde{M}_{i-1}(t)dt = \frac{2}{m(m + 1)}\tilde{s}_2(T_i)$$

$$= \frac{2}{m(m + 1)}\sum_{1\le r\le s\le m} t_{i-m+r}t_{i-m+s}$$

Theorem 6.2. *For any degree* m, *the iQIs* G_2 *are UB. More specifically, for all partitions of* I:

$$\|G_2\|_\infty \leq 5$$

The detailed proof will be given in [78].

6.3. Near-best dQIs

Let us consider the family of dQIs of degree m defined, for the sake of simplicity, on $I = \mathbb{R}$ endowed with an arbitrary non-uniform increasing sequence of knots $T = \{t_i; i \in \mathbb{Z}\}$, by

$$Qf = Q_{p,q}f = \sum_{i\in\mathbb{Z}} \mu_i(f)B_i.$$

Their coefficient functionals depend on $2p + 1$ parameters, with $p \geq m$:

$$\mu_i(f) = \sum_{s=-p}^{p} \lambda_i(s)f(\theta_{i+s}),$$

and they are exact on the space $\mathbb{P}_q$, where $q \leq \min(m, 2p)$. The latter condition is equivalent to $Qe_r = e_r$ for all monomials of degrees $0 \leq r \leq q$. It implies that for all indices i, the parameters $\lambda_i(s)$ satisfy the system of $q + 1$ linear equations:

$$\sum_{s=-p}^{p} \lambda_i(s)\theta_{i+s}^r = \theta_i^{(r)}, \quad 0 \leq r \leq q.$$

The matrix $V_i \in \mathbb{R}^{(q+1)\times(2p+1)}$ of this system, with coefficients $V_i(r, s) = \theta_{i+s}^r$, is a Vandermonde matrix of maximal rank $q + 1$, therefore there are $2p - q$ *free parameters*. Denoting $b_i \in \mathbb{R}^{q+1}$ the vector in the right-hand side, with components $b_i(r) = \theta_i^{(r)}, \quad 0 \leq r \leq q$, we consider the sequence of minimization problems, for $i \in \mathbb{Z}$:

$$\min \|\lambda_i\|_1, \quad V_i\lambda_i = b_i.$$

We have seen in the introduction that $\nu_1^*(Q) = \max_{i\in\mathbb{Z}} \min \|\lambda_i\|_1$ is an upper bound of $\|Q_q\|_\infty$ which is easier to evaluate than the true norm of the dQI.

Theorem 6.3. *The above minimization problems have always solutions, which, in general, are non unique.*
The objective function being convex and the domains being affine subspaces, these classical optimization problems have always solutions, in general non unique.

Example of optimal dQIs are given in [1], [4], [40].

7. Bivariate quadratic spline dQIs on non uniform criss-cross triangulations

At the author's knowledge, the only bivariate box-splines which have been extended to non uniform partitions of the plane are C^1-quadratic box-splines on criss-cross triangulations [20], [62]. Recently, we have constructed a set of B-splines generating the space of quadratic splines on a rectangular domain and we have defined a discrete quasi-interpolant which is exact on $\mathbb{P}_2$ and uniformly bounded independently of the partition [74]–[76].

8. Abbreviations for publishers and journals

Publishers: AP=Academic Press, New-York ; BAS=Bulgarian Academy of Science, Sofia; BV=Birkhäuser-Verlag, Basel; CUP=Cambridge University Press; JWS=John Wiley & Sons, New-York; K=Kluwer, Dordrecht; NH= North-Holland, Amsterdam; NP=Nashboro Press, Brentwood; SV=Springer-Verlag, Berlin; SIAM=Society for Industrial and Applied Mathematics, Philadelphia; VUP=Vanderbilt University Press, Nashville.

Journals: AiCM=Advances in Comput. Mathematics; ATA=Approximation Theory and its Applications (now Analysis in Theory and Applications); CAGD=Computer Aided Geometric Design; JAT=Journal of Approximation Theory; JCAM=Journal of Computational and Applied Mathematics.

Proceedings: AT2=*Approximation Theory II*, G.G. Lorentz, C.K. Chui, L.L. Schumaker (eds), AP 1976; AT4 & AT5=Approximation Theory IV & V, C.K. Chui, L.L. Schumaker, J.D. Ward (eds), AP 1983 and 1986; CMSB=Colloquia Mathematica Soc. Janos Bolyai; CS02=Curve and Surface Fitting (St Malo 2002), A. Cohen, J.L. Merrien and L.L. Schumaker (eds), NP 2003.

Preprints: PI=Prépublications IRMAR, Inst. de Recherche Math. de Rennes.

References

[1] D. Barrera, M.J. Ibañez, P. Sablonnière: Near-best quasi-interpolants on uniform and nonuniform partitions in one and two dimensions. In CS02 (2003), 31–40.

[2] D. Barrera, M.J. Ibañez, P. Sablonnière, D. Sbibih: Near-minimally normed spline quasi-interpolants on uniform partitions. PI 04–12, 2004. (Submitted).

[3] D. Barrera, M.J. Ibañez, P. Sablonnière, D. Sbibih: Near-best quasi-interpolants associated with H-splines on a three directional mesh. PI 04–14, 2004 (Submitted).

[4] D. Barrera, M.J. Ibañez, P. Sablonnière, D. Sbibih: Near-best univariate spline discrete quasi-interpolants on non-uniform partitions. PI 04–15, 2004 (Submitted).

[5] H. Berens, H.J. Schmidt and Y. Xu: Bernstein-Durrmeyer polynomials on a simplex. JAT **68** (1992), 247–261.

[6] H. Berens and Xu: On Bernstein-Durrmeyer polynomials with Jacobi weights. In *Approx. Theory and Functional Anal.*, C.K. Chui (ed.), AP (1991), 25–46.

[7] B.D. Bojanov, H.A. Hakopian, A.A. Sahakian: *Spline functions and multivariate interpolation*, K 1993.

[8] C. de Boor: Splines as linear combinations of B-splines. In AT2 (1976), 1–47.

[9] C. de Boor: Quasi-interpolants and approximation power of multivariate splines. in *Computation of Curves and Surfaces*, W. Dahmen, M. Gasca and C.A. Micchelli (eds), K (1990), 313–345.

[10] C. de Boor: *A practical guide to splines*, SV 2001. (revised edition).

[11] C. de Boor and M.G. Fix: Spline approximation by quasi-interpolants. JAT **8** (1973), 19–54.

[12] C. de Boor, K. Höllig and S. Riemenschneider: *Box-splines*. SV 1992.

[13] D. Braess and Ch. Schwab: Approximation on simplices with respect to weighted Sobolev norms. JAT **103** (2000), 329–337.

[14] C. Brezinski, M. Redivo-Zaglia: *Extrapolation methods, theory and practice*, NH 1992.

[15] P.L. Butzer, M. Schmidt, E.L. Stark, L. Vogt: Central factorial numbers, their main properties and some applications. *Numer. Funct. Anal. and Optimiz.* **10** (1989), 419–488.

[16] P.L. Butzer, M. Schmidt: Central factorial numbers and their role in finite difference calculus and approximation. In *Approximation theory*, J. Szabados and K. Tandori (eds), CMSB **58**, NH(1990), 127–150.

[17] G. Chen, C.K. Chui, M.J. Lai: Construction of real-time spline quasi-interpolation schemes, ATA **4** (1988), 61–75.

[18] W. Chen and Z. Ditzian: Multivariate Durrmeyer-Bernstein operators, in *Israel mathematical conference proceedings, Vol. 4, Conference in honour of A. Jakimowski* (1991), 109–119.

[19] C.K. Chui: *Multivariate splines*. SIAM 1992.

[20] Chui, Schumaker, Wang: Concerning C^1 B-splines on triangulations of nonuniform rectangular partitions. ATA **1** (1984), 11–18.

[21] Z. Ciesielski: Local spline approximation and nonparametric density estimation. In *Constructive theory of functions '87*, BAS (1988), 79–84.

[22] F. Costabile, M.I. Gualtieri, S. Serra, Asymptotic expansions and extrapolation for Bernstein polynomials with applications. *BIT* **36** (1996), 676–687.

[23] C. Dagnino, P. Lamberti: Numerical integration of 2D integrals based on local bivariate C^1 quasi-interpolating splines. Adv. Comput. Math. **8** (1998), 19–31.

[24] M.M. Derriennic: Sur l'approximation des fonctions intégrables sur $[0, 1]$ par des polynômes de Bernstein modifiés. JAT **31**, No 4 (1981), 325–343.

[25] M.M. Derriennic: Polynômes de Bernstein modifiés sur un simplexe T de $\mathbb{R}^l$. Problème des moments. In *Polynômes orthogonaux et applications*, C. Brezinski et al. (eds), LNM 1171, SV (1985), 296–301.

[26] M.M. Derriennic: Polynômes orthogonaux de type Jacobi sur un triangle. *C.R. Acad. Sci. Paris* Ser I, **300** (1985), 471–474.

[27] M.M. Derriennic: On multivariate approximation by Bernstein type polynomials. JAT **45** (1985), 155–166.

[28] M.M. Derriennic: Linear combinations of derivatives of Bernstein type polynomials on a simplex. CMSB **58** (1990), 197–220.

[29] R.A. DeVore, G.G. Lorentz: *Constructive Approximation*. SV 1993.

[30] A.T. Diallo: Rate of convergence of Bernstein quasi-interpolants. Publ. IC/95/295, Int. Centre for Theoretical Physics, Miramare-Trieste, 1995.

[31] Z. Ditzian: Multidimensional Jacobi type Bernstein-Durrmeyer operators. *Acta Sci. Math. (Szeged)* **60** (1995), 225–243.

[32] Z. Ditzian, V. Totik: *Moduli of smoothness.* SV 1987.

[33] J.L. Durrmeyer: Une formule d'inversion de la transformée de Laplace: applications à la théorie des moments. Thèse, Université de Paris, 1967.

[34] H.II. Gonska, J. Meier: Quantitative theorems on approximation by Bernstein-Stancu operators. *Calcolo* **21**(1984), 317–335.

[35] H.H. Gonska, J. Meier: A bibliography on approximation of functions by Bernstein type operators, in AT4, AP (1983), 739–785.

[36] H.H. Gonska, J. Meier: A bibliography on approximation of functions by Bernstein type operators. (suppl. 1986), in AT5, AP (1986), 621–654.

[37] T.N.T. Goodman, H. Oruç, G.M. Phillips: Convexity and generalized Bernstein polynomials. *Proc. of the Edinburgh Math. Soc.* **42** (1999), 179–190.

[38] T.N.T. Goodman and A. Sharma: A modified Bernstein-Schoenberg operator. In *Constructive theory of functions '87*, BAS (1988), 166–173.

[39] T.N.T. Goodman and A. Sharma: A Bernstein type operator on the simplex. *Mathem. Balkanica* **5** (1991), 129–145.

[40] M.J. Ibañez-Pérez: Cuasi-interpolantes spline discretos con norma casi minima: teoria y aplicaciones. Tesis, Univ.de Granada (Sept. 2003).

[41] K. Jetter, J. Stöckler: An identity for multivariate Bernstein polynomials, CAGD **20**(2003), 563–577.

[42] K. Jetter, J. Stöckler: New polynomial preserving operators on simplices. Report Nr 242, University of Dortmund, 2003 (Submitted).

[43] V. Kac, P. Cheung: *Quantum calculus*, SV, New-York, 2002.

[44] Y. Kageyama: Generalization of the left Bernstein quasi-interpolants. JAT **94**, No 2 (1998), 306–329.

[45] Y. Kageyama: A new class of modified Bernstein operators. JAT **101**, No 1 (1999), 121–147.

[46] M.J. Lai: On dual functionals of polynomials in B-form, JAT **67**, No 1 (1991), 19–37.

[47] M.J. Lai: Asymptotic formulae of multivariate Bernstein approximation, JAT **70** No 2 (1992), 229–242.

[48] B.G. Lee, T. Lyche, L.L. Schumaker: Some examples of quasi-interpolants constructed from local spline projectors. In *Mathematical methods for curves and surfaces: Oslo 2000*, T. Lyche and L.L. Schumaker (eds), VUP (2001), 243–252.

[49] G.G. Lorentz: *Bernstein polynomials.* University of Toronto Press, 1953.

[50] T. Lyche and L.L. Schumaker: Local spline approximation methods. JAT **15** (1975), 294–325.

[51] P. Mache and D.H. Mache: Approximation by Bernstein quasi-interpolants. *Numer. Funct. Anal. and Optimiz.* **22**, 1& 2 (2001), 159–175.

[52] J.M. Marsden, I.J. Schoenberg: On variation diminishing spline approximation methods. Mathematica (Cluj) **31** (1966), 61–82.

[53] J.M. Marsden, I.J. Schoenberg: An identity for spline functions with applications to variation diminishing spline approximation. JAT **3** (1970), 7–49.

[54] J.M. Marsden: Operator norm bounds and error bounds for quadratic spline interpolation, In: *Approximation Theory*, Banach Center Publications, vol. **4** (1979), 159–175.

[55] E. Neuman: Moments and Fourier transforms of B-splines. JCAM **7**, 51–62.

[56] H. Oruç and G.M. Phillips: q-Bernstein polynomials and Bézier curves. JCAM **151** (2003), 1–12.

[57] G.M. Phillips: On generalized Bernstein polynomials. In *Numerical Analysis: A.R. Mitchell 75-th birthday Volume*, D.F. Griffiths and G.A. Watson (eds), World Scientific, Singapore (1996), 263–269.

[58] G.M. Phillips: Bernstein polynomials based on the q-integers. Annals of Numer. Math. **4** (1997), 511–518.

[59] G.M. Phillips: *Interpolation and approximation by polynomials*. SV 2003.

[60] M.J.D. Powell: *Approximation theory and methods*. CUP 1981.

[61] P. Sablonnière: Opérateurs de Bernstein-Jacobi. Rapport ANO 37, Université de Lille, 1981 (unpublished).

[62] P. Sablonnière: Bernstein-Bézier methods for the construction of bivariate spline approximants. CAGD **2** (1985), 29–36.

[63] P. Sablonnière: Positive spline operators and orthogonal splines. JAT **52** (1988), 28–42.

[64] P. Sablonnière: Bernstein quasi-interpolants on [0, 1]. In *Multivariate Approximation Theory IV*, C.K. Chui, W. Schempp and K. Zeller (eds), ISNM, Vol. 90, BV (1989), 287–294.

[65] P. Sablonnière: Bernstein quasi-interpolants on a simplex. Meeting *Konstruktive Approximationstheorie*, Oberwolfach (July 30–August 5, 1989). Publ. LANS 21, INSA de Rennes, 1989 (unpublished).

[66] P. Sablonnière: Bernstein type quasi-interpolants. In *Curves and surfaces*, P.J. Laurent, A. Le Méhauté, L.L. Schumaker (eds), AP (1991), 421–426.

[67] P. Sablonnière: A family of Bernstein quasi-interpolants on [0, 1]. ATA **8:3** (1992), 62–76.

[68] P. Sablonnière and D. Sbibih: Spline integral operators exact on polynomials. ATA **10:3** (1994), 56–73.

[69] P. Sablonnière: Quasi-interpolants aasociated with H-splines on a three-direction mesh. JCAM **66** (1996), 433–442.

[70] P. Sablonnière: Representation of quasi-interpolants as differential operators and applications. In *New developments in approximation theory*, M.W. Müller, M.D. Buhmann, D.H. Mache, M. Felten (eds). ISNM Vol. 132, BV (1999), 233–253.

[71] P. Sablonnière: Quasi-interpolantes sobre particiones uniformes, First meeting in Approximation Theory, Ubeda, Spain, July 2000. PI 00-38, 2000.

[72] P. Sablonnière: H-splines and quasi-interpolants on a three-directional mesh. In *Advanced Problems in Constructive Approximation*, M.D. Buhmann and D. Mache (eds), ISNM Vol. 142, BV (2002), 187–201.

[73] P. Sablonnière: On some multivariate quadratic spline quasi-interpolants on bounded domains. In *Modern Developments in Multivariate Approximation*, W. Haussmann et al. (eds), ISNM Vol. 145, BV (2003), 263–278.

[74] P. Sablonnière: BB-coefficients of basic bivariate quadratic splines on rectangular domains with uniform criss-cross triangulations. PI 02-56, 2002.

[75] P. Sablonnière: Quadratic spline quasi-interpolants on bounded domains of $\mathbb{R}^d$, $d = 1, 2, 3$. *Spline and radial functions*, Rend. Sem. Univ. Pol. Torino, Vol. **61** (2003), 61–78.

[76] P. Sablonnière: BB-coefficients of bivariate quadratic B-splines on rectangular domains with non-uniform criss-cross triangulations. PI 03-14, March 2003.

[77] P. Sablonnière: Refinement equation and subdivision algorithm for quadratic B-splines on non-uniform criss-cross triangulations. Int. Conf. on *Wavelets and splines*, St Petersburg, 2003 (submitted). PI 03-35, October 2003.

[78] P. Sablonnière: Near-best univariate spline integral quasi-interpolants on non-uniform partitions. PI 2004 (In preparation).

[79] Th. Sauer: The genuine Bernstein-Durrmeyer operator on a simplex. *Results Math.* **26** (1994), 99–130.

[80] I.J. Schoenberg: *Cardinal spline interpolation*, CBMS-NSF Regional Conference Series in Applied Mathematics, vol. 12, SIAM, Philadelphia 1973.

[81] L.L. Schumaker: *Spline functions: basic theory*, JWS 1981.

[82] A. Sidi: *Practical extrapolation Methods*, CUP 2003.

[83] D.D. Stancu: Approximation properties of a class of linear positive operators. *Studia Univ. Babes-Bolyai* **15** (1970), 31–38.

[84] G. Vlaic: On the approximation of bivariate functions by the Stancu operator for a triangle. Analele Univ. Timişoara **36**, Seria Mat.-Inform.(1998), 149–158.

[85] G. Walz: Asymptotic expansions for multivariate polynomial approximation. JCAM **122** (2000), 317–328.

[86] Wu Zhengchang: Norm of Bernstein left quasi-interpolant operator. JAT **66** (1991), 36–43.

Paul Sablonnière
Centre de Mathématiques
INSA de Rennes
20 Avenue des Buttes de Coësmes, CS 14315
F-35043 Rennes cedex, France
e-mail: `psablonn@insa-rennes.fr`

In a Lecture of the Conference (with J. Szabados)

Trends and Applications in Constructive Approximation
(Eds.) M.G. de Bruin, D.H. Mache & J. Szabados
International Series of Numerical Mathematics, Vol. 151, 247–258
© 2005 Birkhäuser Verlag Basel/Switzerland

A Strong Converse Result for Approximation by Weighted Bernstein Polynomials on the Real Line

Jozsef Szabados

Abstract. We prove that the weighted error of approximation by generalized Bernstein polynomials introduced in [1] is equivalent to the modulus of smoothness of the function. This result is analogous to a well-known theorem of Ditzian and Ivanov [2] for the classical Bernstein polynomials.

Mathematics Subject Classification (2000). primary 41A36, secondary 41A25.

Keywords. Freud weights, weighted modulus of smoothness, Bernstein operator, Markov type inequalities.

1. Introduction

In [1] we introduced some Bernstein-type operators to approximate unbounded functions on the real line. Here we complete those results by giving a Ditzian-Ivanov type theorem proving the complete equivalence of the weighted error and the modulus of smoothness.

In the following c denotes a positive constant which may assume different values in different formulas. Moreover let $\nu \sim \mu$, for ν and μ two quantities depending on some parameters, if $|\nu/\mu|^{\pm 1} \leq c$, with c independent of the parameters.

Let
$$w(x) = \exp(-Q(x)), \qquad x \in (-\infty, +\infty)$$
be a Freud weight, i.e., assume that $Q(x)$ is even and continuous in $\mathbf{R}$, Q'' is continuous in $(0, \infty)$, $Q' > 0$ in $(0, \infty)$ and
$$1 < A \leq \frac{(xQ'(x))'}{Q'(x)} \leq B < \infty, \qquad x \in (0, \infty)$$
(see [4], Theorem 1.1, p. 184).

Research supported by OTKA No. T032872.

Now consider the following class of functions:

$$C_w = \{f \in C(\mathbf{R})) : \lim_{|x| \to \infty} (wf)(x) = 0\},$$

equipped with the norm $\|wf\|_{C_w} := \|wf\| = \sup_{x \in \mathbf{R}} |(wf(x)|$.

We also put $\|wf\|_{[c,d]} = \max_{c \le x \le d} |(wf)(x)|$. For $f \in C_w$, the weighted modulus of smoothness is

$$\omega_2(f,t)_w = \sup_{0 < h \le t} \|w\Delta_h^2 f\|_{[-h^*, h^*]} + \inf_{\ell \in \mathcal{P}_1} \|w(f - \ell)\|_{[t^*, \infty)} + \inf_{\ell \in \mathcal{P}_1} \|w(f - \ell)\|_{(-\infty, -t^*]},$$

(1)

where t^* is defined by $tQ'(t^*) = 1$ (see [3, Definition 11.2.2, p. 182]) and $\mathcal{P}_n$, $n \in \mathbb{N}$, is the set of algebraic polynomials of degree at most n.

Next, define the monotone increasing sequence $\{\lambda_n\}$ by the relations

$$\lambda_n Q'(\lambda_n) = \sqrt{n}, \qquad n = 1, 2, \ldots.$$

(2)

Finally, for every $f \in C_w$ let

$$B_n(f, x) = \sum_{k=0}^{n} p_{n,k}(x) f(x_k)$$

(3)

with

$$p_{n,k}(x) = \frac{1}{2^n} \binom{n}{k} \left(1 + \frac{x}{2\lambda_n}\right)^k \left(1 - \frac{x}{2\lambda_n}\right)^{n-k},$$

(4)

and $x_k = \frac{2k-n}{n} 2\lambda_n$. Further let

$$B_n^*(f, x) = \begin{cases} B_n(f, x) & \text{if } |x| \le \lambda_n, \\ B_n(f, \lambda_n) + B_n'(f, \lambda_n)(x - \lambda_n) & \text{if } x \ge \lambda_n, \\ B_n(f, -\lambda_n) + B_n'(f, -\lambda_n)(x + \lambda_n) & \text{if } x \le -\lambda_n. \end{cases}$$

(5)

Remarks. Note that $B_n^{*'} \in \mathrm{AC}_{\mathrm{loc}}$ and B_n^* is a linear operator, which reproduces linear functions ℓ, i.e., $B_n^*(\ell, x) \equiv \ell(x)$. We could not consider only B_n because its weighted norm is not bounded.

The most natural definition for B_n^* could be (4)–(5) with λ_n instead of $2\lambda_n$, but then the corresponding operator would be unbounded.

2. The result

Theorem 2.1. *If $f \in C_w$ then*

$$\frac{c}{\mu} \omega_2 \left(f, \frac{\lambda_n}{\sqrt{n}}\right)_w \le \| w (f - B_n^*(f))\| + \| w (f - B_{\mu n}^*(f))\| \left(f, \frac{\lambda_n}{\sqrt{n}}\right)_w,$$

(6)

where $\mu > 1$ is a suitable fixed integer depending only on the weight w.

This result, together with the upper estimate

$$\|w(f - B_n^\star(f))\| \le c\omega_2\left(f, \frac{\lambda_n}{\sqrt{n}}\right)_w$$

(see Theorem 1 in [1]) is very complex: it includes direct and strong converse error estimates, reproducing property (for linear functions), saturation classes and order.

3. Lemmas

The proof of Theorem is based on several lemmas.

Lemma 3.1. *We have*

$$1 - \frac{2}{n} \le \frac{\lambda_{n-2}}{\lambda_n} \le 1 - \frac{1}{Bn} \tag{7}$$

and

$$\mu^{\frac{1}{2B}} \le \frac{\lambda_{\mu n}}{\lambda_n} \le \mu^{\frac{1}{2A}} \qquad (\mu \ge 1). \tag{8}$$

Proof. Using (2) and

$$t^{A-1} \le \frac{Q'(tx)}{Q'(x)} \le t^{B-1}, \qquad 0 \le x < \infty, \ 1 \le t < \infty \tag{9}$$

(cf. [4], inequalities (5.3)) with $x = \lambda_{n-2}$ and $t = \lambda_n/\lambda_{n-2}$,

$$\left(\frac{\lambda_n}{\lambda_{n-2}}\right)^{A-1}\sqrt{1 - \frac{2}{n}} \le \frac{\lambda_{n-2}}{\lambda_n} = \frac{Q'(\lambda_n)}{Q'(\lambda_{n-2})}\sqrt{\frac{n-2}{n}} \le \left(\frac{\lambda_n}{\lambda_{n-2}}\right)^{B-1}\sqrt{1 - \frac{2}{n}},$$

i.e.,

$$1 - \frac{2}{n} \le \left(1 - \frac{2}{n}\right)^{\frac{1}{2A}} \le \frac{\lambda_{n-2}}{\lambda_n} \le \left(1 - \frac{2}{n}\right)^{\frac{1}{2B}} \le 1 - \frac{1}{Bn}.$$

On the other hand, using (2) as well as (9) with $t = \lambda_{\mu n}/\lambda_n$ and $x = \lambda_n$ we obtain

$$\frac{1}{\sqrt{\mu}}\left(\frac{\lambda_{\mu n}}{\lambda_n}\right)^{A-1} \le \frac{Q'(\lambda_{\mu n})}{Q'(\lambda_n)}\frac{1}{\sqrt{\mu}} = \frac{\lambda_n}{\lambda_{\mu n}} \le \frac{1}{\sqrt{\mu}}\left(\frac{\lambda_{\mu n}}{\lambda_n}\right)^{B-1}$$

which proves (8) after rearranging.

Denote

$$I_n := \left[-\frac{3}{2}\lambda_n, \frac{3}{2}\lambda_n\right].$$

Lemma 3.2. *We have*

$$\left\|w(x)\sum_{|x-x_k|\ge\delta}\frac{|x - x_k|^i p_{nk}(x)}{w(x_k)}\right\|_{I_n} \le c_i\left(\frac{\lambda_n}{\sqrt{n}}\right)^i\exp\left(-C\frac{n}{\lambda_n^2}\delta^2\right), \qquad i = 0, 1, \dots,$$

where $\delta \ge 0$, $C > 0$ is an absolute constant, and the constants $c_i > 0$ depend only on i.

Proof. By symmetry, it is sufficient to estimate the norm for $0 \le x \le \frac{3}{2}\lambda_n$. In this interval we use the following estimates for the polynomials $p_{nk}(x)$:

$$p_{nk}(x) \le \begin{cases} \frac{c}{\sqrt{n}} \exp\left\{-c\frac{n}{\lambda_n^2}(x - x_k)^2\right\} & \text{if } |x_k| \le \lambda_n + \frac{x}{2}, \\ e^{-cn} & \text{if } |x_k| > \lambda_n + \frac{x}{2}, \end{cases} \qquad 0 \le x \le \frac{3}{2}\lambda_n.$$

These follow from the proof of Lemma 1 in [1]. The other tool we use is

$$\frac{w(x)}{w(x_k)} \le e^{|Q(x) - Q(x_k)|} \le e^{|x - x_k| Q'(2\lambda_n)} \le e^{d|x - x_k|\frac{\sqrt{n}}{\lambda_n}}, \qquad x \in I_n, \ k = 0, \dots, n \tag{10}$$

where we used (2) and applied the relation $Q'(2\lambda_n) \sim Q'(\lambda_n)$, which follows from (9). Thus we obtain

$$w(x) \sum_{|x - x_k| \ge \delta} \frac{|x - x_k|^i p_{nk}(x)}{w(x_k)}$$

$$\le \frac{c}{\sqrt{n}} \sum_{|x - x_k| \ge \delta} |x - x_k|^i \exp\left\{-c\frac{n}{\lambda_n^2}(x - x_k)^2 + d|x - x_k|\frac{\sqrt{n}}{\lambda_n}\right\} + e^{-cn}$$

$$\le \frac{c_1}{\sqrt{n}} \sum_{|x - x_k| \ge \delta} |x - x_k|^i \exp\left\{-C\frac{n}{\lambda_n^2}(x - x_k)^2\right\} + e^{-cn}. \tag{11}$$

Here

$$|x - x_k|^i \sim \delta^i + \left(\frac{j\lambda_n}{n}\right)^i, \qquad i = 0, 1, \dots$$

with some j depending on k. Substituting this into (11) we get

$$w(x) \sum_{|x - x_k| \ge \delta} \frac{|x - x_k|^i p_{nk}(x)}{w(x_k)}$$

$$\le \frac{c}{\sqrt{n}} \exp\left(-C\frac{n}{\lambda_n^2}\delta^2\right) \sum_{j=0}^{\infty} \left[\delta^i + \left(\frac{j\lambda_n}{n}\right)^i\right] \exp\left(-\frac{C^2 j^2}{n}\right) + e^{-cn}. \tag{12}$$

Here

$$\sum_{j=0}^{\infty} \left[\delta^i + \left(\frac{j\lambda_n}{n}\right)^i\right] \exp\left(-\frac{C^2 j^2}{n}\right)$$

$$\le c \int_0^{\infty} \left[\delta^i + \left(\frac{x\lambda_n}{n}\right)^i\right] \exp\left(-\frac{C^2 x^2}{n}\right) dx$$

$$\le c\sqrt{n} \int_0^{\infty} \left[\delta^i + \left(\frac{x\lambda_n}{\sqrt{n}}\right)^i\right] e^{-C^2 x^2} dx \le c_i \sqrt{n} \left[\delta^i + \left(\frac{\lambda_n}{\sqrt{n}}\right)^i\right].$$

Substituting this into (12) we obtain the statement of the lemma.

It is clear from the above proof that we can easily obtain the following "unweighted" version of Lemma 2:

Corollary 3.3. *We have*

$$\left\| \sum_{|x-x_k|\geq\delta} |x-x_k|^i p_{nk}(x) \right\|_{I_n} \leq c_i \left(\frac{\lambda_n}{\sqrt{n}}\right)^i \exp\left(-C\frac{n}{\lambda_n^2}\delta^2\right), \quad i=0,1,\ldots,$$

where $\delta \geq 0$, $C > 0$ is an absolute constant, and the constants $c_i > 0$ depend only on i.

Lemma 3.4. *Let $f \in C_w$. Then*

$$\|wB_n'''(f)\|_{I_n} \leq c\frac{n^{3/2}}{\lambda_n^3}\|wf\|. \tag{13}$$

Proof. Using the identity

$$p_{nk}'(x) = \frac{n}{4\lambda_n^2 - x^2} p_{nk}(x)(x_k - x) \tag{14}$$

(which follows from (4) by an easy computation) successively, and differentiating (3) three times we obtain

$$B_n'(f,x) = \frac{n}{4\lambda_n^2 - x^2} \sum_{k=0}^{n} f(x_k)(x_k - x)p_{nk}(x),$$

$$B_n''(f,x) = -\frac{n}{4\lambda_n^2 - x^2} \sum_{k=0}^{n} f(x_k)p_{nk}(x)$$

$$+\frac{2nx}{(4\lambda_n^2 - x^2)^2} \sum_{k=0}^{n} f(x_k)(x_k - x)p_{nk}(x) + \frac{n^2}{(4\lambda_n^2 - x^2)^2} \sum_{k=0}^{n} f(x_k)(x_k - x)^2 p_{nk}(x),$$

and

$$B_n'''(f,x) = -\frac{4nx}{(4\lambda_n^2 - x^2)^2} \sum_{k=0}^{n} f(x_k)p_{nk}(x) \,|$$

$$+\left[\frac{2n - 3n^2}{(4\lambda_n^2 - x^2)^2} + \frac{8nx^2}{(4\lambda_n^2 - x^2)^3}\right] \sum_{k=0}^{n} f(x_k)(x_k - x)p_{nk}(x)$$

$$+\frac{6n^2 x}{(4\lambda_n^2 - x^2)^3} \sum_{k=0}^{n} f(x_k)(x_k - x)^2 p_{nk}(x) + \frac{n^3}{(4\lambda_n^2 - x^2)^3} \sum_{k=0}^{n} f(x_k)(x_k - x)^3 p_{nk}(x).$$

Multiplying the last relation by $w(x)$ and using Lemma 2 with $\delta = 0$ and $i = 0, 1, 2, 3$ we get

$$\|wB_n'''\|_{I_n} \leq c\|wf\| \left\| \frac{n}{\lambda_n^3} \sum_{k=0}^{n} \frac{p_{nk}(x)}{w(x_k)} + \frac{n^2}{\lambda_n^4} \sum_{k=0}^{n} \frac{|x-x_k|p_{nk}(x)}{w(x_k)} \right.$$

$$\left. +\frac{n^2}{\lambda_n^5} \sum_{k=0}^{n} \frac{|x-x_k|^2 p_{nk}(x)}{w(x_k)} + \frac{n^3}{\lambda_n^5} \sum_{k=0}^{n} \frac{|x-x_k|^3 p_{nk}(x)}{w(x_k)} \right\|_{I_n}$$

$$\leq c\|wf\| \left(\frac{n}{\lambda_n^3} + \frac{n^{3/2}}{\lambda_n^3} + \frac{n}{\lambda_n^3} + \frac{n^{3/2}}{\lambda_n^3}\right) \leq c\frac{n^{3/2}}{\lambda_n^3}\|wf\|.$$

This proves the lemma.

Lemma 3.5. *Let $f \in C_w$. Then*

$$\|wB_n'''\|_{I_n} \le c\frac{n^{5/2}}{\lambda_n^3} \max_{1 \le k \le n-1} [w(x_k)|\Delta^2_{2\lambda_n/n}f(x_k)|].$$

Proof. In order to avoid confusion, in this proof we will use the more precise notation x_{kn} for x_k. Introducing also the notation $p_{nk}(x) \equiv 0$ for $k < 0$ and $k > n$, it is easy to see from (4) that

$$p_{nk}'(x) = \frac{n}{4\lambda_n}\left[p_{n-1,k-1}\left(\frac{\lambda_{n-1}}{\lambda_n}x\right) - p_{n-1,k}\left(\frac{\lambda_{n-1}}{\lambda_n}x\right)\right], \qquad 0 \le k \le n.$$

Differentiating once more and using this formula again we get

$$p_{nk}''(x) = \frac{n(n-1)}{16\lambda_n^2}\left[p_{n-2,k-2}\left(\frac{\lambda_{n-2}}{\lambda_n}x\right) - 2p_{n-2,k-1}\left(\frac{\lambda_{n-2}}{\lambda_n}x\right)\right.$$
$$\left. + p_{n-2,k}\left(\frac{\lambda_{n-2}}{\lambda_n}x\right)\right], \qquad 0 \le k \le n.$$

Hence

$$B_n''(f,x) = \sum_{k=0}^{n} f(x_{kn})p_{nk}''(x) = \frac{n(n-1)}{16\lambda_n^2}\sum_{k=2}^{n-1}\left[f(x_{kn})p_{n-2,k-2}\left(\frac{\lambda_{n-2}}{\lambda_n}x\right)\right.$$
$$\left. -2\sum_{k=0}^{n} f(x_{kn})p_{n-2,k-1}\left(\frac{\lambda_{n-2}}{\lambda_n}x\right) + \sum_{k=0}^{n} f(x_{kn})p_{n-2,k}\left(\frac{\lambda_{n-2}}{\lambda_n}x\right)\right].$$

Changing the running indices in the first and third sums we obtain

$$B_n''(f,x) = \frac{n(n-1)}{16\lambda_n^2}\sum_{k=1}^{n-1}\Delta^2_{2\lambda_n/n}f(x_{kn})p_{n-2,k-1}\left(\frac{\lambda_{n-2}}{\lambda_n}x\right),$$

where we have used the usual notation for the second differences. Differentiating once more and using the identity (14) we get

$$B_n'''(f,x) = \frac{n(n-1)\lambda_{n-2}}{16\lambda_n^3}\sum_{k=1}^{n-1}\Delta^2_{2\lambda_n/n}f(x_{kn})p_{n-2,k-1}'\left(\frac{\lambda_{n-2}}{\lambda_n}x\right)$$

$$= \frac{n(n-1)(n-2)\lambda_{n-2}}{16\lambda_n^3\left(4\lambda_{n-2}^2 - \frac{\lambda_{n-2}^2}{\lambda_n^2}x^2\right)}$$

$$\times \sum_{k=1}^{n-1}\Delta^2_{2\lambda_n/n}f(x_{kn})\left(x_{k-1,n-2} - \frac{\lambda_{n-2}}{\lambda_n}x\right)p_{n-2,k-1}\left(\frac{\lambda_{n-2}}{\lambda_n}x\right).$$

Hence

$$w(x)|B_n'''(f,x)| \le c\frac{n^3}{\lambda_n^4}\max_{1 \le k \le n-1}[w(x_k)\Delta^2_{2\lambda_n/n}f(x_k)|]$$

$$\times \sum_{k=1}^{n-1}\left|x_{k-1,n-2} - \frac{\lambda_{n-2}}{\lambda_n}x\right|p_{n-2,k-1}\left(\frac{\lambda_{n-2}}{\lambda_n}x\right)e^{Q(x_{kn})-Q(x)}, \qquad x \in I_n.$$

Here

$$|Q(x_{kn}) - Q(x)| \leq \left| Q\left(\frac{\lambda_{n-2}}{\lambda_n}x\right) - Q(x_{k-1,n-2}) \right|$$

$$+ \left| Q(x) - Q\left(\frac{\lambda_{n-2}}{\lambda_n}x\right) \right| + |Q(x_{k-1,n-2}) - Q(x_{kn})|, \qquad k = 1, \ldots, n-1.$$

The last two absolute values are easily estimated by the mean value theorem, (9) and (7):

$$0 \leq Q(x) - Q\left(\frac{\lambda_{n-2}}{\lambda_n}x\right) \leq |x|\left(1 - \frac{\lambda_{n-2}}{\lambda_n}\right)Q'(2\lambda_n) \leq c\sqrt{n} \cdot \frac{2}{n} \leq \frac{c}{\sqrt{n}}, \qquad x \in I_n,$$

and by (7) again

$$|Q(x_{kn}) - Q(x_{k-1,n-2})| \leq \left|\frac{(2k-n)2\lambda_n}{n} - \frac{(2k-n)2\lambda_{n-2}}{n-2}\right| \leq \left|\frac{2n}{n-2}\lambda_{n-2} - 2\lambda_n\right|$$

$$= 2\lambda_n\left(\frac{n}{n-2}\frac{\lambda_{n-2}}{\lambda_n} - 1\right) \leq 2\lambda_n\left(\frac{n}{n-2}\frac{Bn-1}{Bn} - 1\right) \leq c\frac{\lambda_n}{n}.$$

Collecting these estimates,

$$|Q(x_{kn}) - Q(x)| \leq \left| Q\left(\frac{\lambda_{n-2}}{\lambda_n}x\right) - Q(x_{k-1,n-2}) \right| + o(1),$$

i.e., Lemma 2 with $i = 1$ and $n - 2$ instead of n yields

$$w(x)|B_n'''(f,x)| \leq c\frac{n^3}{\lambda_n^4}\max_{1 \leq k \leq n-1}[w(x_k)|\Delta_{2\lambda_n/n}f(x_k)|]$$

$$\times \sum_{k=1}^{n-1}\frac{\left|x_{k-1,n-2} - \frac{\lambda_{n-2}}{\lambda_n}x\right| p_{n-2,k-1}\left(\frac{\lambda_{n-2}}{\lambda_n}x\right)}{w(x_{k-1,n-2})}$$

$$\leq c\frac{n^{5/2}}{\lambda_n^3}\max_{1 \leq k \leq n-1}[w(x_k)|\Delta_{2\lambda_n/n}^2 f(x_k)|], \qquad x \in I_n.$$

Lemma 3.6. *Let* $K_n := \left[-\lambda_n - \frac{4\lambda_n}{n}, \lambda_n + \frac{4\lambda_n}{n}\right]$. *If* $f \in C_w$ *and*

$$\|w(f - B_n^\star(f))\| \leq \frac{\lambda_n^2}{n}\|wB_n''(f)\|_{K_n}, \tag{15}$$

then

$$\|wB_n'''(f)\|_{I_n} \leq d\frac{\sqrt{n}}{\lambda_n}\|wB_n''(f)\|_{K_n}. \tag{16}$$

Remark. The absolute constant $d > 0$ will play a special role, this is why the distinguished notation.

Proof. Using

$$\|wB_n(f)\|_{I_n} \leq c\|f\| \tag{17}$$

(this follows from Lemma 2 applied with $i = \delta = 0$), as well as Lemmas 3 and 4,

$$\|wB_n'''(f)\|_{I_n} \leq \|wB_n'''(f - B_n^\star(f))\|_{I_n} + \|wB_n'''(B_n^\star(f))\|_{I_n}$$

$$\leq c\frac{n^{3/2}}{\lambda_n^3}\|w[f - B_n^\star(f)]\| + c\frac{n^{5/2}}{\lambda_n^3}\max_{1\leq k\leq n-1}[w(x_k)|\Delta_{2\lambda_n/n}^2 B_n^\star(f,x_k)|]$$

$$\leq c\frac{\sqrt{n}}{\lambda_n}\|wB_n''(f)\|_{K_n} + c\frac{n^{5/2}}{\lambda_n^3}\max_{|x_k|<\lambda_n+\frac{2\lambda_n}{n}}[w(x_k)|\Delta_{2\lambda_n/n}^2 B_n^\star(f,x_k)|], \qquad (18)$$

since $B_n^\star$ is linear in $\mathbf{R} \setminus J_n$, where

$$J_n := [-\lambda_n, \lambda_n].$$

Here, by symmetry, it suffices to estimate the maximum for $0 \leq x_k < \lambda_n + \frac{2\lambda_n}{n}$. If $m \sim \frac{3}{4}n$ denotes the index for which $x_m \leq \lambda_n < x_{m+1}$, then by the mean value theorem

$$w(x_k)|\Delta_{2\lambda_n/n}^2 B_n^\star(f,x_k)| = w(x_k)|\Delta_{2\lambda_n/n}^2 B_n(f,x_k)| = \frac{4\lambda_n^2}{n^2}w(x_k)|B_n''(f,\xi_k)|$$

$$\leq c\frac{\lambda_n^2}{n^2}\|wB_n''(f)\|_{J_n} \qquad 0 \leq x_k < x_m,$$

where $|x_k - \xi_k| \leq \frac{2\lambda_n}{n}$ which in turn implies $w(x_k) \sim w(\xi_k)$ (cf. (10)). Further by (5)

$$w(x_m)|\Delta_{2\lambda_n/n}^2 B_n^\star(f,x_m)| = w(x_m)|B_n(f,x_{m-1}) - 2B_n(f,x_m) + B_n(f,\lambda_n)$$

$$+(x_{m+1} - \lambda_n)B_n'(f,\lambda_n)| \leq w(x_m)|\Delta_{2\lambda_n/n}^2 B_n(f,x_m)|$$

$$+\frac{1}{2}w(x_m)|B_n''(f,\xi_m)|(x_{m+1} - \lambda_n)^2 \leq c\frac{\lambda_n^2}{n^2}\|wB_n''(f)\|_{K_n}$$

for the same reasons as above. Finally,

$$w(x_{m+1})|\Delta_{2\lambda_n/n}^2 B_n^\star(f,x_{m+1})| = w(x_{m+1})|B_n(f,x_m) - 2[B_n(f,\lambda_n)$$

$$+(x_{m+1} - \lambda_n)B_n'(f,\lambda_n)] + B_n(f,\lambda_n) + (x_{m+2} - \lambda_n)B_n'(f,\lambda_n)|$$

$$\leq w(x_{m+1})\left[|\Delta_{2\lambda_n/n}^2 B_n(f,x_{m+1})| + \frac{1}{2}|B_n''(f,\xi_{m+1})|(x_{m+1} - \lambda_n)^2\right.$$

$$\left.+\frac{1}{2}|B_n''(f,\eta_{m+2})|(x_{m+2} - \lambda_n)^2\right] \leq c\frac{\lambda_n^2}{n^2}\|wB_n''(f)\|_{K_n}.$$

Substituting these estimates into (18) we obtain the statement of the lemma.

4. Proof of Theorem

We know that the modulus of smoothness is equivalent to the corresponding K-functional, i.e.,

$$\omega_2\left(f, \frac{\lambda_n}{\sqrt{n}}\right)_w \sim \inf_{g' \in AC_{\mathrm{loc}}} \left[\|w(f-g)\| + \frac{\lambda_n^2}{n}\|wg''\| \right]$$

(cf. [3], Theorem 11.2.3). Therefore choosing $g = B_n^\star(f)$, it will be sufficient to prove that

$$\|w(f - B_n^\star(f))\| + \|w(f - B_{\mu n}^\star(f))\| \geq c\frac{\lambda_n^2}{n}\|wB_n''(f)\|_{J_n} \left(= c\frac{\lambda_n^2}{n}\|wB_n^{\star''}(f)\| \right).$$

We may assume that condition (15) of Lemma 5 holds (otherwise there is nothing to prove).

Now suppose that the norm $\|wB_n''(f)\|_{K_n}$ is attained at a point $0 \leq y_n \in K_n$, and that $B_n''(f, y_n) > 0$. (If the latter condition does not hold, we can apply the subsequent argument to $-f$ instead of f.) Then (16) implies

$$|B_n'''(f, x)| \leq d\frac{\sqrt{n}}{\lambda_n}\frac{w(y_n)}{w(x)}B_n''(f, y_n) \leq d\frac{\sqrt{n}}{\lambda_n}B_n''(f, y_n), \qquad |x| \leq y_n.$$

This easily yields that

$$B_n''(f, x) \geq \frac{1}{2}B_n''(f, y_n), \qquad x \in L_n := \left[y_n - \frac{\lambda_n}{2d\sqrt{n}}, y_n \right] \cap J_n. \qquad (19)$$

Let

$$z_n := y_n - \frac{\lambda_n}{4d\sqrt{n}};$$

evidently $z_n \in L_n$ for n sufficiently large. Using (17) with μn and $f - B_n^\star(f)$ instead of n and f, respectively, we obtain

$$\|w[f - B_n^\star(f)]\| \geq c\|wB_{\mu n}(B_n^\star(f) - f)\|_{I_{\mu n}} \geq cw(z_n)[B_{\mu n}(B_n^\star(f), z_n) - B_{\mu n}(f, z_n)]$$

$$\geq cw(z_n)B_{\mu n}(B_n^\star(f, x) - B_{\mu n}(f, z_n), z_n) = cw(z_n)B_{\mu n}(B_n^\star(f, x) - B_n(f, z_n), z_n)$$

$$- cw(z_n)|f(z_n) - B_n(f, z_n)| + cw(z_n)|f(z_n) - B_{\mu n}(f, z_n)|$$

$$\geq cw(z_n)B_{\mu n}(B_n^\star(f, x) - B_n(f, z_n), z_n) - c\|w(f - B_n^\star(f))\| - c\|w(f - B_{\mu n}^\star(f))\|,$$

where the integer $\mu > 1$ will be chosen later. Hence

$$\|w(f - B_n^\star(f))\| + \|w(f - B_{\mu n}^\star(f))\| \geq cw(z_n)B_{\mu n}(g_n, z_n) \qquad (20)$$

where

$$g_n(x) := B_n^\star(f, x) - B_n(f, z_n) - (x - z_n)B_n'(f, z_n).$$

(Here we made use of the fact that the operator $B_{\mu n}$ reproduces linear functions.)

Let us estimate $g_n(x)$ on different parts of the real line. First, by using Taylor expansion about z_n,

$$|g_n(x)| = \frac{1}{2}|B_n''(f, \xi_n)|(x - z_n)^2 \leq \|wB_n''(f)\|_{J_n}\frac{(x - z_n)^2}{w(\xi_n)}, \qquad x \in J_n \qquad (21)$$

where $\xi_n \in (x, z_n) \subset J_n$. Hence

$$w(z_n)|g_n(x)| \leq \begin{cases} ||wB_n''(f)||_{J_n}(x - z_n)^2 & \text{if } |x| \leq z_n, \\ ||wB_n''(f)||_{J_n}\dfrac{w(z_n)}{w(x)}(x - z_n)^2 & \text{if } z_n < |x| \leq \lambda_n. \end{cases} \qquad (22)$$

On the other hand, (21) yields by (19)

$$w(z_n)g_n(x) \geq \frac{w(z_n)}{4w(y_n)}||wB_n''(f)||_{J_n}(x - z_n)^2 \geq \frac{1}{4}||wB_n''(f)||_{J_n}(x - z_n)^2, \quad x \in I_n. \tag{23}$$

Finally, using the definition (5), the mean value theorem and (8) we get

$$|g_n(x)| = |B_n(f, \lambda_n) + (x - \lambda_n)B_n'(f, \lambda_n) - B_n(f, z_n) - (x - z_n)B_n'(f, z_n)|$$

$$= (\lambda_n - z_n)|x - \eta_{n,1}| \cdot |B_n''(f, \eta_n)| \leq \frac{(\lambda_n - z_n)(x - z_n)}{w(x)}||wB_n''(f)||_{J_n}, \qquad x \geq \lambda_n, \tag{24}$$

and

$$|g_n(x)| = |B_n(f, -\lambda_n) + (x + \lambda_n)B_n'(f, -\lambda_n) - B_n(f, z_n) - (x - z_n)B_n'(f, z_n)|$$

$$= (\lambda_n + z_n)|x - \eta_{n,2}| \cdot |B_n''(f, \eta_n)| \leq c\mu^{\frac{1}{2A}}\frac{\lambda_n^2}{w(x)}||wB_n''(f)||_{J_n}, \qquad -\lambda_{\mu n} \leq x \leq -\lambda_n, \tag{25}$$

where $z_n < \eta_{n,1}, \eta_{n,2} < \lambda_n$. Using the relation

$$B_{\mu n}\left((x - z_n)^2, z_n\right) = \frac{4\lambda_{\mu n}^2 - z_n^2}{\mu n} \geq \frac{4\lambda_{\mu n}^2 - \lambda_n^2}{\mu n} \geq \frac{3\lambda_n^2}{\mu n}$$

(cf. [5]) we obtain by (23)

$$w(z_n)B_{\mu n}(g_n, z_n) = w(z_n)\left(\sum_{x_{k,\mu n} \in L_n} + \sum_{x_{k,\mu n} \notin L_n}\right)p_{\mu n,k}(z_n)g_n(x_{k,\mu n})$$

$$\geq ||wB_n''(f)||_{J_n}\frac{3\lambda_n^2}{\mu n}$$

$$-c\sum_{x_{k,\mu n} \notin L_n} p_{\mu n,k}(z_n)[||wB_n''(f)||_{J_n}(x_{k,\mu n} - z_n)^2 + w(z_n)|g_n(x_{k,\mu n})|].$$

Here, the Corollary applied with $\delta = \frac{1}{3}|L_n| = \frac{\lambda_n}{12d\sqrt{n}}$ and $i = 2$ yields

$$\sum_{x_{k,\mu n} \notin L_n} p_{\mu n,k}(z_n)(x_{k,\mu n} - z_n)^2 \leq c\frac{\lambda_n^2}{n}\exp\left(-C\frac{\mu n\lambda_n^2}{\lambda_{\mu n}^2 n}\right) \leq c\frac{\lambda_n^2}{n}\exp\left(-C\mu^{1-1/A}\right), \tag{26}$$

where we used (8). Hence

$$w(z_n)B_{\mu n}(g_n, z_n) \geq c||wB_n''(f)||_{J_n}\frac{2\lambda_n^2}{\mu n} - cw(z_n)\sum_{x_{k,\mu n} \notin L_n} p_{\mu n,k}(z_n)|g_n(x_{k,\mu n})|$$

provided the integer μ is large enough.

We still have to give an upper estimate for the last sum. We partition it into three parts.

If $-z_n \leq x_{k,\mu n} \leq y_n - \frac{\lambda_n}{2d\sqrt{n}}$, then we can use the first estimate in (22) and obtain the same inequality as in (26).

Using the second estimate in (22) we obtain by Lemma 2 applied with $i = 2$ and $\delta = \frac{\lambda_n}{4d\sqrt{n}}$,

$$w(z_n) \sum_{\substack{-\lambda_n \leq x_{k,\mu n} \leq \min\left(-z_n, y_n - \frac{\lambda_n}{2d\sqrt{n}}\right) \\ \text{or } y_n \leq x_{k,\mu n} \leq \lambda_n}} p_{\mu n,k}(z_n)|g_n(x_{k,\mu n})|$$

$$\leq c\|wB_n''(f)\|_{J_n} w(z_n) \sum_{|x_{k,\mu n}-z_n| \geq \frac{\lambda_n}{4d\sqrt{n}}} p_{\mu n,k}(z_n) \frac{(x_{k,\mu n} - z_n)^2}{w(x_{k,\mu n})}$$

$$\leq c\|wB_n''(f)\|_{J_n} \frac{\lambda_n^2}{n} \exp\left(-C\frac{\mu n \lambda_n^2}{\lambda_{\mu n}^2 n}\right) \leq c_9\|wB_n''(f)\|_{J_n} \frac{\lambda_n^2}{n} \exp\left(-C\mu^{1-1/A}\right).$$

Now let $x_{k,\mu n} \geq \max(y_n, \lambda_n)$. Then we can apply (24) and Lemma 2 with $i = 1$ and

$$\delta = \max\left(\lambda_n - z_n, \frac{\lambda_n}{4d\sqrt{n}}\right)$$

to get

$$w(z_n) \sum_{x_{k,\mu n} \geq \max(y_n, \lambda_n)} p_{\mu n,k}(z_n)|g_n(x_{k,\mu n})|$$

$$\leq (\lambda_n - z_n)\|wB_n''(f)\|_{J_n} \sum_{x_{k,\mu n}-z_n \geq \delta} \frac{p_{\mu n,k}(z_n)(x_{k,\mu n} - z_n)}{w(x_{k,\mu n})}$$

$$\leq c\|wB_n''(f)\|_{J_n}(\lambda_n - z_n) \frac{\lambda_n}{\sqrt{n}} \exp\left(-C\frac{\mu n}{\lambda_{\mu n}^2} \delta^2\right).$$

Here, if $\delta = \frac{\lambda_n}{4d\sqrt{n}} \geq \lambda_n - z_n$ then this yields

$$w(z_n) \sum_{x_{k,\mu n} \geq \max(y_n, \lambda_n)} p_{\mu n,k}(z_n)|g_n(x_{k,\mu n})| \leq \frac{\lambda_n^2}{n} e^{-C\mu^{1-1/A}}, \tag{27}$$

while if $\delta = \lambda_n - z_n \geq \frac{\lambda_n}{4d\sqrt{n}}$ then

$$w(z_n) \sum_{x_{k,\mu n} \geq \max(y_n, \lambda_n)} p_{\mu n,k}(z_n)|g_n(x_{k,\mu n})| \leq c(\lambda_n - z_n)^2 \exp\left(-C\frac{\mu n}{\lambda_{\mu n}^2}(\lambda_n - z_n)^2\right). \tag{28}$$

But it is easy to see that the function $\varphi(t) = t^2 \exp\left(-C\frac{\mu n}{\lambda_{\mu n}^2} t^2\right)$ attains its maximum at $t = \frac{\lambda_{\mu n}}{\sqrt{C\mu n}}$ and is monotone decreasing after that, whence (28) for sufficiently large μ and $\lambda_n - z_n = \frac{\lambda_n}{4d\sqrt{n}}$ yields the same estimate as in (26).

Finally, if $x_{k,\mu n} \le -\lambda_n$ then (25) and Lemma 2 with $\delta = \lambda_n - \frac{\lambda_n}{4d\sqrt{n}}$ and $i = 0$ yield

$$w(z_n) \sum_{x_{k,\mu n} \le -\lambda_n} p_{\mu n,k}(z_n)|g_n(x_{k,\mu n})| \le c\frac{\lambda_n}{\sqrt{n}} \exp\left(-C\frac{\mu n}{\lambda_{\mu n}^2}\lambda_n^2\right) \le e^{-C\mu^{1-1/A}n}.$$

Collecting all of these estimates we get from (20) the statement of the theorem.

References

[1] B. Della Vecchia, G. Mastroianni and J. Szabados: Weighted approximation of functions on the real line by Bernstein polynomials. *J. Approx. Theory* (accepted).

[2] Z. Ditzian and K.G. Ivanov: Strong converse inequalities. *J. d'Analyse Math.* **61** (1993), 61–111.

[3] Z. Ditzian and V. Totik: *Moduli of smoothness.* Springer Series in Computational Mathematics **9**, Springer-Verlag, New York (1987).

[4] A.L. Levin and D.S. Lubinsky: Christoffel functions, orthogonal polynomials and Nevai's conjecture for Freud weights. *Constr. Approx.* **8** (1992), 463–535.

[5] G.G. Lorentz: *Bernstein Polynomials.* Mathematical Expositions **8**, University of Toronto Press, Toronto (1953).

Jozsef Szabados
Alfréd Rényi Institute of Mathematics
P.O.B. 127
H-1364 Budapest, Hungary
e-mail: szabados@renyi.hu

Trends and Applications in Constructive Approximation
(Eds.) M.G. de Bruin, D.H. Mache & J. Szabados
International Series of Numerical Mathematics, Vol. 151, 259–272
© 2005 Birkhäuser Verlag Basel/Switzerland

Rprop Using the Natural Gradient

Christian Igel, Marc Toussaint and Wan Weishui

Abstract. Gradient-based optimization algorithms are the standard methods
for adapting the weights of neural networks. The natural gradient gives the
steepest descent direction based on a non-Euclidean, from a theoretical point
of view more appropriate metric in the weight space. While the natural gradient has already proven to be advantageous for online learning, we explore
its benefits for batch learning: We empirically compare Rprop (resilient backpropagation), one of the best performing first-order learning algorithms, using
the Euclidean and the non-Euclidean metric, respectively. As batch steepest
descent on the natural gradient is closely related to Levenberg-Marquardt
optimization, we add this method to our comparison.

It turns out that the Rprop algorithm can indeed profit from the natural gradient: the optimization speed measured in terms of weight updates
can increase significantly compared to the original version. Rprop based on
the non-Euclidean metric shows at least similar performance as Levenberg-
Marquardt optimization on the two benchmark problems considered and appears to be slightly more robust. However, in Levenberg-Marquardt optimization and Rprop using the natural gradient computing a weight update requires
cubic time and quadratic space. Further, both methods have additional hyperparameters that are difficult to adjust. In contrast, conventional Rprop has
linear space and time complexity, and its hyperparameters need no difficult
tuning.

1. Introduction

Artificial neural networks such as Multi-Layer Perceptrons (MLPs) have become
standard tools for regression. In essence, an MLP with fixed structure defines
a differentiable mapping from a parameter space $\mathbb{R}^n$ to the space of functions.
Although this mapping is typically not surjective, one can prove that in principle
every continuous function can be well approximated (e.g., see [8], the upper bound
on the approximation error depends on the network structure). For an MLP with
fixed structure, a regression problem reduces to the adaptation of the parameters,
the weights, of the network given the sample data. This process is usually referred
to as learning.

Since MLPs are differentiable, gradient-based adaptation techniques are typically applied to adjust the weights. The earliest and most straightforward adaptation rule, ordinary gradient descent, adapts weights proportional to the partial derivatives of the error functional [1, 17]. Several improvements of this basic adaptation rule have been proposed, some of them based on elaborated heuristics, others on theoretical reconsideration of gradient-based learning. Here we consider three of them, natural gradient descent, resilient backpropagation, and Levenberg-Marquardt optimization. We combine ideas of all three approaches to a new method we call *natural Rprop*.

Resilient backpropagation (Rprop, [10, 15, 16]) is a well-established modification of the ordinary gradient descent. The basic idea is to adjust an individual step size for each parameter to be optimized. These step sizes are not proportional to the partial derivatives but are themselves adapted based on some heuristics. Ordinary gradient descent computes the direction of steepest descent by implicitly assuming a Euclidean metric on the weight space. However, as we are interested in the function corresponding to a weight configuration, a more appropriate metric would take distances in the function space into account. Amari [2] proposed to make use of methods from differential geometry to determine the steepest descent direction, the negative *natural gradient*, based on such a non-Euclidean metric.

Our investigation is based on the following idea: If the natural gradient points in a better descent direction and Rprop improves the ordinary gradient descent, can a combination of both methods profit from the advantages of the two approaches? Can we get best of both worlds, the increased robustness of Rprop and the improved convergence speed due to the decoupling of weight interdependencies when using the right metric?

In order to asses the performance of the new learning approach, we compare it to a standard Rprop algorithm and to Levenberg-Marquardt optimization [12, 13]. We choose Levenberg-Marquardt optimization, because in the function regression scenario that we consider it turns out that natural gradient descent is closely related to Levenberg-Marquardt optimization. In addition, we borough ideas from this classical technique to increase the robustness of the calculation of the steepest descent direction within natural Rprop.

In the next Section we give brief but comprehensive descriptions of Rprop, Levenberg-Marquardt optimization, and the natural gradient. Then we introduce their combination, the natural Rprop. Section 3 presents experimental results on two benchmark problems. Since the Levenberg-Marquardt algorithm as well as the natural Rprop introduce new hyperparameters, we particularly look at the robustness of the algorithms in terms of the choices of these parameters. We conclude by discussing the results in Section 4.

2. Background

We understand a feed-forward MLP with n_{in} inputs and n_{out} outputs as a function $f(\,\cdot\,;w) : \ \mathbb{R}^{n_{\text{in}}} \to \mathbb{R}^{n_{\text{out}}}$ describing some input-output behavior. The function is parameterized by a weight vector $\vec{w} \in W = \mathbb{R}^n$ as indicated by the notation. For instance, $y = f(x;w)$ is the output of the network for a given input x and weights w. To address the components of the weight vector w we use upper indices $(w^1, \ldots, w^n)$ to clarify that they are contra-variant and not co-variant (this will become more apparent and relevant when describing the natural gradient). As an error measure $E(\vec{w})$ to be minimized we assume, in the following, the mean squared error (MSE) on a batch $\{(\vec{x}_1, \vec{y}_1), \ldots, (\vec{x}_P, \vec{y}_P)\} \in (\mathbb{R}^{n_{\text{in}}} \times \mathbb{R}^{n_{\text{out}}})^P$ of sample data points. For simplicity, we restrict ourselves to $n_{\text{out}} = 1$ and define

$$E(\vec{w}) = \frac{1}{P} \sum_{p=1}^{P} ||f(\vec{x}_p; \vec{w}) - \vec{y}_p||^2 \ .$$

In the remainder of this section, we describe four methods for gradient-based minimization of $E(w)$: we review Rprop, natural gradient descent, and Levenberg-Marquardt optimization, and introduce a new approach, natural Rprop.

2.1. Resilient backpropagation

The Rprop algorithms are among the best performing first-order batch learning methods for neural networks with arbitrary topology [9, 10, 15, 16]. They are

- very fast and accurate (e.g., compared to conjugate gradient methods, Quick-prop etc.),
- very robust in terms of the choices of their hyperparameters,
- first-order methods, therefore time and space complexity scales linearly with the number of parameters to be optimized,
- only dependent on the sign of the partial derivatives of the objective function and not on their amount, therefore they are suitable for applications where the gradient is numerically estimated or the objective function is noisy, and
- easy to implement and not very sensitive to numerical problems.

In the following, we describe the Rprop variant with improved backtracking introduced in [9]. The Rprop algorithms are iterative optimization methods. Let t denote the current iteration (epoch). In epoch t, each weight is changed according to

$$w^i(t+1) = w^i(t) - \text{sign}\left(\frac{\partial E(t)}{\partial w^i}\right) \cdot \Delta^i(t) \ .$$

The direction of the change depends on the sign of the partial derivative, but is independent of its amount. The individual step sizes $\Delta^i(t)$ are adapted based on changes of sign of the partial derivatives of $E(w)$ w.r.t. the corresponding weight: If $\frac{\partial E(t-1)}{\partial w^i} \cdot \frac{\partial E(t)}{\partial w^i} > 0$ then $\Delta^i(t)$ is increased by a factor $\eta^+ > 1$, otherwise $\Delta^i(t)$ is decreased by multiplication with $\eta^- \in]0, 1[$. Additionally, some Rprop methods implement weight-backtracking. That is, they partially retract "unfavorable" previous steps. Whether a weight change was "unfavorable" is decided by a heuristic. We use an improved version of the original algorithm called iRprop$^+$, which is described in pseudo-code in Table 1. The difference compared to the original Rprop proposed in [16] is that the weight-backtracking heuristic considers both the evolution of the partial derivatives and the overall error. For a comparison of iRprop$^+$ with other Rprop variants and a detailed description of the algorithms the reader is referred to [10].

for each w^i **do**

    if $\frac{\partial E(t-1)}{\partial w^i} \cdot \frac{\partial E(t)}{\partial w^i} > 0$ **then**

$$\Delta^i(t) = \min\left(\Delta^i(t-1) \cdot \eta^+, \Delta_{\max}\right)$$

$$w^i(t+1) = w^i(t) - \text{sign}\left(\frac{\partial E(t)}{\partial w^i}\right) \cdot \Delta^i(t)$$

    elseif $\frac{\partial E(t-1)}{\partial w^i} \cdot \frac{\partial E(t)}{\partial w^i} < 0$ **then**

$$\Delta^i(t) = \max\left(\Delta^i(t-1) \cdot \eta^-, \Delta_{\min}\right)$$

        if $E(t) > E(t-1)$ **then** $w^i(t+1) = w^i(t-1)$

$$\frac{\partial E(t)}{\partial w^i} := 0$$

    elseif $\frac{\partial E(t-1)}{\partial w^i} \cdot \frac{\partial E(t)}{\partial w^i} = 0$ **then**

$$w^i(t+1) = w^i(t) - \text{sign}\left(\frac{\partial E(t)}{\partial w^i}\right) \cdot \Delta^i(t)$$

    fi

od

TABLE 1. The iRprop$^+$ algorithm with improved weight-backtracking scheme as proposed in [10].

2.2. Levenberg-Marquardt optimization

Levenberg-Marquardt optimization [6, 12, 13] is based on the idea that, to minimize the error functional $E(\vec{w})$, one should find weights such that the derivatives $\frac{\partial E(\vec{w})}{\partial w^i}$ vanish. This search can be realized with a Newton step on an approximation of the error functional as follows. Consider the linear approximation of $f(\vec{x}; \vec{w})$ around $\vec{w}(t)$,

$$\hat{f}(\vec{x}; \vec{w}) = f(\vec{x}; \vec{w}(t)) + \sum_{j=1}^{n} [\vec{w}^j - \vec{w}^j(t)] \frac{\partial f(\vec{x}; \vec{w}(t))}{\partial w^j} .$$

Substituting $\hat{f}$ for f in the MSE gives a new error function $\hat{E}(\vec{w})$ with gradient

$$
\frac{\partial \hat{E}(\vec{w})}{\partial w^i} = \frac{2}{P} \sum_{p=1}^{P} [\hat{f}(\vec{x}_p; \vec{w}) - \vec{y}_p] \frac{\partial \hat{f}(\vec{x}_p; \vec{w})}{\partial w^i}
$$

$$
= \frac{\partial E(\vec{w}(t))}{\partial w^i} + \sum_j A_{ij}(w(t)) \left[\vec{w}^j - \vec{w}^j(t) \right] . \tag{1}
$$

Here the $n \times n$ matrix $A_{ij}(w)$ has the entries

$$
A_{ij}(w) = \frac{2}{P} \sum_{p=1}^{P} \frac{\partial f(\vec{x}_p; \vec{w})}{\partial w^i} \frac{\partial f(\vec{x}_p; \vec{w})}{\partial w^j} .
$$

Setting (1) to zero (i.e., doing a Newton step on $\hat{E}$) leads to the weight update

$$
\vec{w}^i(t+1) = \vec{w}^i(t) - \sum_{j=1}^{n} A^{ij}(w(t)) \frac{\partial E(\vec{w}(t))}{\partial w^j} . \tag{2}
$$

Here A^{ij} is the inverse matrix of A_{ij}. This weight update would lead to an optimum if $\hat{E}(\vec{w}) = E(\vec{w})$. This is in general not the case and the weight update rule (2) is only reasonable close to a minimum. Therefore, the idea is to automatically blend between (2) and standard steepest descent:

$$
\vec{w}^i(t+1) = \vec{w}^i(t) - \sum_{j=1}^{n} [A_{ij} + \lambda \mathbf{I}_{ij}]^{-1} \frac{\partial E(\vec{w}(t))}{\partial w^j} ,
$$

where the parameter $\lambda > 0$ allows soft switching between the two strategies. A large λ corresponds to simple gradient descent. There are several heuristics to adapt λ. We use the most common one to decrease λ by multiplication with $\lambda_- \in]0, 1[$ if the error decreased, and to increase it by multiplication with $\lambda_+ > 1$ (usually $\lambda_- = \lambda_+^{-1}$), otherwise. A drawback of Levenberg-Marquardt optimization is that the choice of λ_0 (the initial value for λ), λ_-, and λ_+ is crucial for the performance of the algorithm.

2.3. Natural gradient descent

Basically, natural gradient descent is steepest descent with a non-Euclidean metric on the parameter space. Two simple facts motivate the natural gradient: First, the steepest descent direction generally depends on the choice of metric on the parameter space – this is very often neglected in standard textbooks describing gradient descent. See Figure 1a) for an illustration. Second, there are good arguments to assume a non-Euclidean metric on the parameter space: Generally, there exists no a priori reason why the Euclidean metric on the parameter space should be a preferential distance measure between solutions. In fact, in the case of function regression, one typically assumes a canonical distance measures *on the function space*, like the mean squared error, or a likelihood measure on the space of distributions, which translate to non-trivial metrics on the parameter space, see

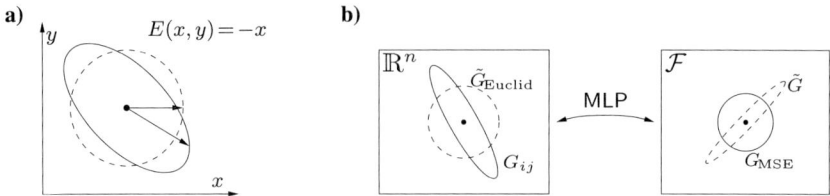

FIGURE 1. **a)** The steepest descent direction of a functional $E(x, y)$ depends on the metric: The ellipses mark the set of vectors of unit length for the Euclidean metric (dashed circle) and the metric $\vec{G} = \begin{pmatrix} 3/4 & 1/4 \\ 1/4 & 3/4 \end{pmatrix}$ (ellipse). For the ellipse, the unit length vector that decreases $E(x, y) = -x$ the most is not pointing directly to the right, it is generally given by equation (3). **b)** An MLP establishes a differentiable relation between the weight space $W = \mathbb{R}^n$ and the manifold of functions $\mathcal{F}$. A canonical distance measure $\vec{G}_{\mathrm{MSE}}$ on the function space induces a non-Euclidean metric G_{ij} on the weight space w.r.t. which steepest descent should be performed. In contrast, if a Euclidean metric $\tilde{\vec{G}}_{\mathrm{Euclid}}$ is presumed on the weight space, this generally leads to a "non-diagonal" metric $\tilde{\vec{G}}$ on the function space. Typically, a non-diagonal metric on the function space is undesirable because this leads to negative inference and cross-talk; e.g., during online learning, if one functional components is trained, others are untrained according to the off-diagonal entries of the metric [18]. Using the natural gradient avoids this effect of catastrophic forgetting during online learning.

Figure 1b). Amari [2, 5, 4] was the first to realize the implications of these facts in the case of gradient-based adaptation of MLPs. In the following, we give a simple derivation of the natural gradient.

An MLP represents a differentiable mapping f from the parameter space $W = \mathbb{R}^n$ to the manifold $\mathcal{F}$ of functions. We write $f : \vec{w} \in W \mapsto f(\,\cdot\,; \vec{w}) \in \mathcal{F}$. Let d be a distance measure on $\mathcal{F}$. Here we assume that $d(f, h)$ is the mean squared distance on a batch $\{x_1, \dots, x_P\}$ of data points between two functions f and h,

$$d(f, h) = \frac{1}{P} \sum_{p=1}^{P} [f(\vec{x}_p) - h(\vec{x}_p)]^2 \;.$$

The pull-back of this metric onto the parameter space is, by definition,

$$G_{ij}(\vec{w}) = \frac{1}{P} \sum_{p=1}^{P} \frac{\partial f(\vec{x}_p; \vec{w})}{\partial w^i} \frac{\partial f(\vec{x}_p; \vec{w})}{\partial w^j} \;.$$

The meaning of this metric G_{ij} on W is: if we measure distances in W using G_{ij}, then these distances are guaranteed to be equal to the mean squared distance when measured on the function space $\mathcal{F}$.[1] Further, if we determine the steepest descent direction in W using the metric G_{ij} we can be sure to find the direction in which the mean squared error on $\mathcal{F}$ decreases fastest – which is generally not true when using the Euclidean metric!

The steepest descent direction of a functional $E(\vec{w})$ over W is given as the vector $\vec{\delta}$ with components

$$\delta^j = \sum_{i=1}^{n} G^{ij}(\vec{w}) \frac{\partial E(\vec{w})}{\partial w^i} \; . \tag{3}$$

Here G^{ij} is the inverse matrix of G_{ij}. (Upper indices denote so-called contra-variant components.) Thus, in summary, natural gradient descent with learning rate $\eta > 0$ reads

$$\vec{w}^i(t+1) = \vec{w}^i(t) - \eta \sum_{j=1}^{n} G^{ij}(\vec{w}(t)) \frac{\partial E(\vec{w}(t))}{\partial w^j} \; . \tag{4}$$

There exists an online version of the natural gradient [5] that approximates the inverse natural metric G^{ij} on the fly and reduces the negative effects of co-inference during online learning (cf. Figure 1b).

Comparing Levenberg-Marquardt adaptation (2) with batch natural gradient (4) descent we find that for $\lambda = 0$ they are equivalent since $A_{ij} = G_{ij}$ (in the case of the mean squared distance $d(f,h)$ on $\mathcal{F}$). This fact has previously been observed by [3, 7, 11]. A small difference is the robustness term for $\lambda \neq 0$. Note that for different distance measures on $\mathcal{F}$, generally $\vec{A}_{ij} = \vec{G}_{ij}$ does not hold.

2.4. Natural Rprop

Rprop is a batch gradient-based learning algorithm that overcomes the problems of standard gradient descent by automatically adjusting individual step sizes. The natural gradient points in a direction that is more appropriate for steepest descent optimization. Now, the question arises whether it can be beneficial to combine natural gradient descent with the heuristics of Rprop. Recall that one of the main features of Rprop is that the update step sizes depend only on the signs of the gradient. Since the metric $\vec{G}$ and also its inverse $\vec{G}^{-1}$ are always positive definite, a vector transformed by $\vec{G}$ changes its direction by up to 90°. The angle between ordinary and natural gradient descent directions can also be up to 90° (which also becomes apparent from Figure 1a). Thus, the signs can generally change when replacing the ordinary gradient by the natural gradient in the Rprop algorithm and, therefore, adaptation behavior changes.

We hence propose to combine iRprop[+], the natural gradient, and the robust-ness term $\lambda \mathbf{I}_{ij}$ of Levenberg-Marquardt simply by replacing the ordinary gradient

---

[1] More precisely, if we measure the distance between w_1 and w_2 in W by the length of the geodesic w.r.t. $G_{ij}(w)$, then this distance is guaranteed to be equal to the mean squared distance $d(f_1, f_2)$ between the two corresponding functions $f_1 = f(\cdot\,; \vec{w}_1)$ and $f_2 = f(\cdot\,; \vec{w}_2)$ in $\mathcal{F}$.

$\frac{\partial E(\vec{w})}{\partial w^i}$ by the robust natural gradient

$$\left[G_{ij} + \lambda \operatorname{trace}(G_{ij})\,\mathbf{I}_{ij}\right]^{-1} \frac{\partial E(\vec{w})}{\partial w^i}$$

within the iRprop$^+$algorithm. As in Levenberg-Marquardt optimization, the parameter $\lambda \in \mathbb{R}^+$ blends between the natural gradient and the ordinary gradient by adding a weighted unity matrix $\mathbf{I}_{ij}$. Additionally, we weight the term proportional to the trace of $\vec{G}$ such that the blending becomes relative w.r.t. the orders of the Eigenvalues of $\vec{G}$. We use the same update rule as before, λ is reduced by multiplication with $\lambda_- \in\,]0,1[$ if the error decreased and is set to $\lambda \leftarrow \lambda \cdot \lambda_+$, otherwise (usually $\lambda_- = \lambda_+^{-1}$). We call this new algorithm *natural Rprop*.

	space	time
iRprop$^+$	$\mathcal{O}(n)$	$\mathcal{O}(n)$
Levenberg-Marquardt optimization	$\mathcal{O}(n^2)$	$\mathcal{O}(n^3)$
natural gradient descent / natural iRprop$^+$	$\mathcal{O}(n^2)$	$\mathcal{O}(n^3)$

TABLE 2. The complexity of the three algorithms w.r.t. space and time. The number of weights in the MLP is denoted by n.

Table 2 displays the complexity of iRprop$^+$, the Levenberg-Marquardt algorithm, and natural gradient descent / natural Rprop w.r.t. space and time. Both, Levenberg-Marquardt and the natural gradient require the storage and inversion of a $n \times n$-matrix, with n being the number of weights, and this dominates the cost of these algorithms leading to cubic time complexity and quadratic space complexity. In contrast, iRprop$^+$needs only linear time and space to update weight by weight separately and to store the weights and the step size for each weight.

3. Experiments

First, the two benchmark problems, sunspot prediction and extended XOR, are introduced. Then we describe the experimental setup. Finally, the empirical results are presented.

3.1. Benchmark problems

The goal of the sunspot prediction task is to reproduce the time series of the average number of sunspots observed per year, see Figure 2. The data are available from http://sidc.oma.be. The average number of spots from the years $t-1$, $t-2$, $t-4$, and $t-8$ are given to predict the value for the year t. The training set contains 289 patterns. The first year to predict is 1708. The input values are normalized between 0.2 and 0.8.

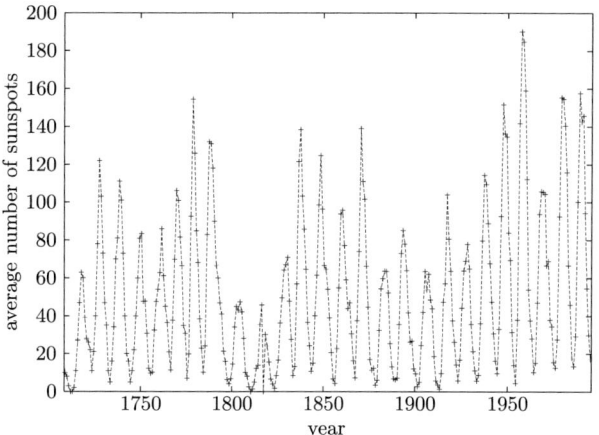

FIGURE 2. Time series of average number of sunspots observed per year.

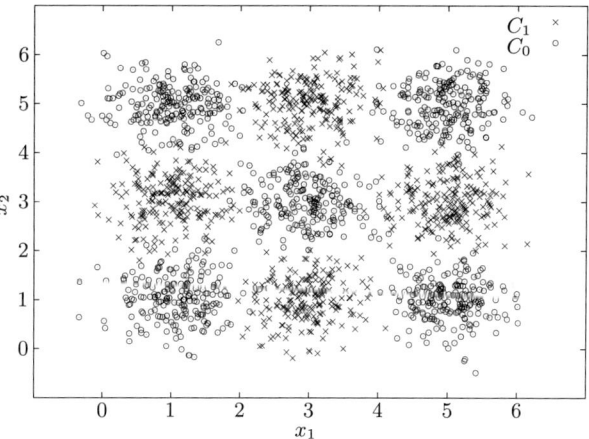

FIGURE 3. The extended XOR problem.

The extended XOR task, see Figure 3, is an artificial classification benchmark [14]. The 1800 training patterns $(\vec{x}, c) \in \mathbb{R}^2 \times \{0, 1\}$ are sampled from

$$p(\vec{x}, c) = \frac{1}{|C_0| + |C_1|} \sum_{\vec{\mu} \in C_c} \frac{1}{\sqrt{2\pi}\sigma} e^{-(\vec{x} - \vec{\mu})^2/(2\sigma^2)} \ ,$$

with $C_0 = \{(1, 5), (5, 5), (3, 3), (1, 1), (5, 1)\}$, $C_1 = \{(3, 5), (1, 3), (5, 3), (3, 1)\}$, and variance $\sigma^2 = 0.2$.

We want to evaluate the optimization speed of different learning algorithms. Thus, in both benchmark problems we just consider the task of learning the sample data and do not consider the import issue of generalization.

3.2. Experiments

We compare the standard iRprop$^+$, natural iRprop$^+$ (i.e., iRprop$^+$ using the natural gradient), and LevenbergMarquardt optimization on the two benchmark problems. The Rprop parameters are set to default values $\eta^+ = 1.2$, $\eta^- = 0.5$, $\Delta_{\min} = 0$, $\Delta_{\max} = 50$, and $\Delta_0 = 0.01$. For natural iRprop$^+$ and Levenberg-Marquardt optimization, we test all combinations of $\lambda_0 \in \{0.01, 0.1, 1, 10\}$ and $\lambda_+^{-1} = \lambda_- \in \{0.5, 0.6, \ldots, 0.9\}$. For every optimization method and parameter setting 20 trials starting from random initializations of the weights are performed; the 20 different initializations are the same for every algorithm and parameter combination. That is, a total of 840 NNs are trained. The number of iterations (learning epochs) is set to 5000.

For the sunspot prediction problem, a 4-10-1 NN architecture without shortcut connections is chosen. The 10 hidden neurons have sigmoidal transfer functions, the logistic / Fermi function $f(x) = 1/(1 + e^{-x})$, and the output neuron is linear. For the extended XOR, we use a 2-12-1 architecture without shortcut connections and only sigmoidal transfer functions. These architectures have not been tailored to the problems (and hence the absolute results are far from being optimal).

3.3. Results

The results are summarized in Figures 4 and 5: Shown are the error trajectories for the 20 parameter combinations averaged over the 20 trials for Levenberg-Marquardt optimization and natural iRprop$^+$, respectively. The parameter settings yielding the lowest final error on average on the extended XOR problem were $\lambda_0 = 0.01$ and $\lambda_- = 0.5$ for Levenberg-Marquardt optimization and $\lambda_0 = 0.1$ and $\lambda_- = 0.8$ for natural iRprop$^+$. On the the sunspot prediction task, the best results were obtained for $\lambda_0 = 10, \lambda_- = 0.9$ and $\lambda_0 = 0.01, \lambda_- = 0.8$, respectively. The results corresponding to these parameter settings are compared to the standard Rprop in the lowest plots in Figures 4 and 5. The differences between the error values of those three curves in the final iteration are pairwise statistically significant (Wilcoxon rank sum test, $p < .001$). The results show:

- The performance of natural iRprop$^+$ and LevenbergMarquardt optimization strongly depends on the choice of λ_0 and λ_-. The "best" values for λ_0 and λ_- are task-dependent. However, natural iRprop$^+$ appears to be more robust.
- For an appropriate parameter setting, Levenberg-Marquardt optimization and natural iRprop$^+$ clearly outperform the standard iRprop$^+$. However, the latter has lower computational complexity and does not depend on critical parameters such as λ_0 and λ_-.
- For one task, the Levenberg-Marquardt method yielded the best final solutions (averaged over 20 trials), for the other iRprop$^+$ using the natural gradient. In both problems, the iRprop$^+$ combined with the natural gradient seems to be slightly faster in the early iterations.

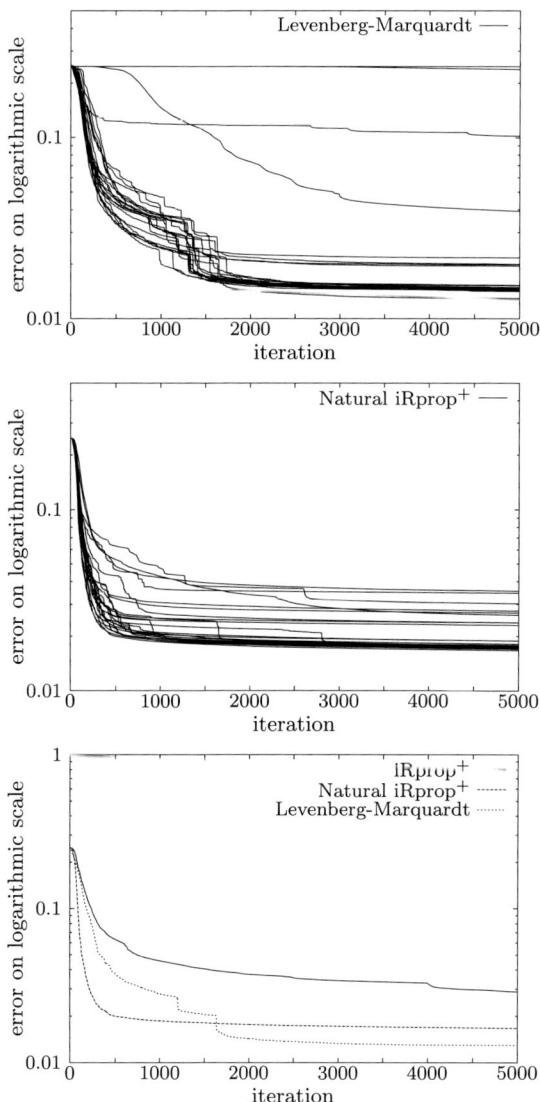

FIGURE 4. Results for the extended XOR problem. The upper plots show the results for the different settings of λ_0 and λ_- averaged over 20 trials for iRprop$^+$ using natural gradient and Levenberg-Marquardt optimization, respectively. The lower plot shows the averaged trajectories for the parameters resulting in the lowest final error in each case compared to standard iRprop$^+$.

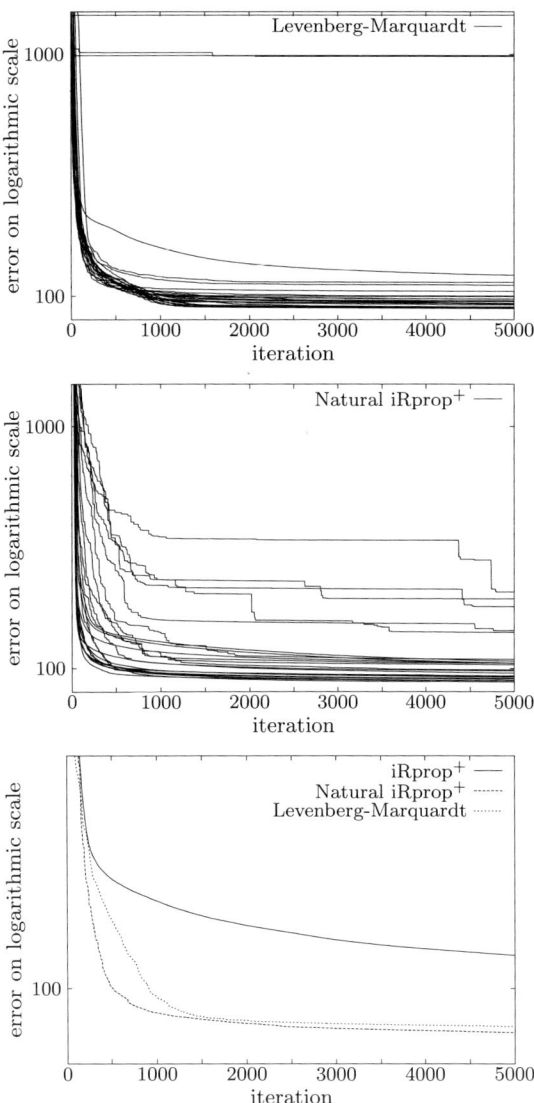

FIGURE 5. Results for the sunspots prediction problem. The upper plots show the results for the different settings of λ_0 and λ_- averaged over 20 trials for natural iRprop$^+$ and Levenberg-Marquardt optimization, respectively. The lower plot shows the averaged trajectories for the parameters resulting in the lowest final error in each case compared to standard iRprop$^+$.

4. Conclusions

In this study, we compared Levenberg-Marquardt optimization (which can be regarded as some kind of batch natural gradient learning in our scenario), iRprop$^+$, and iRprop$^+$ using the natural gradient (natural iRprop$^+$) for optimizing the weights of feed-forward neural networks. It turned out that the Rprop algorithm can indeed profit from using the natural gradient, although the updates done by Rprop are not collinear with the (natural) gradient direction. Natural iRprop$^+$ shows similar performance as Levenberg-Marquardt optimization on two test problems. The results indicate that natural iRprop$^+$ is a little bit faster in the early stages of optimization and more robust in terms of the choices of the parameters λ_0, λ_-, and λ_+. However, a more extensive empirical investigation is needed to substantiate these findings. The standard iRprop$^+$ algorithm is slower than the other methods with appropriate parameters for λ_0, λ_-. Still, these parameters are problem dependent, they considerably influence the performance of the methods, and they are not easy to adjust. Further, the computational costs of each optimization step grow from linear to cubic when replacing Rprop with one of the other two methods. Hence, we conclude that some Rprop algorithm is still the batch-learning method of choice.

Acknowledgments

This work was supported by the DFG, grant Solesys-II SCHO 336/5-2.

References

[1] S. Amari. A theory of adaptive pattern classifiers. *IEEE Transactions on Electronic Computers*, 16(3):299–307, 1967.

[2] S. Amari. Natural gradient works efficiently in learning. *Neural Computation*, 10:251–276, 1998.

[3] S. Amari and S.C. Douglas. Why natural gradient? In *Proceedings of the 1998 IEEE International Conference Acoustics, Speech, Signal Processing (ICASSP 1998)*, volume II, pages 1213–1216, 1998.

[4] S. Amari and H. Nagaoka. *Methods of Information Geometry*. Number 191 in Translations of Mathematical Monographs. American Mathematical Society and Oxford University Press, 2000.

[5] S. Amari, H. Park, and K. Fukumizu. Adaptive method of realizing natural gradient learning for multilayer perceptrons. *Neural Computation*, 12:1399–1409, 2000.

[6] M.T. Hagan and M.B. Menhaj. Training feedforward networks with the marquardt algorithm. *IEEE Transactions on Neural Networks*, 5(6):989–993, 1994.

[7] T. Heskes. On "natural" learning and pruning in multi-layered perceptrons. *Neural Computation*, 12(4):881–901, 2000.

[8] K. Hornik, M. Stinchcombe, and H. White. Multilayer feedforward networks are universal approximators. *Neural Networks*, 2(5):359–366, 1989.

[9] C. Igel and M. Hüsken. Improving the Rprop learning algorithm. In H. Bothe and R. Rojas, editors, *Proceedings of the Second International ICSC Symposium on Neural Computation (NC 2000)*, pages 115–121. ICSC Academic Press, 2000.

[10] C. Igel and M. Hüsken. Empirical evaluation of the improved Rprop learning algorithm. *Neurocomputing*, 50(C):105–123, 2003.

[11] Y. LeCun, L. Bottou, G.B. Orr, and K.-R. Müller. Efficient backprop. In G.B. Orr and K.-R. Müller, editors, *Neural Networks: Tricks of the Trade*, number 1524 in LNCS, chapter 1, pages 9–50. Springer-Verlag, 1998.

[12] K. Levenberg. A method for the solution of certain non-linear problems in least squares. *Quarterly Journal of Applied Mathematics*, 2(2):164–168, 1944.

[13] D. Marquardt. An algorithm for least-squares estimation of nonlinear parameters. *Journal of the Society for Industrial and Applied Mathematics*, 11(2):431–441, 1963.

[14] H. Park, S. Amari, and K. Fukumizu. Adaptive natural gradient learning algorithms for various stochastic models. *Neural Networks*, 13(7):755–764, 2000.

[15] M. Riedmiller. Advanced supervised learning in multi-layer perceptrons – From backpropagation to adaptive learning algorithms. *Computer Standards and Interfaces*, 16(5):265–278, 1994.

[16] M. Riedmiller and H. Braun. A direct adaptive method for faster backpropagation learning: The RPROP algorithm. In E.H. Ruspini, editor, *Proceedings of the IEEE International Conference on Neural Networks*, pages 586–591. IEEE Press, 1993.

[17] D.E. Rumelhart, G.E. Hinton, and R.J. Williams. Learning internal representations by error backpropagation. In D.E. Rumelhart, J.L. McClelland, and the PDP Research Group, editors, *Parallel Distributed Processing: Explorations in the Microstructure of Cognition*, volume 1, pages 318–362. MIT Press, 1986.

[18] M. Toussaint. On model selection and the disability of neural networks to decompose tasks. In *Proceedings of the International Joint Conference on Neural Networks (IJCNN 2002)*, pages 245–250, 2002.

Christian Igel, Marc Toussaint and Wan Weishui
Institut für Neuroinformatik
Ruhr-Universität Bochum
D-44780 Bochum, Germany
e-mail: {christian.igel,marc.toussaint}@neuroinformatik.rub.de

Trends and Applications in Constructive Approximation
(Eds.) M.G. de Bruin, D.H. Mache & J. Szabados
International Series of Numerical Mathematics, Vol. 151, 273–281
© 2005 Birkhäuser Verlag Basel/Switzerland

Some Properties of Quasi-orthogonal and Para-orthogonal Polynomials. A Survey

Biancamaria Della Vecchia, Giuseppe Mastroianni and Peter Vértesi

Dedicated to the memory of Professor Ambikeshwar Sharma

1. Introduction and preliminaries

1.1. This paper deals with quasi-orthogonal and para-orthogonal polynomials defined on $[-1, 1]$ and on the unit circle line, respectively, where the corresponding weight is constructed from moduli of continuity. Fairly general theorems are proved; their special cases were obtained in some previous papers. Many definitions and ideas are in our paper [2]. The detailed proofs will appear soon.

1.2. Everywhere below $\mathbb{C}$ is the complex plane, $\mathbb{R} = (-\infty, \infty)$, $\mathbb{N} = \{1, 2, \ldots\}$, $\mathbb{Z}_+ = \{0, 1, 2, \ldots\}$. Throughout this paper $c, c_1, c_2, \ldots$ denote positive constants; they may take different values even in subsequent formulae. It will always be clear what variables and indices the constants are independent of. If F and G are two expressions depending on some variables then we write

$$F \sim G \text{ if } |FG^{-1}| \leq c_1 \text{ and } |F^{-1}G| \leq c_2$$

uniformly for the variables in consideration.

$L^p[a, b]$ denotes the set of functions F such that

$$\|F\|_{L^p[a,b]} := \left\{ \int_a^b |F(t)|^p dt \right\}^{1/p} \qquad \text{if } 0 < p < \infty \qquad (1)$$

$$\|F\|_\infty := \operatorname*{ess\,sup}_{a \leq t \leq b} |F(t)| \qquad \text{if } p = \infty \qquad (2)$$

is finite. If $p \geq 1$, this is a norm; for $0 < p < 1$ its pth power defines a metric in $L^p[a, b]$.

By a *modulus of continuity* we mean a nondecreasing, continuous semiadditive function $\omega(\delta)$ on $[0, \infty)$ with $\omega(0) = 0$ ($\omega \in$ MC, shortly). If, in addition,

$$\omega(\delta) + \omega(\eta) \le 2\omega\left(\frac{\delta}{2} + \frac{\eta}{2}\right) \quad \text{for any } \delta, \eta \ge 0,$$

then $\omega(\delta)$ is a *concave* modulus of continuity. In the latter case $\delta/\omega(\delta)$ is nondecreasing for $\delta \ge 0$. We define the *modulus of continuity of f in L^p* (where L^p stands for $L^p[0, 2\pi]$) as

$$\omega(f, \delta)_p = \sup_{|\lambda| \le \delta} \|f(\lambda + \cdot) - f(\lambda)\|_p.$$

1.3. We call a function $\mathcal{W}(\vartheta)$ a *trigonometric weight* ($\mathcal{W} \in TW$) if it is 2π-periodic, measurable, $\mathcal{W}(\vartheta) \ge 0$ ($\vartheta \in \mathbb{R}$) and $\int_0^{2\pi} W > 0$. If, in addition, $\mathcal{W} \in L^1$ we call the functions $\{\Phi_n(\mathcal{W}, z)\}_{n=0}^\infty$ the (unique) system of orthonormal polynomials (ONP) on the unit circle line $\Gamma_1 := \partial D$ with respect to $\mathcal{W}$ if

$$\Phi_n(\mathcal{W}, z) = \kappa_n(\mathcal{W})z^n + \text{lower degree terms}, \quad \kappa_n(\mathcal{W}) > 0, \ n \in \mathbb{Z}_+, \tag{3}$$

$$\frac{1}{2\pi} \int_0^{2\pi} \Phi_n(\mathcal{W}, z)\overline{\Phi_m(\mathcal{W}, z)}\mathcal{W}(\vartheta)d\vartheta = \delta_{n,m}, \quad m, n \in \mathbb{Z}_+, \ z = \exp(i\vartheta). \tag{4}$$

We say that a point ϑ_0 is *regular* for $\mathcal{W} \in TW$ if for some $\varepsilon > 0$ the essential suprema of $\mathcal{W}$ and $1/\mathcal{W}$ on $[\vartheta_0 - \varepsilon, \vartheta_0 + \varepsilon]$ are finite. Otherwise, $\mathcal{W}$ has a *singularity* at ϑ_0.

Similarly, $w(x) \in AW$ (*algebraic weight*) if w is measurable in $[-1, 1]$, $w(x) \ge 0$, $w(x) \ne 0$ ($x \in [-1, 1]$) and $\int_{-1}^1 w > 0$. If, in addition, $w \in L_1$ (L_1 stands for $L^1[-1, 1]$), the corresponding ONP $\{p_n(w, x)\}_{n=0}^\infty$ on $[-1, 1]$ are uniquely defined by

$$p_n(w, x) = \gamma_n(w)x^n + \text{lower degree terms}, \quad \gamma_n(w) > 0, \ n \in \mathbb{Z}_+, \tag{5}$$

$$\int_{-1}^1 p_n(w, x)p_m(w, x)w(x)dx = \delta_{n,m}, \quad m, n \in \mathbb{Z}_+. \tag{6}$$

We define the singular and regular points of w as we did for $\mathcal{W}$.

G. Szegö [1, 11.5] established a close connection between certain $\{\Phi_n\}$ and $\{p_n\}$. Let $w \in AW \cap L_1$. If the (even) $\mathcal{W} \in TW$ is defined by

$$\mathcal{W}(\vartheta) = w(\cos \vartheta)|\sin \vartheta|, \quad x = \cos \vartheta, \ x \in [-1, 1], \tag{7}$$

then with $\varphi(x) = \sqrt{1 - x^2}$, we have $a_{2n-1}(\mathcal{W}) := \kappa_{2n}^{-1}(\mathcal{W})\Phi_{2n}(\mathcal{W}, 0)$, and

$$\begin{cases} p_n\left(w, \frac{1}{2}\left(z + \frac{1}{z}\right)\right) = \dfrac{z^{-n}\Phi_{2n}(\mathcal{W}, z) + z^n\Phi_{2n}(\mathcal{W}, z^{-1})}{\sqrt{2\pi}\{1 + a_{2n-1}(\mathcal{W})\}^{1/2}}, \\[3mm] p_{n-1}\left(w\varphi, \frac{1}{2}\left(z + \frac{1}{z}\right)\right) = \dfrac{z^{-n}\Phi_{2n}(\mathcal{W}, z) - z^n\Phi_{2n}(\mathcal{W}, z^{-1})}{\sqrt{2\pi}\{1 - a_{2n-1}(\mathcal{W})\}^{1/2}\frac{1}{2}(z - 1/z)}, \\[3mm] \hspace{6cm} n \in \mathbb{Z}_+, \ z \in \mathbb{C}. \end{cases} \tag{8}$$

Note that $2^{-1}(z + z^{-1}) = \cos\vartheta = x \in [-1,1]$ iff $z = \exp(i\vartheta)$; at the same time $2^{-1}(z - z^{-1}) = i\sin\vartheta$.

Definition 1.1. The trigonometric weight $\mathcal{W}(\vartheta)$ is a generalized trigonometric Jacobi weight ($\mathcal{W} \in GTJ$ shortly), iff

$$\mathcal{W}(\vartheta) = h(\vartheta) \prod_{r=0}^{m+1} w_r \left(\sin \frac{|\vartheta - \tau_r|}{2} \right), \quad \vartheta \in [-\pi, \pi], \tag{9}$$

where

$$-\pi < \tau_0 < \tau_1 < \cdots < \tau_m < \tau_{m+1} \le \pi, \tag{10}$$

$$\tau_{-1} = \tau_{m+1} - 2\pi, \ \tau_{m+2} = \tau_0 + 2\pi,$$

$$w_r(\delta) = \prod_{s=1}^{\ell_r} \{w_{rs}(\delta)\}^{\alpha(r,s)}, \tag{11}$$

$m, \ell_r \in \mathbb{N}$, $\alpha(r,s) \in \mathbb{R}$, $w_{rs}(\delta)$ are moduli of continuity ($s = 1, 2, \ldots, \ell_r$, $r = 0, 1, \ldots, m+1$) and the function h satisfies

$$h(\vartheta) \ge 0, \ h \text{ and } \frac{1}{h} \in L^\infty. \tag{12}$$

(i) If $\mathcal{W} \in GTJ$ and all w_{rs} are *concave* moduli of continuity then we say that $\mathcal{W} \in GTJ1$.

(ii) We say the $\mathcal{W} \in GTJ2$ if $\mathcal{W} \in GTJ$ and

$$\int_0^\delta w_r(\tau)d\tau = O(\delta w_r(\delta)), \ \delta \to +0 \ (r = 0, 1, \ldots, m+1).^1 \tag{13}$$

(iii) If $\mathcal{W} \in GTJ$ and

$$\omega(h, \delta)_\infty \delta^{-1} \in L^1[0,1] \text{ or } \omega(h, \delta)_2 = O(\sqrt{\delta}), \ \delta \to 0,$$

then $\mathcal{W} \in GTJ3$.

Remark 1.2. Generally we combine the above properties. E.g., we write that $\mathcal{W} \in GTJ13$ iff $\mathcal{W} \in GTJ1 \cap GTJ3$.

Definition 1.3. Let $w(x) \in AW$ be defined by

$$w(x) = H(x)(\sqrt{1-x^2})^{-1} w_0(\sqrt{1-x}) w_{m+1}(\sqrt{1+x}) \prod_{r=1}^m w_r(|x - t_r|). \tag{14}$$

If with $h(\vartheta) := H(\cos\vartheta)$, $t_r = \cos\tau_r \in (-1,1)$ ($1 \le r \le m$) the corresponding expressions $h(\vartheta)$, $\{\tau_r\}$ and $\{w_r(\delta)\}$ satisfy (10)–(12), then $w \in GJ$ (generalized Jacobi weight). The definition of $GJ1$, $GJ2$, etc., are similar.

Definition 1.4. Let $w \in GJ$. Then $\mathcal{W}(w, \vartheta) := \mathcal{W}(\vartheta) = w(\cos\vartheta)|\sin\vartheta|$, $\vartheta \in \mathbb{R}$, is called a generalized *even* trigonometric Jacobi weight ($\mathcal{W} \in GETJ$). The definitions of $GETJ1$, etc., are analogous.

---

[1] Notice that (13) is trivial for any w_r if $\delta \ge \delta_0$.

Remarks 1.5.

1. It is easy to see that the weights $(1-x)^\alpha(1+x)^\beta$, $(1-x)^\alpha(1+x)^\beta|x|^\gamma$, $\alpha, \beta, \gamma > -1$ are from $GJ123$. Also, $(1+|x|)(1-x)^\alpha \in GJ123$ and $(x+2)(1+x)^\beta \in GJ123$ $(\alpha, \beta > -1)$.

2. Further it is not difficult to verify the following relations (see [2, pp. 328–329]):

$$x^\Gamma \left(\log \frac{1}{x}\right)^\gamma \in GMC2, \ \Gamma > -1, \ \gamma \in \mathbb{R}, \ 0 \le x \le x_0;$$

$$x^\Gamma \left(\log_k \frac{1}{x}\right)^\gamma \in GMC2, \ \Gamma > -1, \ \gamma \in \mathbb{R}, \ 0 \le x \le x_0$$

and

$$x^\Gamma \left(\log \frac{1}{x}\right)\left(\log\log \frac{1}{x}\right)^\gamma \in GMC2, \ \Gamma > -1, \ \gamma \in \mathbb{R}, \ 0 \le x \le x_0.$$

Investigation of other combinations are left to the interested reader.

3. By $(1/(\log 1/x)^\gamma)' = \alpha/(x(\log 1/x)^{\alpha+1}) := \alpha u(x)$, whence, if $\alpha > 0$,

$$\int_0^\delta \frac{1}{x(\log 1/x)^{\alpha+1}} dx = \frac{1}{\alpha}\frac{1}{(\log 1/\delta)^\alpha} \ne O(\delta u(\delta)),$$

i.e., if $w(x) = u(e/x) = e|x|^{-1}(\log e/|x|)^{-\alpha-1}$, then $w \in L_1$ but $w \notin GJ2$.

2. Quasi-orthogonal polynomials on $[-1, 1]$

2.1. Let $w \in AW \cap L_1$. According to Marcell Riesz, we may investigate the expression

$$\psi_n(x, w, \rho_n) := p_{n-1}(w, \rho_n)p_n(w, x) - p_{n-1}(w, x)p_n(w, \rho_n), \ n \in \mathbb{N}, \qquad (15)$$

where $\rho_n \in \mathbb{R}$ is arbitrary fixed. M. Riesz proved about 1920: All the zeros of $\psi_n(x, \rho_n)$ (with respect to x) are real and simple. Moreover, at least $n-1$ zeros lie in $(-1, 1)$ (cf. G. Freud [3 ,I/3 and p. 53]; here and later we use some obvious short notations).

Let us denote the zeros of ψ_n by

$$\xi_{1n}(\rho_n) > \xi_{2n}(\rho_n) > \cdots > \xi_{n^*,n}(\rho_n), \ n \ge 1, \qquad (16)$$

where $n^* = \deg\psi_n(x, \rho_n)$. By definition, $n^* = n$ if $p_{n-1}(\rho_n) \ne 0$; otherwise $n^* = n-1$; moreover, by (16), ρ_n itself is one of the roots of ψ_n.

A fundamental property of ψ_n is that $\psi_n(x; \xi_{in}(\rho_n))$ and $\psi_n(x; \xi_{kn}(\rho_n))$ have exactly the same roots (namely, $\xi_{1n}(\rho_n), \xi_{2n}(\rho_n), \ldots, \xi_{n^*,n}(\rho_n)$) if $1 \le i, \ k \le n^*$ (see [3, p. 21]).

Another important property of ψ_n is the *quasi-orthogonality*, namely the relation

$$\int_{-1}^1 \psi_n(x, w, \rho_n)x^k w(x) dx = 0, \ \text{if } 1 \le k \le n^* - 1 \qquad (17)$$

(see [3 ,p. 20]).

2.2. Let us define the n. Christoffel function for $w \in AW \cap L_1$ by

$$\lambda_n(w, x) := \min_{P \in \mathcal{P}_{n-1}} \int_{-1}^{1} \left| \frac{P(t)}{P(x)} \right|^2 w(t) dt, \; x \in \mathbb{R}. \tag{18}$$

Then, if $m = n + n^* - 2$, for any $P \in \mathcal{P}_m$

$$\int_{-1}^{1} P(x)w(x)dx = \sum_{k=1}^{n^*} \lambda_n(w, \xi_{kn}(\rho_n)) P(\xi_{kn}(\rho_n)), \tag{19}$$

where, as above, $\xi_{kn}(\rho_n)$ are the roots of $\psi_n(x, \rho_n)$ ([3, p. 31]). This quadrature formula (which is a generalization of the well-known Gauss quadrature) shows that properties of $\{\xi_{kn}\}$ are of fundamental importance. Here we quote a result from [3, p. 111] .

Let $w \in AW \cap L_1$ and suppose that $0 < m \le w(x) \le M$ if $x \in [a, b] \subset [-1, 1]$. Then

$$\frac{c_1(\varepsilon)}{n} \le \xi_{kn}(\rho_n) - \xi_{k+1,n}(\rho_n) \le \frac{c_2(\varepsilon)}{n}, \; n \ge 2 \tag{20}$$

whenever $\xi_k, \xi_{k+1} \in [a + \varepsilon, b - \varepsilon]$. Above, $\varepsilon > 0$, fixed, $0 < c_1(\varepsilon) < c_2(\varepsilon)$ are constants.

Remark 2.1. All the definitions and statements of the previous Part 2 originally were said with a *fixed* ρ (instead of ρ_n). However, it is easy to see that they are valid using a varying sequence $\{\rho_n\}$, too.

3. New results concerning $\psi_n(w, \rho_n)$

3.1. Throughout this part we consider some generalizations of the corresponding results proved in [2] using the roots of $p_n(w)$.

Only in part 3.1 let

$$1 \ge \xi_{1n}(\rho_n) > \xi_{2n}(\rho_n) > \cdots > \xi_{n_1,n}(\rho_n) \ge -1, \; n \in \mathbb{N} \tag{21}$$

denote those roots of ψ_n which are in $[-1, 1]$ (compare to formula (16)). As we stated in Part 2, $n_1 = n_1(n) \ge n - 1$.

We state (cf. [2, Theorem 3.2])

Theorem 3.1. *Let $w \in GJ2$. Then for the roots $\xi_{kn}(\rho_n) = \cos \vartheta_{kn}(\rho_n)$ $(1 \le k \le n_1)$ of $\psi_n(x, w, \rho_n)$ situated in $[-1, 1]$, we have*

$$\vartheta_{k+1,n}(\rho_n) - \vartheta_{kn}(\rho_n) \sim \frac{1}{n}, \; 1 \le k \le n_1 - 1, \; n \ge 3, \tag{22}$$

uniformly in k, ρ_n and n.

With $|x - \xi_{jn}(\rho_n)| = \min_{1 \le k \le n_1} |x - \xi_{kn}(\rho_n)|$, $j = j(n, x)$, we state (cf. [3, Theorem 3.3])

Theorem 3.2. *Let $w \in GJ123$. Then for $n \geq 1$*

$$|\psi'_n(\xi_{kn}(\rho_n), w, \xi_{kn}(\rho_n))| \sim \frac{1}{\lambda_n(w, \xi_{kn}(\rho_n))}, \quad 1 \leq k \leq n_1, \tag{23}$$

$$|\psi_n(x, w, \xi_{jn}(\rho_n))| \sim \frac{1}{\lambda_n(w, \xi_{jn}(\rho_n))}|x - \xi_{jn}(\rho_n)|, \quad x \in [-1, 1], \tag{24}$$

$$\vartheta_{k+1,n}(\rho_n) - \sigma_{k,n-1} \sim \sigma_{k,n-1} - \vartheta_{kn}(\rho_n) \sim \frac{1}{n}, \quad 1 \leq k \leq n_1 - 1, \tag{25}$$

where $y_{k,n-1} = \cos \sigma_{k,n-1}$ $(1 \leq k \leq n_1 - 1)$ stand for those roots (in decreasing order) of $\psi'_n(x, w, \rho_n)$ which are in $(\xi_{n_1,n}(\rho_n), \xi_{1n}(\rho_n))$. The estimations (23)–(25) are uniform in x, k, j, ρ_n and n, respectively.

3.2. Before stating some generalization of formula (19) we introduce the function class GTJ4. $\mathcal{V} \in GTJ4$ if $\mathcal{V} = U/W$, where both U and W from GTJ, moreover they have only positive exponents (cf. (11)). Moreover if W is defined by (9), then

$$\int_0^\delta \frac{1}{w_r(\tau)}d\tau = O\left(\frac{\delta}{w_r(\delta)}\right), \quad \delta \to +0, \quad r = 0, 1, \ldots, m+1.$$

If moreover $U, W \in GETJ$, then we say that $\mathcal{V} \in GETJ4$. Let $v \in GJ$ be of the form (14) (formed by $v_0, v_1, \ldots, v_{m+1}$). Then $|\sin \vartheta|v(\cos \vartheta) = \mathcal{V}(v)(\vartheta) \in GETJ$. If $\mathcal{V}(v) \in GETJ4$, too, then we can say that $v \in GJ4$.

Example 3.3. *Let $\omega(\delta) = |\log \delta|^{-\alpha}$, $\eta(\delta) = \delta^\beta$ $(\alpha \in \mathbb{R}, 0 < \beta < 1, \delta > 0)$. Then $\mathcal{V}(\vartheta) = \omega(\sin|\vartheta - \pi|/2)/\eta(\sin(|\vartheta|/2)) \in GTJ4$. On the other hand, $\mathcal{V}(\vartheta) = \sin(|\sin \vartheta|/2)/\sin(|\vartheta|/2) \in GTJ2\backslash GTJ4$.*

Theorem 3.4. *Let $w \in GJ2$ and $v \in GJ4$. Then for any fixed positive integer ℓ and $1 \leq p < \infty$*

$$\sum_{k=1}^{n^*} \lambda_n(v, \xi_{kn}(\rho_n))|P(\xi_{kn}(\rho_n))|^p \leq c \int_{-1}^1 |P(x)|^p v(x)dx, \quad n \geq 2, \tag{26}$$

whenever $P \in \mathcal{P}_{\ell n}$. Here c does not depend on n and P.

The "converse" inequality is as follows.

Theorem 3.5. *Let $w \in GJ123$, further v, $V^{-p} \in GJ2$, finally $V^q \in GJ24$. Here $V = \sqrt{w\varphi}v^{-1/p}$, $1/p + 1/q = 1$, $1 < p < \infty$. Then for any $p \in \mathcal{P}_{n^*-1}$*

$$\int_{-1}^1 |P(x)|^p v(x)dx \leq c \sum_{k=1}^{n^*} \lambda_n(v, \xi_{kn}(\rho_n))|P(\xi_{kn}(\rho_n))|^p, \quad n \geq 2. \tag{27}$$

We may emphasize that (as before) in formulas (26) and (27) $\xi_{kn}(\rho_n)$ $(1 \leq k \leq n^*)$ are the roots of $\psi_n(x, w, \rho_n)$ (cf. (16)).

Remark 3.6. *Theorems 3.4 and 3.5 generalize [2, Theorems 3.6 and 3.7]. Other statements can be obtained using [2, Part 3.2].*

4. Para-orthogonal polynomials on the unit circle

4.1. Szegö's fundamental relations (8) create an obvious correspondence between the orthonormal polynomials $p_n(w)$ and $\Phi_n(\mathcal{W})$ (if $\mathcal{W}$ is even). However, while the zeros o f $p_n(w)$ are in $(-1, 1)$ and they are simple, in contrast, the zeros of $\Phi_n(\mathcal{W})$ are *inside* of the unit circle (for any $\mathcal{W} \in T\mathcal{W}$). In the simplest case when $\mathcal{W}(\vartheta) = 1$, $\Phi_n(\mathcal{W}, z) = z^n$, i.e., all the zeros are located at the origin that is "far away" from the support of the measure. This simple fact results that there are *no* quadrature-type formulas based on the roots of $\Phi_n(\mathcal{W})$ (compare to (20), (26) and (27) where $\psi_n(w, \rho_n) = c_n p_n(w)$ or $\psi_n(w, \rho_n) = d_n p_{n-1}(w)$, supposing that $p_n(w, \rho_n) = 0$ or $p_{n-1}(w, \rho_n) = 0$, respectively). However, using the polynomial

$$B_n(z, \mathcal{W}, u_n) := \Phi_n(\mathcal{W}, z) + w_n \Phi_n^*(\mathcal{W}, z), \ |u_n| = 1, \ n \geq 1, \tag{28}$$

where $w_n = -\Phi_n(\mathcal{W}, u_n)/\Phi_n^*(\mathcal{W}, u_n)$,

one can prove that B_n is a polynomial of degree exactly n, its zeros are simple and lie *on* the unit circle line; moreover the parameter $u_n = \exp(i\delta_n)$ is one of them ($\mathcal{W} \in T\mathcal{W} \cap L^1$). We denote these roots by $\zeta_{kn}(u_n) = \exp(i\eta_{kn}(u_n))$, $1 \leq k \leq n$, according to

$$\delta_n = \eta_{1n}(u_n) < \eta_{2n}(u_n) < \cdots < \eta_{nn}(u_n) < 2\pi + \delta_n, \ n \geq 1. \tag{29}$$

As before, one can prove that $B_n(\mathcal{W}, u_n)$ and $B_n(\mathcal{W}, \zeta_{kn}(u_n))$ $(1 \leq k \leq n)$ have the *same* roots. (The proofs of these facts are in the exhausting survey of W.B. Jones, O. Njåstad and W.J. Thorn [4]).

4.2. The "para-orthogonality" means that

$$\int_0^{2\pi} B_n(z, \mathcal{W}, u_n) z^{-k} \mathcal{W}(\vartheta) d\vartheta = 0, \ z - \exp(i\vartheta), \ 1 \leq k \leq n - 1, \tag{30}$$

while formula (19) can be replaced by

$$\frac{1}{2\pi} \int_0^{2\pi} P_{n-1}(z) \overline{Q_{n-1}(z)} \mathcal{W}(\vartheta) d\vartheta =$$
$$\sum_{k=1}^n \mu_n(\mathcal{W}, \zeta_{kn}(u_n)) P_{n-1}(\zeta_{kn}(u_n)) \overline{Q_{n-1}(\zeta_{kn}(u_n))}, \tag{31}$$

$z = \exp(i\vartheta)$, $P_{n-1}, Q_{n-1} \in \mathcal{P}_{n-1}$, otherwise arbitrary; moreover

$$\mu_n(\mathcal{W}, z) := \min_{P \in \mathcal{P}_{n-1}} \frac{1}{2\pi} \int_0^{2\pi} \frac{|P(\exp(i\vartheta)|^2}{|P(z)|^2} \mathcal{W}(\vartheta) d\vartheta, \ z \in \mathbb{C}, \ n \geq 1, \tag{32}$$

the corresponding n. Christoffel function (see [4, (6.1) and Part 7] and [5, (10)]).

In his recent paper, L. Golinskii proved (among others) as follows (cf. [5, Theorem 5]).

If $0 < A \leq W(\vartheta) \leq B$, then with $\zeta_{1n}(u_n) = \zeta_{n+1,n}(u_n) = u_n$

$$\frac{4}{n-1} \sqrt{\frac{A}{B}} \leq |\zeta_{k+1,n}(u_n) - \zeta_{kn}(u_n)| \leq \frac{4\pi B}{An}, \ 1 \leq k \leq n, \ n \geq 2. \tag{33}$$

5. New results concerning $B_n(\mathcal{W}, u_n)$

5.1. We intend to get result analogous to those in Part 3. Because of the obvious correspondences sometimes we only indicate the analogous statements.

Theorem 5.1. *Let* $\mathcal{W} \in GTJ2$. *Then*

$$\zeta_{k+1,n}(u_n) - \zeta_{kn}(u_n) \sim \frac{1}{n}, \ 1 \le k \le n, \ n \ge 2 \tag{34}$$

uniformly in k *,*u_n *and* n.

Theorem 5.2. *Let* $\mathcal{W} \in GTJ123$. *Then for* $n \ge 1$

$$|B'_n(\zeta_{kn}(u_n), W, \zeta_{kn}(u_n))| \sim \frac{1}{\mu_n(\mathcal{W}, \xi_{kn}(u_n))}, \ 1 \le k \le n, \tag{35}$$

$$|B_n(z, W, \zeta_{jn}(u_n))| \sim \frac{|z - \zeta_{jn}(u_n)|}{\mu_n(\mathcal{W}, \zeta_{kn}(u_n))}, \ |z| = 1. \tag{36}$$

5.2. Theorems 3.4 and 24 may be replaced by

Theorem 5.3. *Let* $\mathcal{W} \in GTJ2$ *and* $\mathcal{V} \in GTJ4$. *Then for any fixed positive integer* ℓ *and* $1 \le p < \infty$

$$\sum_{k=1}^{n} \mu_n(\mathcal{V}, \zeta_{kn}(u_n))|P(\zeta_{kn}(u_n))|^p \le c \int_0^{2\pi} |P(\exp(i\vartheta))|^p \mathcal{V}(\vartheta)d\vartheta \tag{37}$$

for any $P \in \mathcal{P}_{\ell n}$. *Here* c *does not depend on* n *and* P.

Theorem 5.4. *Let* $\mathcal{W} \in GTJ123$, $\mathcal{V}$, $U^{-p} \in GTJ2$, $U^q \in GTJ4$ *where* $U = \sqrt{\mathcal{W}}\mathcal{V}^{-1/p}$, $1/p + 1/q = 1$, $1 < p < \infty$. *Then for any* $P \in \mathcal{P}_{n-1}$

$$\int_0^{2\pi} |P(z)|^p \mathcal{V}(\vartheta)d\vartheta \le c \sum_{k=1}^{n} \mu_n(\mathcal{V}, \zeta_{kn}(u_n))|P(\xi_{kn}(u_n))|^p, \ z = \exp(i\vartheta). \tag{38}$$

6. Final remarks

The introduction of the quasi-orthogonal polynomials is due to M. Riesz in a serious of papers written in 1921-1923.

The word "para-orthogonality" was coined in the paper of W.B. Jones, O. Njastad and J. Thron [4] in 1989. The idea in both cases is that we investigate the polynomial which formed from the numerator of the corresponding Christoffel-Darboux formula.

We may mention that in 1963 G. Szegö [6] wrote a paper on certain bi-orthogonal systems which have a strong connection with the para-orthogonal polynomials (see P. Gonzáles and M. Camacho in [7]).

Acknowledgement

Research supported by OTKA Nos. T 037299 and T 047132.

References

[1] G. Szegö, *Orthogonal Polynomials*, AMS Coll. Publ., Vol. 23 (Providence, R.I. 1967)

[2] G. Mastroianni, P. Vértesi, Some applications of generalized Jacobi weights. *Acta Math. Hungar.* **77** (1977), 323–357.

[3] G. Freud, *Orthogonal Polynomials*, Akadémiai Kiadó (Budapest, 1971).

[4] W.B. Jones, O. Njåstad, W.J. Thron, Moment theory, orthogonal polynomials, quadrature, and continued fractions associated with the unit circle. *Bull. London Math. Soc.* **21** (1989), 113–152.

[5] L. Golinskii, Quadrature formulas and zeros of para-orthogonal polynomials on the unit circle. *Acta Math. Hungar.* **96** (2002), 169–186.

[6] G. Szegö, On bi-orthogonal systems of trigonometric polynomials, *Magyar Tnd. Akad. Kutadó Int. Közl* **8** (1963), 255–273.

[7] P. Gonzáles-Vera and M. Camacho, A note on para-orthogonality and biorthogonality, *Det Kongelige Norske Videnskabers Selskab* **3** (1992) 1–16.

Biancamaria Della Vecchia
Dipartimento di Matematica
Università di Roma "La Sapienza"
Piazzale Aldo Moro 2
I-00185 Roma (Italy)

and

Istituto per le Applicazioni del Calcolo 'M. Picone' CNR
Sezione di Napoli
Via Pietro Castellino 111
I-80131 Napoli (Italy)
e-mail: biancamaria.dellavecchia@uniroma1.it

Giuseppe Mastroianni
Dipartimento di Matematica
Università della Basilicata
Contrada Macchia Romana
I-85100 Potenza (Italy)
e-mail: mastroianni@unibas.it

Peter Vértesi
Alfréd Rényi Institute of Mathematics
P.O.B. 127
H-1364 Budapest (Hungary)
e-mail: vertesi@renyi.hu

M. de Bruin, D.H. Mache and R. Beatson

At the German Mining Museum & Institute for
Research into History of Mining in Bochum

International Series of Numerical Mathematics

Edited by
Hoffmann, K.-H., Technische Universität München, Germany
Mittelmann, H.D. , Arizona State University, Tempe, USA

International Series of Numerical Mathematics is open to all aspects
of numerical mathematics. Some of the topics of particular interest
include free boundary value problems for differential equations, phase transitions, problems of optimal control and optimization, other
nonlinear phenomena in analysis, nonlinear partial differential equations, efficient solution methods, bifurcation problems and approximation theory. When possible, the topic of each volume is discussed
from three different angles, namely those of mathematical modeling,
mathematical analysis, and numerical case studies.

Your Specialized Publisher in Mathematics
Birkhäuser

For orders originating from all over the world
except USA/Canada/Latin America:

Birkhäuser Verlag AG
c/o Springer GmbH & Co
Haberstrasse 7
D-69126 Heidelberg
Fax: +49 / 6221 / 345 4 229
e-mail: birkhauser@springer.de
http://www.birkhauser.ch

For orders originating in the
USA/Canada/Latin America:

Birkhäuser
333 Meadowland Parkway
USA-Secaucus
NJ 07094-2491
Fax: +1 201 348 4505
e-mail: orders@birkhauser.com

Operator Theory: Advances and Applications

Your Specialized Publisher in Mathematics
Birkhäuser

For orders originating from all over the world except USA/Canada/Latin America:

Birkhäuser Verlag AG
c/o Springer GmbH & Co
Haberstrasse 7
D-69126 Heidelberg
Fax: +49 / 6221 / 345 4 229
e-mail: birkhauser@springer.de
http://www.birkhauser.ch

For orders originating in the USA/Canada/Latin America:

Birkhäuser
333 Meadowland Parkway
USA-Secaucus
NJ 07094-2491
Fax: +1 201 348 4505
e-mail: orders@birkhauser.com

Edited by
Gohberg, I., School of Mathematical Sciences, Tel Aviv University, Ramat Aviv, Israel

This series is devoted to the publication of current research in operator theory, with particular emphasis on applications to classical analysis and the theory of integral equations, as well as to numerical analysis, mathematical physics and mathematical methods in electrical engineering.